Lizard Social Behavior

Lizard Social Behavior

Edited by

Stanley F. Fox,

J. Kelly McCoy,

and

Troy A. Baird

The Johns Hopkins University Press

Baltimore and London

© 2003 The Johns Hopkins University Press
All rights reserved. Published 2003
Printed in the United States of America on acid-free paper
9 8 7 6 5 4 3 2 1

The Johns Hopkins University Press
2715 North Charles Street
Baltimore, Maryland 21218-4363

www.press.jhu.edu

Library of Congress Cataloging-in-Publication Data

Lizard social behavior / edited by Stanley F. Fox, J. Kelly McCoy,
and Troy A. Baird.
     p. cm
Includes bibliographical references and index.
   ISBN 0-8018-6893-9 (hardcover : alk. paper)
     1. Lizards—Behavior. 2. Social behavior in animals. I. Fox,
Stanley F., 1946– II. McCoy, J. Kelly. III. Baird, Troy A.
QL666.L2 L59 2002
597.95'156—dc21   2001004066

A catalog record for this book is available from the British Library.

Contents

III. VARIATION AMONG SPECIES

Color plates follow page 210.

Preface

For more than three decades now, many zoologists have chosen lizards as model systems with which to address questions that are central to ecological and evolutionary theory. During this time, contemporary ecological and evolutionary research conducted on lizard model systems prompted three important symposia at international meetings. All three symposia resulted in published volumes that have become essential references not only for those specifically interested in lizard ecology, but for those studying a wide range of ecological and evolutionary issues. Although all these volumes included chapters discussing aspects of behavior in lizards, none focused specifically on the evolution of social behavior. The need for such a volume emphasizing behavior is acute because sexual selection and its influence on social and mating systems, spacing behavior, sexual dimorphism, and sexual dichromatism have become major foci in behavioral ecology, and contemporary studies on lizards that examine adaptive variation in lizard social behavior are yielding exciting results in this area.

The large number of high-quality and taxonomically diverse studies that use lizards to explore the nexus between social behavior and evolutionary ecology inspired us to organize a symposium at the 1999 joint meetings of the American Society of Ichthyologists and Herpetologists, the Herpetologists' League, and the Society for the Study of Amphibians and Reptiles, held at Penn State University, and to produce a volume from the contributions made in that symposium. Our objective for this volume, however, is to appeal to an audience that is broader than lizard behavioral ecologists. To achieve this goal, in each chapter we use lizards as models to address hypotheses that are central to important theory in sexual selection and the evolution of social behavior. We begin each chapter by reviewing the pertinent general literature for developing the theoretical basis for one or more hypotheses. We then present data on the specific lizard system(s) that address these theoretical issues. The chapters are arranged in three sections that reflect the three primary levels at which behavioral ecologists examine adaptive variation in social behavior: individual variation within populations, variation among different populations of the same species, and variation among species. To emphasize the important relationships between work on saurians and other vertebrate taxa, each section is introduced by an eminent behavioral ecologist who has made major contributions to the study of adaptive variation in social behavior at one of these levels of inquiry, but working with other vertebrate taxa. It is our hope that this volume will continue the spirit and impact of the three previous volumes on lizard ecology, and perhaps play a small part in showcasing lizards as model organisms for the study of social behavior of animals.

Each chapter was read critically by at least two outside reviewers, after which we revised them to address the reviewers' comments and suggestions. We wish to express our sincere appreciation to the following colleagues for their careful and constructive reviews: Allison Abell, Robin Andrews, Michael Bull, John Carothers, Emilio Civantos, Doug Eifler, John Godwin, Monique Halloy, Jerry Husak, Nora Ibargüengoytía, Tom Jenssen, Rufus Johnstone, Mark Jordan, Rosemary Knapp, Don Miles, Michael C. Moore, Anders Møller, Mats Olsson, Gad Perry, Charles C. Peterson, Stan Rand, Doug Ruby, Judy Stamps, and Richard Tokarz. We thank the Herpetologists' League and the American Society of Ichthyologists and Herpetologists for providing a forum for Variation in Lizard Behavior: Individuals, Populations, and Species at the 1999 annual meetings held at Penn State University, State College. Funding for the symposium participants was provided by the Herpetologists' League and the National Science Foundation (IBN#9816994). Funding for preparation of the manuscript was provided by the College of Graduate Studies and Research at the University of Central Oklahoma, the Faculty Development Program at Angelo State University, and the College of Arts and Sciences at Oklahoma State University.

We also wish to thank our wives for putting up with the prolonged strain and tension of putting together such a multiauthored volume, and for helping us to retain our sanity. Thank you Susana Perea, Therese McCoy, and Teresa Baird; thank goodness we humans are pair-bonded, social creatures.

Contributors

TROY A. BAIRD, Department of Biology, University of Central Oklahoma, Edmond, tbaird@ucok.edu

GEORGE W. BARLOW, Department of Integrative Biology and Museum of Vertebrate Zoology, University of California, Berkeley, barlowgw@socrates. berkeley.edu

PHILIP W. BATEMAN, Department of Zoology and Entomology, University of Pretoria, Pretoria, South Africa, pwbateman@zoology.up.ac.za

MARGUERITE BUTLER, Department of Integrative Biology and Museum of Vertebrate Zoology, University of California, Berkeley, marguerite@mac.com

WILLIAM E. COOPER, JR., Department of Biology, Indiana University–Purdue University at Fort Wayne, Fort Wayne, Indiana, cooperw@ipfw.edu

STANLEY F. FOX, Department of Zoology, Oklahoma State University, Stillwater, foxstan@okstate.edu

PAUL J. GIER, Department of Biology, Huntingdon College, Montgomery, Alabama, pgier@huntingdon.edu

MASAMI HASEGAWA, Department of Ecological Science, Natural History Museum and Institute, Chiba, Japan, PX1M-HSGW@asahi-net.or.jp

DIANA K. HEWS, Department of Life Sciences, Indiana State University, Terre Haute, lshews@scifac.indstate.edu

JONATHAN B. LOSOS, Department of Biology, Washington University, St. Louis, Missouri, losos@biology.wustl.edu

PETER MARLER, Animal Communication Lab, Section of Neurobiology, Physiology and Behavior, University of California, Davis, prmarler@ucdavis.edu

J. KELLY MCCOY, Department of Biology, Angelo State University, San Angelo, Texas, kelly.mccoy@angelo.edu

KENNETH A. NAGY, Department of Organismic Biology, Ecology, and Evolution, University of California, Los Angeles, kennagy@biology.ucla.edu

GORDON H. ORIANS, Department of Zoology, University of Washington, Seattle, blackbird@u.washington.edu

VANESSA S. QUINN, Department of Life Sciences, Indiana State University, Terre Haute, lsrquinn@scifac.indstate.edu

THOMAS W. SCHOENER, Section of Evolution and Ecology, University of California, Davis, twschoener@ucdavis.edu

PAUL A. SHIPMAN, Department of Zoology, Oklahoma State University, Stillwater, shipman@cowboy.net

BARRY SINERVO, Department of Ecology and Evolutionary Biology, University of California, Santa Cruz, sinervo@biology.ucsc.edu

CHRIS L. SLOAN, Department of Biology, University of Central Oklahoma, Edmond, chris.sloan@worldnet.att.net

HEIDI M. SNELL, Charles Darwin Research Station, Puerto Ayora, Isla Santa Cruz, Galápagos, Ecuador, hsnell@fcdarwin.org.ec

HOWARD L. SNELL, Department of Biology, University of New Mexico, Albuquerque, and Charles Darwin Research Station, Puerto Ayora, Isla Santa Cruz, Galápagos, Ecuador, hsnell@mail.unm.edu

PAUL A. STONE, Department of Biology, University of Central Oklahoma, Edmond, pstone@ucok.edu

DUSTI K. TIMANUS, Department of Biology, University of Central Oklahoma, Edmond, dusti-sloan@ouhsc.edu

MARTIN J. WHITING, School of Animal, Plant and Environmental Sciences, University of the Witwatersrand, South Africa, martin@gecko.biol.wits.ac.za

KELLY R. ZAMUDIO, Department of Ecology and Evolutionary Biology, Cornell University, Ithaca, New York, krz2@cornell.edu

The Evolutionary Study of Social Behavior and the Role of Lizards as Model Organisms

Stanley F. Fox, J. Kelly McCoy, and Troy A. Baird

Historically, much of the research conducted to test hypotheses about the evolution of animal social behavior has focused on endotherms and fishes. Relatively recently, lizards have emerged to play a more prominent role as model organisms for field studies that test theory in behavioral ecology. Sexual selection has come to be a central principle in the evolution of social and reproductive behavior, and over the last decade investigators have increasingly used lizards to conduct studies in this area. A powerful approach is to document and investigate the causes of variation in social behavior as tests of sexual selection theory. Our objective for this volume is to bring together a collection of articles describing recent work by contemporary behavioral ecologists utilizing field studies of lizards to investigate the evolution of social behavior, particularly by intra- and intersexual selection. We have chosen contributors and organized the book to showcase the adaptive value of variation in social behavior at three levels of ecological organization: among individuals within populations, among different populations of the same species, and among different species of lizards. Studies at all three levels have proven to be effective paradigms with which to test hypotheses about the role of sexual selection in the evolution of social and mating tactics, spacing patterns, and associated coloration and morphological variation. In part 1, we present studies that focus on within-population variation in the behavior of individuals that illustrate the different ontogenetic trajectories of males and females and the evolution and maintenance of alternative male reproductive tactics in sexually selected species, the trade-offs between social behavior and the critical need to evade predators as a test of optimal escape theory, and the importance of sexually selected male coloration patterns as social badges. In part 2, we provide investigations focusing on variation among populations of the same species that reveal to what extent local environmental characteristics such as heterogeneity in the distribution and structure of habitat, demography, and predation pressure act as strong selective agents that differentially influence the strength of sexual selection in disparate habitats. In part 3, we include studies that compare species of lizards and how they differ in social behavior and development of sexual dimorphism and dichromatism as a consequence of both environmental conditions and phylogeny.

Lizards are uniquely positioned as model organisms in which to address questions about the adaptive value of social behavior owing to their tractability as subjects of field research, and their phylogenetic position as ectothermic amniotes. Because lizards lack the distinct larval phase characteristic of amphibians, juveniles and adults do not necessarily occupy markedly different habitats; both adults and juveniles have similar ecological requirements for survival. Most lizards rely heavily on the same visual and chemosensory modalities utilized by higher vertebrates, but they also show differences. Unlike endothermic birds and mammals, lizards, like ectotherms, are intimately tied to thermal aspects of their environment. Whereas food and predation have been the primary influential forces in the evolution of social behavior of birds and mammals, food, predation, as well as thermal environment have played prominently in the evolution of social behavior of lizards. Even more than fish, amphibians, and other reptiles, lizards exhibit a rich repertoire of thermoregulatory behavior. Often their social behavior is constrained (or released) by thermal conditions, but sometimes social interactions take precedence and lead to risky thermal exposure, but modulated by predation and food. Another difference is the widespread lack of parental care in lizards. Because parental care is not displayed in most lizards, there is less potential for learning to shape the behavior of neonates than in most birds and mammals. Therefore, lizards at once provide a more simplified, but also more complex, amniote model in which to test many of the important theoretical advances that have been proposed to explain the adaptive value of variation in social and reproductive behavior of birds and mammals.

A primary factor driving the utility of lizards as model organisms for studies of the evolution of social behavior is their low vagility relative to that of other vertebrates. Unlike birds, which disperse via flight, and aquatic vertebrates with larval or juvenile stages that are dispersed by water currents, many lizards display a high degree of lifelong site philopatry. Philopatry can have profound consequences for the evolution of social behavior, because successful behavioral tactics are strongly influenced by selective forces operating on localized spatial scales, and these selective forces often show spatial and temporal heterogeneity. Essentially, individual lizards may be "caught" in particular locales because they are prevented from sampling numerous habitat patches by their low vagility, the biophysical constraints imposed by ectothermy, and the costs associated with dispersal. Because there is high potential for heterogeneity in the environmental and demographic conditions both across and within localized populations, limited dispersal in lizards results in a rich landscape in which to address the evolutionary forces influencing behavioral variation among individuals within populations, among different populations of the same species, and among related species subjected to different selective pressures.

From a practical standpoint, lizards are outstanding organisms with which to test hypotheses using field studies. Many species occupy open, unobstructed habitats where they are highly visible. Mostly they are diurnal, carrying out their easily observed social behavior throughout the day. Lizards are generally indifferent to the presence of human observers and appear to behave normally under observation. Coupled with their strong site fidelity, this makes it possible for observers to record the location, outcome, participants of social interactions, and behavior shown by known individuals over long stretches of time. Lizards are also amenable to a variety of experimental manipulations in the field such as changes in levels of threat from predators; modification of coloration patterns that serve as social signals; and alteration of the distribution and abundance of important resources such as food, perch sites, or refugia. Owing to their relative ease of capture, free-living lizards can be measured repeatedly to monitor changes in reproductive state, body condition, and growth in body size and body structures that might relate importantly to aspects of their social behavior. Blood samples can be obtained from the suborbital sinuses of most lizards without harming them, and because lizards have nucleated erythrocytes, blood is a ready source of DNA that can be subjected to state-of-the-art molecular techniques to establish parentage and genetic relatedness of animals observed in the field. Because free-living lizards can often be captured and bled quickly, levels of steroid hormones can also be measured with confidence using radioimmunoassays. The relationship between hormones and social behavior has become better understood by using castration, implants, and even external hormone patches in free-living lizards. Together, the ability to apply these techniques to free-living animals residing under different environmental and social conditions has allowed the application of powerful empirical approaches in lizard studies for testing a wide range of contemporary hypotheses.

We hope that the theoretical approaches and practical techniques as exemplified in this book will spur more advances in the study of variation in animal social behavior. We are convinced that lizards will continue to play a definitive role in such advances and hope that our book will convince others.

Variation among Individuals

Introduction

Peter Marler

It is a salutary reminder of the generality of evolutionary concepts of behavioral organization, social behavior, and sexual selection that one can move so freely across taxa and find phylogenetically remote organisms conforming to similar principles. Studies of the social behavior of birds have an obvious bearing on our understanding of mammalian behavior, and the same is true of reptiles and amphibians, and even fish. Nevertheless, many will ask, why lizards?

At several points in his thought-provoking essay, "Competition and the Niche," Schoener (1977) spells out some of the characteristics that make lizards such suitable subjects for the empirical testing of a wide range of theories and problems in behavioral and population biology, some classical, some only recently conceived. He acknowledges the critical role of observational studies of birds, by pioneers such as Lack, Crook, and Orians, in generating new concepts about the adaptive significance of variation in competitive and other kinds of behavior. But then he goes on to make a compelling case that lizards are, in some respects, better suited than birds as a source of basic information about how and why the social behavior of animals is so variable, not only between species, but intraspecifically as well (cf. the introduction to this volume). Schoener (p. 37) notes that "lizards are, in general, closer to the 'modal' animal than birds; they are terrestrial, quadrupedal, poikilothermic, relatively slow-growing, and lack parental care," all features that hold promise of relative tractability in dealing with issues that call for theoretical, quantitative analysis. In fact, lizard enthusiasts, including several contributors to this volume, are prone to speak of their subjects as "model organisms" (e.g., Huey, Pianka and Schoener, 1983; see the introduction to this volume).

Lizards "sometimes surpass birds in conspicuousness, abundance, and ease of study," and are "sometimes denser than one per square meter." Close scrutiny is often possible over long periods of time. Further, many species can be easily captured, marked, and released with apparently no increase in wariness of humans. Some anoles, for example, can literally be "plucked like fruits" from their nocturnal sleeping perches on peripheral leaves and twigs (Schoener 1977, p. 36). The fact that many lizards are predators, and subjected to predation by other vertebrates (see chapters 4-6 and 10), adds up to a good case for lizards as ideal subjects for observational studies of behavior, ecology, and demography to which theoretical propositions can be submitted for empirical testing.

Sometime in the 1960s, word of this potential began to diffuse through evolutionary and organismal biology, and the stage was set for an explosion of studies to the point that, in some domains, lizards, along with other reptiles, as well as some amphibia, have surpassed birds as subjects of first choice.

The consequences have been pervasive. Where would our appreciation of the great potential diversity of the endocrinological controls of reproductive behavior, the range of proximate constraints on reproduction, or the mysteries of parthenogenesis be without lizard studies (Licht and Gorman 1970; Crews 1975, 1979; Crews and Williams 1977; Crews, Gustafson and Tokarz 1983; Sinervo and Licht 1991b)? Unraveling the complex proximate and ultimate determinants of behavioral and morphological polymorphisms, including the throat color "ensembles" with which lizards are unusually well endowed, and beginning to understand the many intricacies of alternative mating strategies, owe much to students of lizard behavior and ontogeny (Fox 1975, 1978; Moore and Thompson 1990; Moore 1991; Thompson and Moore 1991b; Cooper and Greenberg 1992; Rand 1992; see chapters 1 and 3). Experimental investigations of the visual signaling behavior of lizards have provided exemplary models for how to approach the analysis of animal signals and badges (see chapter 2) and their role in species recognition (Hunsaker 1962; Carpenter and Ferguson 1977; Williams and Rand 1977). The many efforts to submit modern concepts of sexual selection, the evolution of social organization, and the dynamics of evolutionary change have been profoundly influenced by the intensive, long-term field studies of lizard enthusiasts (Blair 1960; Milstead 1961; Rand 1964; Pianka 1966, 1973; Tinkle 1967b; Schoener 1968; Williams

1969, 1972; Pianka and Pianka 1976). Modern views on territorial behavior, its ontogeny, and the role of social factors in territorial establishment and the use of living space have been strongly influenced by the work of Stamps and colleagues, to say nothing of pioneering studies of display behavior (Stamps 1973, 1978, 1994; Stamps and Barlow 1973).

The first four chapters of this book continue the tradition established by these pioneers, combining observations and experiments to unravel the intricacies of social behavior and the use of living space within natural lizard populations. In chapter 1, Troy Baird, Dusti Timanus, and Chris Sloan present a thorough characterization of behavioral variation in collared lizards by testing predictions of sexual selection theory, exploring the influence of ontogeny and environment on behavioral variation in both sexes, and examining the role of morphological and behavioral attributes in mating success. In chapter 2, Martin Whiting, Kenneth Nagy, and Philip Bateman use male flat lizards, whose coloration rivals that of the most brightly plumed birds, to experimentally evaluate contemporary interpretations of the evolution of social badges. In chapter 3, Kelly Zamudio and Barry Sinervo couple modern molecular techniques with field studies on social behavior to examine the evolution of fixed alternative male mating strategies. They propose a general ecological and social framework for the evaluation of lizard social systems, which is perhaps applicable to other taxa, as well. Finally, in chapter 4, William Cooper, Jr. employs an optimality model approach to interpret the results of clever field experiments (by himself and others) designed to examine theoretical trade-offs between the necessity to evade predators and the need for social and mating interactions.

As in all productive research, answering one set of questions leads inevitably to the emergence of others, with ever-increasing diversity and complexity. No issue has bedeviled behavioral science more than the distinction between proximate, mechanistic causation and ultimate, evolutionary causation, made so difficult to handle because the two issues are closely linked and interdependent in some respects, and independent in others. In recommending lizards for their tractability as subjects for field experimentation, Schoener (1977, p. 39) draws a bead on the proximate-ultimate distinction by reminding us that "experimental perturbation is better suited to provide information on the mechanisms maintaining differences than on the original causes of such differences. Failure to make this distinction is responsible for much of the present controversy on whether ecological overlap or its absence implies competition. The answer depends in part on whether we mean past or present competition." He is focusing here on the issue of niche overlap, pointing out that "genuine resource overlap between resource-limited species must imply competition, but its absence may imply an effective, evolutionary cure has been found for past competition." There is a message of profound importance here for all who seek to test theories of social evolution. To be fully meaningful, investigations that strive to explain the past must eventually be linked and compatible with a thorough understanding of the present. If ultimate causes accomplish evolutionary change by modifying proximate causes, it behooves us to devote equal attention to both.

Another theme due for more attention is that of phenotypic plasticity (West-Eberhard 1989; Schlichting and Pigliucci 1998; Marler 1999). We are still all too ready to assume, implicitly if not explicitly, that a given genotype encodes instructions for a single trajectory toward a given phenotype, to be perturbed and diverted to varying degrees by environmental influences. In fact, the more typical case is for the provision of multiple pathways. The developing organism is guided toward one or the other by the particular patterns of environmental stimulation it encounters as an individual, to which its responsiveness is often specific and genetically preordained. Thus, demonstrations of variability in a phenotype may have no place in an antigenetic argument, whether our aim is to explore the ontogeny of a behavioral trait, or whether we are focused on its evolutionary history. This is a caveat worth bearing in mind if we assume that some populations possess "genetically-based mating strategies," and others lack them (chapter 3). The very communality across organisms in how changes in social organization correlate with similar changes in their environments, noted at the beginning of this introduction, would seem to argue in favor of shared genetic mechanisms. However impatient some may be with the nature-nurture controversy, it is still with us, and must be addressed more aggressively in the future if progress is to be made. What role lizard studies are destined to play in this adventure is a question for posterity.

CHAPTER 1

Intra- and Intersexual Variation in Social Behavior

Effects of Ontogeny, Phenotype, Resources, and Season

Troy A. Baird, Dusti K. Timanus, and Chris L. Sloan

Sexual selection is a powerful agent influencing the evolution of social behavior and mating systems in many phyletically diverse vertebrates (Vehrencamp and Bradbury 1984; Anderson 1994; Arnold and Duvall 1994; Wade 1995). It is widely appreciated that sexual selection can result in marked differences in the social behavior exhibited by adult males and females (Trivers 1972), and, more recently, it has become apparent that sexual selection may also be responsible for significant behavioral variation within each sex (Caro and Bateson 1986; Moore and Thompson 1990; Moore 1991; Gross 1996). Especially in males, sexual selection may result in asymmetries in both behavioral and morphological attributes that are important in competitive contests for mates and resources, and mating preferences by females (Andersson 1994). Finally, temporal variation in the frequency and intensity of sexually selected social behavior, such as intrasexual competition and courtship, may reflect the changing costs and benefits of these behavior patterns during the course of the breeding season (Ruby 1978; Orians and Wittenberger 1991; Stamps 1994).

Detailed field studies of populations in which the gender, age, size, and reproductive condition of individuals are known have been especially powerful in elucidating behavioral variation within populations and the functional advantages for individuals that adopt different behavior patterns (e.g., Clutton-Brock, Guiness, and Albon 1982; Davies 1992; Sinervo and Lively 1996). The objective of this chapter is to use observational and experimental data recorded on free-ranging collared lizards of known gender, age, size, reproductive condition, and social history to test predictions of several hypotheses that stem from theoretical and empirical studies of sexual selection. We review selected literature to develop each hypothesis that we test and then outline the expected consequences for the social behavior of individuals. In the next section, we use our data from field studies on eastern collared lizards, *Crotaphytus collaris,* to address these hypotheses. This chapter focuses primarily on the action of intrasexual selection within one population of collared lizards. Chapter 5 considers variation in the environmental potential for both intra- and intersexual selection among three populations of this species.

The most obvious consequence of sexual selection for variation in social behavior is that adult males and females often are selected to adopt fundamentally different tactics to maximize reproductive success (Trivers 1972; Vehrencamp and Bradbury 1984). Most basically, differences in how the sexes maximize reproductive success stem from different patterns of parental investment (Trivers 1972) together with differential rates of gamete production (Bayliss 1981). Female parental investment per gamete is relatively high, and the reproductive success of individual females is generally limited more by the costs of producing and caring for eggs and offspring than by access to males (Bateman 1948; Williams 1966; Trivers 1972; Alexander and Borgia 1979; Andersson 1994). Although females may increase their fitness by mating preferentially with males that possess certain attributes (Fisher 1930; Trivers 1972), females likely gain less from mating with multiple partners than males do, and whether or not mature females reproduce is usually not limited by access to males. Rather than engaging in intrasexual competition for males, females usually maximize efforts to secure the resources that they require for production and survival of eggs and offspring (e.g., Hoffman 1983; Baird and Liley 1989). Maximizing parental investment sometimes involves competing intrasexually for resources and/or choosing a male mate that controls high-quality resources (e.g., Howard 1978b; Searcy 1979; Jones 1981; Yasukawa 1981).

In contrast with females, male parental investment/gamete is relatively low, male fertility is usually not limited by the number of sperm that they can produce, and males often maximize reproductive success by competing to mate with as many females as possible (Trivers 1972; Andersson 1994). As a consequence, the reproductive success of a given male may be highly dependent on his ability to outcompete other males and/or attract females, and variance in the mating success of males is often high relative to that of females (Bateman 1948). Intrasexual male competition for mates often results in selection for agonistic social behavior patterns (Le Boeuf 1974; Otte and Stayman 1979; Shine 1979;

Warner 1984; Hews 1990), establishment of hierarchical social rankings (Dewsbury 1982; Ellis 1995), defense of space and resources (Stamps 1994), and exaggerated morphological traits (Geist 1978; Eberhard 1979; Clutton-Brock, Guiness, and Albon 1982). Polygynous social organizations characterized by male defense of breeding territories and pronounced asymmetries in male mating success evolve when the ability of males to compete for and attract females varies and the distribution of resources and/or females favors mate monopolization by some males (Emlen and Oring 1977). Because asymmetries in male mating success can be pronounced in territorial polygynous species, these systems have proven fruitful for examining phenotypic correlates of high male mating success (Le Boeuf 1974; Trivers 1976; Borgia 1979; Warner and Hoffman 1980; Stamps 1983b; Arak 1988; Jarman 1988).

Polygynous social organizations and territorial defense are common in diurnal lizards (Stamps 1983b, 1994), and sexual selection has been implicated in the evolution of social behavior in several species (e.g., Ruby 1984; Vitt and Cooper 1985a; Carpenter 1995b; Marler et al. 1995; McCoy 1995; Wikelski, Carbone, and Trillmich 1996; Olsson and Madsen 1998). We tested the hypothesis that the social behavior of adult male and female collared lizards has evolved by intrasexual selection by comparing the types of behavior exhibited by each sex, the frequencies of agonistic behavior patterns, and the extent to which adult males and females compete with consexuals for mates or resources. If male collared lizards compete for access to females through territoriality, then they should maximize behavior associated with advertisement and defense of space from other males. By contrast, if the reproductive success of females is limited by the energy and nutrients necessary to produce eggs, then females are expected to compete intrasexually for access to food and/or other resources, and the intensity of intrasexual female competition should depend on the availability of resources. We tested these predictions by comparing the social behavior of female collared lizards with that of males, and by comparing female social and reproductive behavior, growth, survival, and responses to experimental manipulation of social conditions in two populations in which access to resources varied owing to substantial differences in habitat structure.

In species with polygynous mating systems, there is considerable evidence from diverse taxa that male mating success may be skewed in favor of males with certain behavioral and/or morphological attributes including large or high-quality territories, frequent or intense displays, large body size or peak physical condition, and bright coloration (e.g.,

Clutton-Brock, Guiness, and Albon 1982; Ruby 1984; Kodric-Brown 1985; Kennedy et al. 1987; Andersson 1994; Kodric-Brown 1995). Males that possess such characteristics may be more successful because they prevail in intrasexual agonistic encounters for females or necessary resources, are more attractive to females, or both. Asymmetries in male mating success may result when females prefer to mate with males that display certain attributes either because such traits signal superior male genetic qualities (Zahavi 1975; Hamilton and Zuk 1982) or because female bias for male attributes unrelated to genetic quality result in strong directional selection for these male traits (Fisher 1930). Female preferences for males displaying certain traits have also been supported with empirical studies in a wide range of vertebrate taxa (e.g., Sullivan 1983; Borgia 1985; Andersson 1986; Basolo 1990; Baird, Fox, and McCoy 1997). We examined which attributes promote high mating success in adult male collared lizards by recording detailed behavioral and morphological data on males in the field and using multivariate analyses to examine correlates of annual male mating success that we estimated from observations on courtship interactions among known males and females.

Although sexual selection often favors aggressive tactics by some individuals (usually larger and older) to maintain high social status and dominate mates or resources, younger and smaller individuals that are not physically equipped to obtain matings by direct competition may be selected to adopt alternative social and mating tactics (Caro and Bateson 1986; Moore 1991; Gross 1996). Alternative mating tactics sometimes evolve in females (Eadie and Fryxell 1992). However, the evolution of alternative mating tactics appears most pronounced among males because intense intrasexual competition for females is more common among males (reviewed by Moore [1991] and Gross [1996]). When a relatively few high-ranking males are able to monopolize the majority of mating opportunities through agonistic activity, lower-ranking males either may defer reproductive efforts until they can successfully compete in agonistic contests or may adopt one or more behavior patterns that circumvent the aggressive tactics of higher-ranking males (reviewed by Lott [1991]). Deferred male reproductive efforts and/or the evolution of alternative male tactics appears especially likely when dispersal ability is reduced, limiting opportunities for low-ranking males to move to areas where they can acquire mates immediately by adopting aggressive tactics (Shapiro 1991; Warner 1991; see chapter 3).

Alternative male mating behavior appears to fall into two categories depending on whether or not individuals can switch between tactics

(Moore 1991; Gross 1996). In some cases, the different social and mating tactics are two or three fixed alternatives (Gross and Charnov 1980; Gross 1982, 1983; Mason and Crews 1985; Moore and Thompson 1990; Sinervo and Lively 1996). In species in which alternatives are fixed, males become sexually mature as one phenotype, and the alternative phenotypes develop distinct and immutable morphological traits and behavioral tactics. By contrast, it may be more common for alternative male phenotypes to be plastic (Howard 1978a; Orians 1980; Beletsky, Orians, and Wingfield 1989, 1990; Cardwell and Liley 1991). In species with plastic alternative male phenotypes, young individuals generally adopt subordinate or sneaking tactics when they first become reproductively mature, and they later switch to socially dominant tactics when larger and older. While exhibiting subordinate behavior, young males may display coloration that is less intense than that of aggressively dominant males (Howard 1978a) and, in some cases, may even mimic the appearance of females (Gross 1982, 1983; Warner 1984).

We tested the hypothesis that alternative social tactics are displayed by both male and female collared lizards by examining the behavior of individuals of different ages and sizes. Moreover, we tested whether the alternate social tactics of individuals are fixed or plastic by conducting field experiments involving controlled manipulation of the social composition of collared lizard groups.

Competition for mates among polygynous males usually involves defense of territories (*sensu* Wilson 1975; Stamps 1983b, 1994), which almost always carries costs (Brown 1964; Brown and Orians 1970; Davies and Houston 1984; Carpenter 1987; Hixon 1987), such as the time and energy required to patrol and defend territory borders (Moore and Marler 1987; Marler and Moore 1988, 1989, 1991). Territorial defense may also heighten predation risk owing to increased conspicuousness during advertisement (e.g., Tuttle and Ryan 1981; Ryan, Tuttle, and Rand 1982), increase the risk of injury during agonistic contests (e.g., Clutton-Brock et al. 1979), and detract from maintenance activities such as foraging (Wittenberger 1981). Given the potential costs of spatial defense, it is reasonable to hypothesize that behavior patterns that reduce one or more costs may be favored by selection (e.g., Krebs 1982; Fox and Baird 1992). Although territorial behavior has a long, rich history of research by behavioral ecologists, relatively little is known about the temporal dynamics of agonistic activity over the period that territories are established and defended (Stamps 1994; but see Hews [1993] and Jenssen, Greenberg, and Hovde [1995]). In cases in which

males are polygynous through territorial defense, in addition to repelling same-sex competitors, males must court and mate with females
when their eggs are receptive for fertilization. Therefore, to be successful, territorial males must effectively budget their social energies between the defense of territories and courtship.

A common expectation for the temporal pattern of agonistic activity among territorial males is that the intensity of agonism will be highest at the beginning of the breeding season when territory boundaries
are being established and become less frequent and intense later when
males become more involved in courtship and mating (Nolan 1978;
Stamps 1990). This pattern is supported by field data on some species;
however, these appear to all be cases in which individuals are establishing territories in new areas either because they disperse at the conclusion of breeding and there is little interseasonal fidelity to the same
areas, or because they are species that survive only for 1 yr. A different
pattern of seasonal variation in territorial activity might be expected in
species with low dispersal ability that remain on the breeding grounds
during the nonreproductive season, display a high level of interseason
fidelity to the same territories, and have a breeding season that is long
enough for females to produce multiple clutches of eggs. High interseasonal fidelity to the same breeding areas may allow males to reestablish
occupancy of territories without frequent and intense agonism at the
beginning of the season. Furthermore, establishment of territories may
not involve high levels of aggression if numerous males arrive at (or
emerge onto) the breeding grounds simultaneously (Stamps 1992). Establishment of territories with only low levels of agonism could allow
males to direct more of their social energies early in the season toward
courtship. When females produce multiple clutches, early courtship and
insemination may facilitate production of additional eggs that male territory owners likely would also have the opportunity to fertilize. We
tested the hypotheses that adult male collared lizards establish occupancy of territories without frequent agonism and instead spend more
of their social energy courting females early in the season by comparing
the frequency of territorial and courtship activities during six observation periods distributed throughout the breeding season.

Although females are generally less aggressive than males, vertebrate
female social organizations that are mediated through intrasexual aggression ranging from social dominance hierarchies (Robertson 1972;
Stamps 1983b; Rodda 1992) to defense of exclusive territories (Stamps
1973, 1978, 1983b; Ruby and Dunham 1987; Baird and Liley 1989;

Mahrt 1998a, 1998b) have been described. In many vertebrates, female aggression is associated with defense of eggs, offspring, and nests. If defense of offspring is the function of intrasexual aggression, then females should be most aggressive after they have laid their eggs, and aggression should be most frequent near nesting locations. Alternatively, females may be more aggressive after oviposition not because they are defending progeny and/or nests, but because high levels of physical activity are more cost-effective at this time. In female iguanians, clutches may weigh up to 50% of total body mass (Schwarzkopf and Shine 1992), and the eggs become increasingly heavy prior to oviposition. Because aggressive behavior usually involves rapid locomotory behavior, the energetic cost of defense may be prohibitively high when females are gravid. Increased body mass impairs locomotion of gravid females in some lizard species (Cooper et al. 1990). Therefore, selection may favor female aggression following oviposition because it is more cost-effective after eggs have been laid. Moreover, because the construction of nests and oviposition probably detract from a female's ability to defend resources, a high level of aggression following egg laying may be important for females to reestablish social dominance over their home ranges. Female aggression is expected to be highest following oviposition whether aggression functions in the defense of eggs, or because agonism is less costly following oviposition. However, defense of eggs predicts that aggression should occur near nests, whereas agonistic social encounters should be independent of nesting locations if higher aggression following oviposition occurs as a consequence of reduced energetic costs. We tested these hypotheses by mapping the social activities of females on a daily basis, prior to and following oviposition, in individuals for which we knew the locations of their nests.

Empirical Studies
Study Animal and General Field Techniques

To test the hypotheses we have described, we conducted field studies on the eastern collared lizard, *C. collaris* (Sauria, Crotaphytidae) in central Oklahoma. This species is a diurnal iguanian that is restricted to exposed rocky substrata throughout the central and southwestern United States (Fitch 1956). Collared lizards are sit-and-wait predators that use visual cues to detect their prey, which are primarily arthropods (Blair and Blair 1941; Best and Pfaffenberger 1987; Husak and McCoy 2000). In central Oklahoma, collared lizards emerge from hibernation in late March or early April. Intrasexual and intersexual social and reproductive activity occurs from late April through July 15, during which females produce and lay one to three clutches of eggs (Baird, Sloan, and Timanus 2001). Following oviposition of the last clutches in late June and early July, social interaction and general activity gradually diminish until adult lizards become inactive in late August or early September. Emergence of hatchlings begins in late July, and continues into October until temperatures become too cold for lizards to be active.

Crotaphytus collaris is an excellent model system in which to conduct field studies on the influence of sexual selection on the social behavior of individuals of both sexes and of different ages. Because numerous lizards are restricted to relatively small patches of exposed rock, both competitive intrasexual and courtship interactions can be observed directly. Neither sex invests parentally beyond the production of fertilized eggs. Therefore, the reproductive success of males is expected to be limited by factors that influence access to females, whereas female reproductive success should be limited primarily by access to a dependable food supply that provides sufficient energy and nutrients to manufacture eggs (Rostker 1983).

We conducted most of our studies on lizards occupying three patches of granite boulders that were imported to construct the flood-control spillway at the Arcadia Lake Dam (AL) located on State Highway 66, 9.6 km east of Edmond, Oklahoma (Baird, Acree, and Sloan 1996). We conducted additional studies of female lizards on a separate population (Hranitz and Baird 2000) living on five patches of sandstone at the Morningside Farms Ranch (MS) 2.6 km north of Arcadia, Oklahoma. In 1991, we first mapped study sites by measuring distance and angles among mapping markers, and later, in 1994, using global positioning satellite measurements that are accurate to the nearest meter. We conducted mark-recapture studies on lizards from their hatchling seasons (late July through October), throughout their lives at AL from 1991 through 1999, and from 1993 through 1997 at MS. By monitoring known individuals through time, we know the ages of almost all individuals at both sites. We toe-clipped individual lizards for permanent identification and gave unique color markings by applying spots of acrylic paint to the dorsum. At each capture, we recorded the snout-vent length (SVL) and body mass, and for females we palpated the abdomen to monitor the development of eggs.

We recorded behavioral data on scale-drawn maps of rock patches over which we established mapping markers in 10-m grids. We recorded space use and social behavior during daily censuses and focal observations of known individuals (Baird, Acree, and Sloan 1996; Baird and Timanus 1998; Sloan and Baird 1999). Daily censuses involved recording the static locations of all emergent lizards while observers walked a routine path around study sites. Focal observations involved tracing the movement path of one lizard for 20 min during which the location, outcome, and participants in all social behavior (described subsequently) were recorded on maps. From several focal observations per individual (for details see Baird, Acree, and Sloan [1996], Baird and Timanus [1998], Timanus [1999], and Baird, Sloan, and Timanus [2001]), we calculated the hourly frequency of behavior patterns (or travel) per individual by dividing the cumulative number of acts by the total observation time. We recorded varying numbers of focal observations on different numbers of lizard subjects to accomplish the specific objectives of each study; these are given in the next sections. Unless indicated otherwise, we made statistical comparisons using t-tests and parametric analysis of variance (ANOVA) when data met the requirements for parametric procedures, and Mann-Whitney U-tests and Kruskal-Wallis ANOVA when data did not meet the requirements for parametric statistics.

Contrasts in Social Behavior of Adult Males and Females

The social behavior of adult male and female collared lizards differed in several ways that are consistent with expectations of sexual selection theory. Differences between the sexes were first apparent in the earlier emergence from hibernacula by males. To examine emergence dates quantitatively, from 1993 to 1997 we regressed the cumulative percentage of lizards captured that were male against first capture date during the first 10 d that collared lizards were emerging. For this analysis, we used females that had an SVL ≥ 65 mm and males that had an SVL ≥ 70 mm because lizards of these sizes are reproductively mature during their first (yearling) season. We captured $>95\%$ of all emergent lizards sighted, and although it is possible that we failed to detect some emergent lizards, it is unlikely that these instances were biased sexually. A statistically significant negative relationship (slope $= -5.57$, $t = 12.2$, $P < 0.001$) between the percentage of lizards captured that were male and later collection date indicates that male collared lizards emerged earlier than females.

Both the type and relative frequency of social behavior differed between adult males and females in a pattern consistent with expectations. By the beginning of their second year, most (76 of 79 from 1991 to 1998)

males exhibited strongly developed territorial behavior. Males adver-
tised territories primarily by high rates of patrol, and by performing visually transmitted displays when they were distant (>5 m) from other
lizards (Baird, Acree, and Sloan 1996). During patrol behavior, territorial males typically moved among a series of rock perches and stopped
for short periods to perform displays. The most prominent displays
consisted of lateral compression of the torso while elevating on all four
legs and simultaneously extending the brightly colored dewlap. While in
this position, males performed push-ups by alternately straightening
and flexing the legs, which elevated and lowered the torso (Carpenter
1978b; Rostker 1983). More subtle displays involved vertical bobbing of
the head while the torso was held stationary, or brief extension of the
dewlap without lateral torso compression. Occasionally, males displayed
by walking in a circular or figure-eight pattern one to two body lengths
in total diameter, while remaining on a single perch.

Less frequently (Table 1-1) (Baird, Acree, and Sloan 1996), intrasexual male aggression involved proximal encounters consisting of two
males exchanging a series of displays or more intense acts while in close
(within 1 m) proximity to one another. Encounters began with an approach in which one male moved toward another in a deliberate fashion, which resulted in the opponent either yielding or standing his
ground and confronting the approaching male. Chases ensued when the
recipient of an approach retreated, and the approaching male pursued
at a high rate of speed, sometimes running bipedally. Long chases continued for up to 60 m, but most were terminated by the pursuing male
after a shorter distance. Chases were sometimes terminated when the
pursued male (particularly smaller yearlings) sought refuge under
rocks. When approached males did not yield, both males performed lateral displays and push-ups while maintaining close proximity to their
opponent, and sometimes moving in a circular pattern around one another (Baird, Acree, and Sloan 1996; Baird and Timanus 1998; Baird,
Slaon, and Timanus 2001). During such standoffs, males occasionally
attacked by lunging toward their opponent with their jaws opened wide
to bite. Bites sometimes produce skin lacerations and fractures of the
jaw or leg bones. Such escalated aggression was rare, however, because
competitively inferior adult males (and smaller yearling males) usually
avoided potentially damaging encounters by fleeing.

Although females in our populations also exhibited intrasexual agonism and some researchers have reported that female collared lizards
defend territories (Yedlin and Ferguson 1973; Rostker 1983), our studies

Table 1-1 Mean (± 1 SE) rate of travel (patrol in adult males, m/h), home range (females) or territory (males) area (m²), hourly frequencies of displays distant (>5 m) from other lizards, and proximal encounters in which adult male and female lizards moved to within 1 m of one another from 1991 through 1998 at AL and MS study sites. Data for 1991–92 are from Baird, Acree, and Sloan (1996), for 1994–97 from Baird and Sloan (unpublished data), and for 1997–98 from Timanus (1999). Statistical comparisons between sexes were performed in which data on both sexes were available from the same observation periods, and for all years pooled.

	Females					Males				
Year/Site	N	Travel (m/h)	Area (m²)	Hourly Displays	Hourly Encounters	N	Patrol (m/h)	Area (m²)	Hourly Displays	Hourly Encounters
1991–92, AL	23	19.5 ± 3.0	358 ± 37	0.93 ± 0.51	0.24 ± 0.07	15	75.9 ± 7.4*	1,311 ± 175**	71.5 ± 12.8*	0.36 ± 0.10
1995–96, AL	27	18.5 ± 2.7	183 ± 25	0.72 ± 0.21	0.18 ± 0.06	—	—	—	—	—
1994–97, MS	27	34.0 ± 4.9	140 ± 21	0.98 ± 0.23	0.41 ± 0.13	—	—	—	—	—
1997, AL	5	18.6 ± 3.0	304 ± 94	0.60 ± 0.47	0.20 ± 0.12	14	71.3 ± 6.5**	1,229 ± 246**	100.6 ± 10.9**	0.20 ± 0.05
1998, AL	—	—	—	—	—	15	57.0 ± 6.6	1,360 ± 218	92.0 ± 10.5	0.23 ± 0.07
All data pooled	82	26.1 ± 2.2	224 ± 19	0.86 ± 0.17	0.27 ± 0.05	44	67.9 ± 4.1**	1,302 ± 121**	87.0 ± 6.8**	0.27 ± 0.05

*$p < 0.05$.
**$p < 0.01$.

show that intrasexual agonistic behavior patterns of females were clearly different from those of territorial males. Female travel differed from male patrol in that females did not give frequent displays while distant from conspecifics (Table 1-1) (Baird, Acree, and Sloan 1996). In particular, females did not perform the distinct push-ups that were performed very frequently by patrolling males. Instead, intrasexual female aggression much more frequently took the form of proximal encounters, with >70% of female agonistic acts occurring during proximal encounters compared with only 18% in adult males (Fig. 1-1). Encounters began with an approach, which, like that in males, often resulted in chases if the recipient female yielded. However, if the recipient female stood her ground, the two opponents exchanged displays consisting of lateral torso compression, extension of the dewlap, and staccato bobbing of the head. Most encounters ended without further escalation; however, some escalated to biting and grabbing onto one another violently (Sloan 1997; Baird and Sloan, unpublished data).

Quantitative examination of the frequencies of agonistic behavior clearly indicates that adult males had much higher rates of patrol and advertisement displays than did females. We made statistical comparisons between adult females (those producing and laying at least one clutch of eggs) and adult males (2 yr and older) separately for two periods (1991–92 and 1997) for which we recorded data on both sexes, and then using pooled data from 8 yr and two populations. Adult males patrolled throughout their territories at 2.6 to 3.9 times (P values < 0.05) the rate that females traveled their home ranges (Table 1-1). Not only did males travel at a higher rate, but their average rate of display in some cases was more than two orders of magnitude higher (P values < 0.05) than that of females (Table 1-1). By contrast, intrasexual proximal encounters between males were infrequent relative to displays, and the frequencies of intrasexual proximal encounters by the two sexes were not different in any of the comparisons. Adult males also utilized larger areas than females. On average, in 1991–92 and 1997, adult males used areas that were 3.7 and 4.0 times larger (P values < 0.01), respectively, than those of females, and pooling data from all years revealed that males used 5.8 times the area used by females (P < 0.01) (Table 1-1). Furthermore, patterns of intrasexual overlap were clearly different between adult males and females. In 1991–92 and 1997, the average percentage of overlap between neighboring males (expressed as the percentage of each male territory) was only 15.8% (SE = 4.3) and 8.7 (SE = 3.0), respectively, whereas overlap between adjacent females for

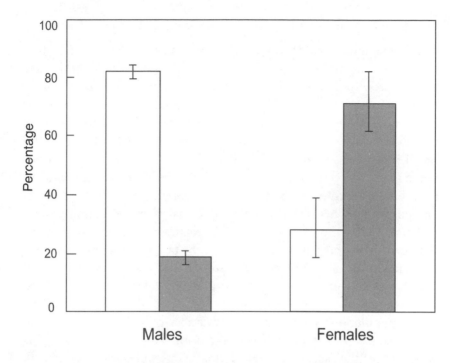

Figure 1-1 Mean percentage (± 1 SE) of total intrasexual agonistic acts performed distant (>5 m) from other lizards (□) and during proximal encounters (■) in adult males and
females during 1991–92. Data from Baird, Acree, and Sloan (1996).

these same periods was 40.0% (SE = 2.4) and 30.6% (SE = 3.8). When
we pooled these data, we found that the mean overlap between adjacent
males (12.6%, SE = 2.7) was less than one-third ($P < 0.0001$) that between adjacent females (39.4%, SE = 2.0).

Higher rates of patrol and advertisement over relatively large areas
that overlapped little with the areas patrolled by other adult males are
consistent with the hypothesis that adult males are polygynous through
territorial defense. By contrast, females occupied smaller areas, did not
patrol like males, and agonistic behavior was much more restricted to
proximal encounters. The behavior of females relative to that of males,
therefore, was also consistent with the hypothesis that females compete
intrasexually for one or more resources. In the following sections, we
further examine the distribution of mating success among polygynous
adult males and phenotypic correlates of high male mating success and
test the hypothesis that intrasexual female aggression functions in resource competition.

Courtship and Phenotypic Correlates of Adult Male Mating Success

The home ranges of reproductively active adult females were either completely or partially overlapped by the territories of one to three adult males. In 1991–92, for example, the home ranges of 47 of 81 (58%) AL females were overlapped by one adult male, whereas the home ranges of 25 (31%) and 9 (11%) females were partially overlapped by two and three adult males, respectively (Baird, Acree, and Sloan 1996). On average, adult males initiated 1.78 courtship encounters/h (SE = 0.22) with female residents. Courtship involved males approaching to within one body length while perfoming lateral displays and push-ups. When approached by adult males, females almost always responded by giving one or more displays, including extension of the dewlap, raising the proximal portion of the tail, lateral compression of the torso, and walking or hopping around the male using a stiffened gait. Although some of these behavioral patterns have been interpreted as female rejection of males (Mosely 1963; Yedlin and Ferguson 1973), in our populations, females almost always gave these displays prior to engaging in a variety of courtship behavior lasting 1–30 min during which either the male or the female maintained physical contact with his or her partner. Our observations, therefore, are not consistent with the hypothesis that these female displays signal rejection of males.

Courtship behavior involved physical contact consisting of either lizard sitting on or superimposing part of his or her body across the dorsum of his or her partner, the two lizards sitting together while touching side to side or snout to snout, or the lizards turning circles in tandem while maintaining contact. During some courtship encounters, males grasped females by the loose skin of the neck and attempted to juxtapose their vents with those of females, presumably to copulate. Attempts at copulation often resulted in the two lizards moving into rock crevices where we could not observe them without disturbing their activities. Therefore, we could not reliably determine whether mounting resulted in successful intromission, and because the home ranges of some females were overlapped by more than one adult male territory, we could not use intersexual overlap patterns alone to assign mating partnerships. Instead, we assigned mating partnerships using the relative frequencies of courtship activities among individual males and females (discussed later). Using this approach, we estimated that annual mating success of adult males ranged from one to seven females (Baird, Acree, and Sloan 1996; Timanus 1999).

To examine patterns of variation in adult male behavioral and morphological traits and their possible relationships with estimated male mating success, we recorded social behavior during focal observations on males ($n = 39$) of known size and age. Because the estimated mating success of adult males far exceeded that of yearlings (Baird, Acree, and Sloan 1996), we focused only on adult males for these analyses. To examine male phenotypic variation, we measured SVL, tail length, body mass, head width and length, frequency of display, rate of patrol, and territory area and performed principal components analysis (PCA) on these variables using a correlation matrix (Timanus 1999). PCA reduces the dimensions of the measured variables to a smaller number of synthetic components that are linear combinations of the original variables (Digby and Gower 1986). Only components with eigenvalues >1.0 were included because smaller eigenvalues account for a negligible amount of variance relative to any variable (Digby and Gower 1986). We also used Munsell charts (Baird, Fox, and McCoy 1997) to measure the hue, value, and chroma of six body areas on males in the field in full sunshine (lizard body temperatures = 33–38°C) and examined variation using a separate PCA analysis. We could not include the color measurements in a single PCA with the size and behavioral variables because this would have violated the requirement that the number of data points (individual males) be less than or equal to the number of variables measured.

To estimate annual male mating success, we determined the number of courtship encounters by all adult males with each female (those that produced at least one clutch), and then used the proportion accounted for by each male as his probablity of inseminating individual females (after Clutton-Brock, Guiness, and Albon [1982]). We then summed these probablilities to estimate the number of female mates for each male (Baird, Acree, and Sloan 1996; Timanus 1999).

The high loading variables from PCA of our color data failed to reveal any clear patterns with respect to hue and/or brightness in particular body areas as important correlates of mating success (Timanus 1999). By contrast, PCA of the body size and behavioral variables constructed two synthetic factors (eigenvalues >1.0) that explained a total of 64.7% of the male variability defined by the eight variables. Moreover, the five body size variables all loaded highest together on factor 1, whereas all three behavioral variables related to male use of space loaded highest together on factor 2. To examine the relative importance of body size and space use as factors that influence the ability of males to acquire matings, we divided males into low and high mating success groups using values from the original data set, and then graphed the corresponding body size/space-use scores (Fig. 1-2). The majority of males with high mating success estimates had high space-use scores and males with low mating success had low space-use scores (Fig. 1-2), whereas no such pattern was apparent for males with high versus low body size scores (Timanus 1999). Moreover, mating success estimates for large males were not statistically different from those of small males, whereas the mating success of males with high space-use scores were significantly higher than those of males with low space-use scores, even after the effects of body size were removed using analysis of covariance (ANCOVA) (Timanus 1999). To understand which body size and behavioral variables associated with space use promote high male mating success, we used the univariate methods of Endler (1986) to calculate measures of directional and stabilizing or disruptive selection. The results of these analyses revealed a statistically significant selection differential only for directional selection on the frequency of displays by adult males (Timanus 1999).

Our data strongly suggest that high male success is explained by variation in behavioral parameters associated with territorial defense rather than body size parameters. All of the adult males examined were large enough and old enough to have obtained territories; there was relatively little variation in size among these males, with the SVLs of the smallest adults being 85% of the largest males; and there was no directional selection on any of the body size variables. By contrast, the display and patrol rates of the least active males were <19% of the most active males, and calculation of selection differentials revealed significant directional selection on male display rate. Because the PCA space-use factor consisted of three variables that were highly intercorrelated statistically, males may achieve high mating success by maximizing any or all of these behavioral parameters.

Controlling a large territory may promote high mating success if the number of females that males can monopolize is positively correlated with male territory area. However, this model assumes that females are distributed equally homogeneously over all habitat patches where males control territories. If, instead, females were more concentrated in some areas than in others, like they were at AL, some males might be able to consort with large numbers of females by controlling relatively small areas. Variation in the spatial distribution of females, therefore, may confound the expected positive relationship between male territory area

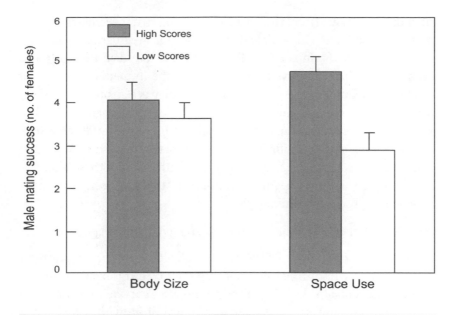

Figure 1-2   Mean (± 1 SE) estimates of annual mating success in males with high and low PCA body size scores and high and low space-use scores. Data from Timanus (1999).

and mating success such that even though territory area loaded high on the PCA space use factor, selection differentials did not reveal significant directional selection for large territories. Maintenance of exclusive territories requires owners to patrol territory boundaries. Therefore, because maintenance of a large area requires a high rate of patrol, the strong correlation between these two variables is not surprising, and high patrol rate as a determinant of high mating success is sensitive to the same potential confound as territory area.

Although the rate that males performed displays was statistically correlated with rate of patrol, having a large territory and a high patrol rate does not necessarily require a high rate of display. For example, it is theoretically possible for males to display at a high rate while remaining within a relatively small territory. Males could also maintain a large territory without high rates of display if they immediately attacked intruders, rather than displaying from a distance. Frequent attacks on intruders would almost certainly carry very high energetic (Marler and Moore 1991; Marler et al. 1995) and perhaps injury and/or predation costs, probably selecting against such a tactic. Displays that are visually conspicuous from a distance likely function to discourage intrusions by competitor males without the high costs of frequent attacks. However, such displays may also play a role in attracting mates. Females may be more likely to mate with a male that they see on a regular basis displaying at a high rate (Stamps 1977a; Ruby 1981, 1984; Kodric-Brown 1995), even though these displays do not occur in the immediate context of courtship encounters; the rate that males display is a strong correlate of reproductive success in guppies (Bischoff, Gold, and Rubenstein 1985). Although successful collared lizard males often have large territories and high rates of patrol, our finding that display rate was the only variable under direction selection is consistent with the hypothesis that frequent display may be the most important attribute promoting high male mating success in this species.

Influence of Resource Availability on Social Behavior of Females

We tested the hypothesis that intrasexual female aggression functions in competition for resources by comparing rates of intrasexual aggressive encounters, female social organization, growth, survival, reproductive effort, and responses to experimental removal of individual females at AL and MS, where the availability of elevated foraging perches and refugia varied markedly. At AL, lizards occupied relatively large (0.5- to 2-m greatest diameter), loosely piled boulders that were used to construct the banks of the flood-control spillways behind and adjacent to the dam (Baird, Acree, and Sloan 1996). The interface of these boulder patches with the surrounding grass is nearly a straight edge, and virtually every rock along the interface is an elevated perch from which lizards can scan the vegetation for arthropod prey. Sticky-trap sampling revealed that potential arthropod prey were 4 to 38 times more abundant along this interface than within rock patches. On average, AL females were perched along the interface during nearly 80% of censuses and females made many foraging strikes into the grass from these rocky perches (Baird and Sloan, unpublished data). Because the boulders at AL are loosely piled in layers, there are also abundant crevices and interstitial spaces that the females used as refugia to which they could readily retreat no matter where they were perched. By contrast, at MS lizards occupied natural outcrops and washes of sandstone each of which had an irregular interface with the surrounding grassland. The abundance of arthropods in sticky traps along the rock-grass interface at MS was similar to that at AL. However, at MS the average number of elevated perches adjacent to the rock-grass interface was only 3% that available at AL. Loose rocks with crevices that the lizards used as refugia were also significantly more scarce at MS than at AL (Baird and Sloan, unpublished data).

Females at MS initiated aggressive encounters more than twice as frequently as AL females. Also, although our sample size was small, introduction of tethered consexual intruders 5 m from females elicited aggressive responses in 5 of 6 trials at MS compared with only 4 of 17 trials at AL (Baird, Acree, and Sloan 1996). These differences prompted us to compare further the social organization of females at the two sites. We calculated a dominance index (DI) as the percentage of total encounters won (female chased or displaced her opponent that yielded) against each opponent with which a given female was observed to interact. The percentage of total encounters won was then averaged over all pairs of interacting females to yield an average dominance ranking (Hoffman 1985; Sloan 1997). Fifty percent of the AL females occupying adjacent home ranges were never observed to interact aggressively, whereas only 22% of adjacent MS females did not interact. Most (90%) of the encounters at MS resulted in decisive winners and losers. Encounters among AL females, however, usually involved only low-intensity displays rather than chases resulting in clear displacement of an opponent. Together, the low frequency and indecisive nature of the aggressive encounters at AL strongly suggest that neighboring females did not have clearly different social rankings (Baird and Sloan, unpublished data).

By contrast, calculation of DIs for MS females revealed asymmetries in social rank influenced by body size and age among the females occupying adjacent home ranges on the same patches of rock habitat (social groups). DIs within social groups were clearly discontinuous, with the DIs of dominant females in three groups at least twice those of subordinate females, and 30% higher in the other social group (Sloan 1997). Dominant females initiated aggressive encounters nearly six times more frequently than subordinate females. Adult females (2 yr and older) were almost always dominant over yearling (first-year) females, with the average DI of adults significantly higher than that of yearlings. In >60% of encounters between adults, encounters escalated to attacks and biting, whereas encounters between adults and yearlings involved physical contact in only 12% of encounters because yearlings usually fled immediately. In all social groups with only one adult female, that female had the highest DI. In social groups containing more than one dominant female, each had a DI that was high relative to those of subordinate females (Sloan 1997).

Overlaps between the home ranges of adjacent subordinate females as well as those between dominant and subordinate females were each significantly higher than the 18% average overlap between adjacent dominant females (Sloan 1997). Dominant females also more frequently had a core area within their home ranges where a minimum of 63% (range = 63–80%) of census sightings were concentrated within <30% of the total area used (Sloan 1997). All eight of the dominant females had core areas, whereas only 2 of 11 subordinate yearlings held a core area. The frequency of foraging strikes by dominant females with core areas was significantly higher than that of females without core areas. The locations of core areas, however, did not appear to be related only to the location of prey, because females struck prey significantly more frequently outside than within their core areas (Sloan 1997).

Because aggression was more frequent and intense at MS and aggression may carry energetic and survival costs (Moore and Marler 1987; Marler and Moore 1989, 1991), we tested the hypothesis that annual growth, survival, and reproductive output were lower in MS than in AL females. For these comparisons, we collected data over the entire lives of AL females that were yearlings in 1994 and 1995. There were fewer lizards at MS and we stopped studies in 1997. Therefore, we monitored all MS females that were yearlings from 1993 to 1996 until they died. We calculated annual growth rates for yearlings (MS, $n = 17$; AL, $n = 36$) as the change in SVL between the first measurement during year 1 and the first measurement during year 2, and for adults (MS, $n = 6$; AL, $n = 29$) as the first SVL measured during year 2 and the last year of the female's life, divided by life span in years. We determined the annual number of clutches by palpating the abdomen for oviductal eggs at least every 21 d (usually every 5–14 d) in yearling (MS, $n = 30$; AL, $n = 56$) and adult (MS, $n = 14$; AL, $n = 46$) females and dividing the total number of clutches in adults by life span in years. For estimates of survival, we concluded that females had died only when individuals that we had monitored during a previous season were not captured on the same or adjacent sites for at least one full subsequent season. Because survival may be influenced by predation as well as competition costs, we also recorded tail loss as an index of predator activity at the two sites. We compared female life span at MS ($n = 56$) and AL ($n = 84$) by determining age at mortality.

Mean annual growth rate of AL yearling females was significantly higher than that of MS yearlings, whereas growth rates of adults appeared similar (Fig. 1-3A), although small sample size at MS ($n = 6$) owing to high mortality precluded statistical comparison. The number of clutches produced annually was similar in both yearlings and adult

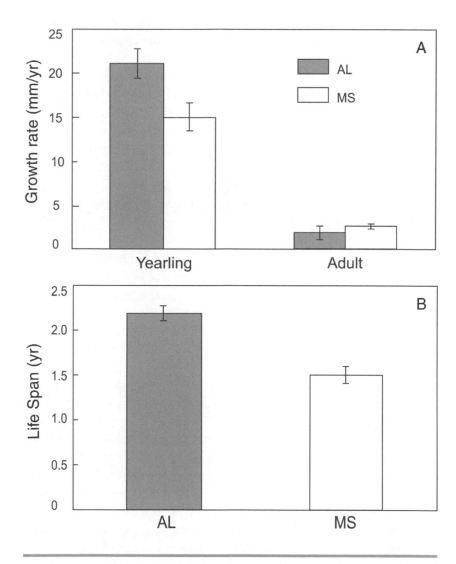

Figure 1-3 (A) Mean (± 1 SE) growth in yearling and adult females at AL and MS. (B) Mean (± 1 SE) life span in AL and MS females. Data from Baird and Sloan (unpublished data).

females. By contrast, the mean life span of AL females was significantly longer than that of MS females, whereas the frequency of tail loss at MS (2 of 31, 6.5%) and AL (8 of 79, 10%) was similar (Baird and Sloan, unpublished data).

We further tested the hypotheses that female collared lizards compete for resources, and that intrasexual competition was more intense at MS by removing individual females with high-quality home ranges and recording the behavior of adjacent females. Specifically, experiments ($n = 7$ trials at MS, $n = 8$ trials at AL) involved recording the social and spatial behavior of females during focal observations ($n = 6$ focals/female/experimental phase) prior to manipulation when we had removed one female from a social group and held her off the site for 4 d, and then again when this female was returned to her home range (Sloan 1997; Baird and Sloan, unpublished data). If females were to compete for resources, then at both AL and MS we expected one or more remaining females to begin using the home range of the removed individuals, and if intrasexual competition were more intense at MS, then we expected a more pronounced response to removals at this site than at AL. At MS, we removed the female in each group with the highest dominance score, whereas at AL where, females did not display a hierarchy of social rankings, we removed one of the largest and oldest females having several neighbors. During removal phases, some females made obvious spatial changes within the first 2 d, whereas others did not shift their space use. In all experiments, responding females were neighbors of the removed female. At MS, 91% of responding females were yearlings, whereas at AL, 60% of the responding females were yearlings (Baird and Sloan, unpublished data).

We performed two-way repeated measures ANOVA with treatment (preremoval, removal, replacement) and site (MS, AL) as main effects, and the frequency of aggressive encounters and the percentage of focal traces within the removed female's home range (separate tests) as the independent variables. Treatment did not have a significant effect on the frequency of encounters (Table 1-2). However, on average for both sites combined, percentage of travel within the removed female's home range increased significantly during the removal, and then decreased again when the removed female was replaced. Site had a significant effect on the frequency of encounters, with levels of aggression higher at MS, whereas percentage of travel within the removed female's home range did not vary as a function of site (Table 1-2). There were no statistically significant interactions between treatment and site for either variable (Baird and Sloan, unpublished data).

Higher levels of aggression at MS than at AL, the presence of an age-dependent social hierarchy at MS when no such social structure was evident at AL, and more pronounced aggressive behavior during removal trials at MS are all consistent with expectations of sexual selection theory that intrasexual aggression in female collared lizards functions in the competitive acquisition of resources. One possible explanation of

Table 1-2 Influence of female removals on mean (± 1 SE) percentage of focal traces of responding females overlapping removed female's home range (% Travel) and the mean hourly frequency (± 1 SE) of intrasexual encounters at the AL and MS sites. Data from Sloan (1997) and Baird and Sloan (unpublished data).

	AL (n = 10)			MS (n = 11)		
	Pre-removal	Removal	Replacement	Pre-removal	Removal	Replacement
% Travel	14.8 ± 8.3	59.8 ± 5.5	29.3 ± 10.9	25.2 ± 6.7	73.0 ± 6.0	42.6 ± 12.2
Encounters/h	0	0.05 ± 0.05	0.06 ± 0.06	0.10 ± 0.07	0.41 ± 0.16	0.15 ± 0.08

the site difference in the intensity of competition is that arthropod prey were less abundant at MS. However, the number of arthropods captured in sticky traps was similar at the two sites, indicating that the abundance of prey was not the explanation. Rather, we suggest that rock perches from which females can simultaneously meet all their requirements for fitness were limiting for females at MS. The types of rocks that were favored by dominant individuals and usurped during many of the removals suggest that preferred perches shared three features. Favored perches were located near the rock-grass interface and were elevated enough that they provided a good vantage point from which to scan for mobile insects. Perch rocks also had crevices beneath them, which allowed female occupants to take refuge quickly without traveling far. Such perches also promoted behavioral thermoregulation because females had ready access to both sun and shade. We suggest that the frequency of aggression was lower at AL because rocks with all these qualities were extremely abundant. By contrast, perches with these qualities were more limited at MS, promoting higher levels of agonism that results in a female social organization that is more structured by age-dependent social asymmetries (Baird and Sloan, unpublished data).

Ontogenetic Variation in Territorial and Courtship Behavior of Males

The social behavior of collared lizard males undergoes a striking ontogenetic transformation that is independent of the onset of sexual maturity. Males ranged in size from 60 to 80 mm SVL when they emerged for their first full activity season (yearling season). The rate of growth in yearling males was high ($\bar{X}$ growth rate $= 0.23$ mm/d, SE $= 0.01$, $n = 80$) until the end of their yearling season, by which time SVL had reached at least 98 mm. At the beginning of their second year, male growth rate ($\bar{X} = 0.02$ mm/d, SE $= 0.003$, $n = 65$) decreased ($P < 0.001$). Ninety-four percent (32 of 34) of yearling males with an SVL ≥ 72 mm from which we collected smears of the fluids emitted from their extruded hemipenes had mature spermatozoa, indicating that sexual maturity was achieved during the first year (Baird and Timanus 1998). Our findings agree with those of Trauth (1979) for male collared lizards in Arkansas.

Behavioral data collected in 1991–92 and 1998 on yearling and adult males showed that yearlings did not exhibit the conspicuous patrol and display that were the most conspicuous agonistic behavioral patterns in adult males (Baird, Acree, and Sloan 1996). The average rates of displays by yearling males were only 12.9 and 7.9% ($P < 0.001$) of the rates of adult males in 1991–92 and 1998, respectively (Tables 1-1 and 1-3).

Table 1-3 Mean (± 1 SE) rate of patrol, display, and percentage of yearling home ranges within adult male territories. Data for 1991–92 from Baird, Acree, and Sloan (1996).

Year/Number of Territories	Patrol (m/h)	Displays/h	Yearling Home Ranges Inside Adult Male Territories (%)
1991–92 (n = 13)	27.6 ± 6.8	9.2 ± 3.0	49.0 ± 4.1
1998 (n = 19)	35.7 ± 4.7	7.3 ± 2.2	88.4 ± 2.2

Yearling males also patrolled at significantly lower rates than adult males in both years (Tables 1-1 and 1-3). Yearling male home ranges highly overlapped the territories of one or more adult males (Fig. 1-4), with on average at least one-half of the area that they used being within the territories of one or more adult males (Table 1-3). These data indicate that yearling males adopted "stealthy" subordinate social tactics that probably functioned to minimize aggression while remaining within adult male territories. Furthermore, in addition to the advertisement of territory ownership to competitor males, the marked differences in the behavior of the two age classes suggest that frequent patrol and display by adults may also function to suppress social activity by sexually mature yearlings.

To test the hypothesis that the subordinate social tactics of sexually mature yearlings were a response to the frequent patrol and display activity of adult males, we recorded the social behavior of sexually mature yearling males when adult males were present, when we had removed all adult males from study plots for 4 d, and then when we replaced adult males on their territories (Baird and Timanus 1998). Relative to the pre-removal period, yearlings showed higher rates of patrol and display in response to removal (Fig. 1-5). Yearlings began to increase these behavioral patterns during the first 2 d of removal (Rem1), and this trend intensified during the last two removal days (Rem2). When adult males were replaced, both patrol and display frequency by yearlings abruptly decreased to levels similar to those prior to removals (Baird and Timanus 1998). These results demonstrate that sexually mature yearling males are capable of rapidly adopting the social behavior patterns that characterize territorial defense by adults, and that the subordinate behavior displayed by yearlings is a consequence of social inhibition by larger males.

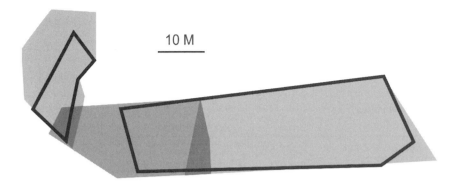

Figure 1-4 Relative locations and overlap of adult male territories (shaded polygons, $n = 3$) and yearling male home ranges (solid lines, $n = 2$) from one rock patch at AL in 1997. Note the low overlap between adjacent adult male territories and the high overlap between adult male territories and yearling male home ranges.

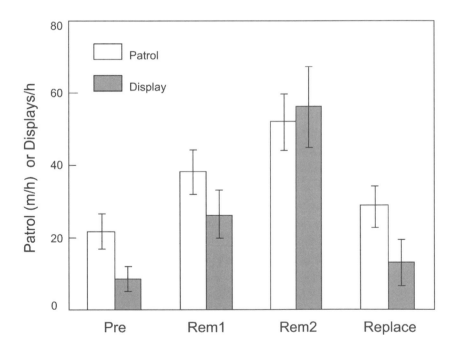

Figure 1-5 Mean (± 1 SE) hourly rates of patrol and display by sexually mature yearling males prior to removal of adults (Pre); during days 1 to 2 (Rem1) and 3 to 4 (Rem2), when adult males were removed; and when adult males were replaced (Replace) on their territories. Data from Baird and Timanus (1998).

The marked changes in yearling male behavior just described were elicited by removing all adult males from study plots, which entirely eliminated the inhibitory effect of advertisement by territorial adults. Although each year up to three territorial males died during individual study seasons, these males disappeared one at a time, and we did not observe the simultaneous disappearance of more than one adult male from the same habitat patch like that simulated in the experiments we have discussed. Therefore, although yearling males rapidly assumed territorial behavior patterns when all adults were removed experimentally, these experiments did not address whether the death of individual adult males can result in earlier opportunities for yearlings to acquire territories.

To address this issue, we recorded observations on the behavior of yearling males 1–6 and 14–20 d after five instances when a single adult male died from natural causes. The results suggest that opportunities for acquisition of territories by yearlings depended on the proximity of the nearest neighboring territorial adult(s), and the extent to which adjacent adults expanded into the area made vacant by the removal. In three of these single male removals, the nearest adult male was at least 70 m away, and no adult male moved into the vacated area. One of four yearling residents in trial 1 (Fig. 1-6A) and one of three yearling residents in trial 2 (Fig. 1-6B) markedly increased their hourly display rates in the 6 d following the removal. The responding yearling was the smallest one present in trial 1 and the largest one in trial 2. In trial 1, the display rate of the responding male diminished during days 14–20 postremoval (but was still higher than preremoval), whereas a second yearling increased his display activity during days 14–20 (Fig. 1-6A). In trial 2, the display rate of the first responding male continued to increase during postremoval days 14–20, and a second male showed a moderate increase during this period (Fig. 1-6B). In trial 3, in which no adult males were present, the only resident yearling increased his display rate over four times during days 1–6, and over six times during days 14–20 (Fig. 1-6C). By contrast, in trials 4 and 5, one or more adjacent adult males expanded into the former territory of the removed male. In both of these trials, only one yearling was present, and neither increased his display rate appreciably during either days 1–6 or 14–20 postremoval. Indeed, in trial 4, the preremoval display rate of the yearling decreased following removal (Fig. 1-6D).

Although it is premature to draw conclusions on the basis of only five trials, thus far these results suggest that the potential for yearlings to acquire a territory may be influenced by the economics governing

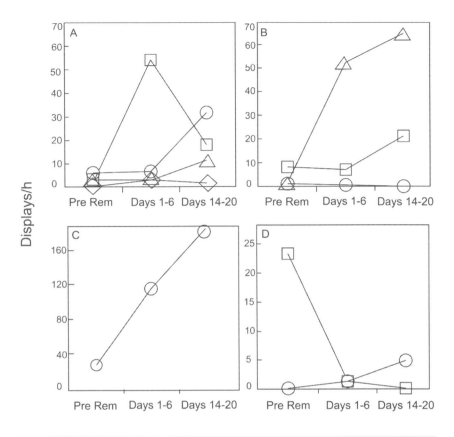

Figure 1-6 Mean (± 1 SE) hourly display by yearling males prior to (Pre Rem) and 1–6 and 14–20 d following removal of a single adult male. (A–C) Data for three separate trials in which there were four, three, and one yearling males (individual yearlings indicated by different symbols) present, respectively, and no adult males present after the removal; (D) data from two trials, each involving one yearling, in which an adjacent adult male expanded into the area made vacant by the removal.

expansion of territories by the nearest territorial males. If the borders of one or more neighboring male territories are nearby, like in trials 4 and 5, it may be economically feasible for adults to expand their existing territory boundaries to usurp the vacant area (see similarly, Baird and Liley [1989]). The potential for expansion by neighboring adult males is likely also influenced by the size and location of existing territories of adult males relative to the vacated area, as well as the number of neighboring competitors contending for the expanded territory because these factors will impact the temporal and energetic costs of defense (Hixon 1987). The outcomes of trials 4 and 5 suggest that as long as one larger adult moves in, the potential for acquisition of a territory by yearlings is inhibited.

By contrast, if the nearest neighboring adult males are farther from the site of the removal, then the temporal and energetic costs of territorial expansion and the potential cost of losing female mates may prohibit adult males from expanding their territories, allowing one or more yearling males to establish a territory. Consistent with this hypothesis, in trials 1–3, which resulted in marked display rate changes by a yearling, the territories of the nearest adult males were distant. Furthermore, the territory geometries in relation to the topographies of the habitats in two of these trials were such that the vacant areas were not readily visible from the neighboring males' territories. Therefore, not only would it have been costly for neighboring adults to expand, but they might not even have detected that a territorial male had disappeared. The marked responses of yearling males in trials 1–3 suggest that under conditions in which it is not economically feasible for adult males to expand, it may be possible for yearling males to acquire territories.

Subordinate social tactics in yearling collared lizard males are also apparent because the average rate of courtship encounters by yearlings was only one-fourth that of adult males (Table 1-4). Females were also less receptive to courtship by yearling males. We considered females moving away or fleeing from males to indicate rejection, whereas receptivity was indicated by females remaining with a male and engaging in mutual courtship behavior (described earlier). Using avoidance as the criterion for rejection, females were significantly less likely to be receptive to approaches initiated by yearlings than by adult males (Table 1-4). As a consequence of the low frequency of courtship and frequent rejection by females, estimates of mating success for yearling males were significantly lower (only 12%) than those for adult males with territories (Baird, Acree, and Sloan 1996).

Even though yearling males courted infrequently and were rejected more frequently, yearlings could possibly achieve higher than expected mating success with even a small number of successful copulations if they produced large quantities of sperm relative to that of adult males. If collared lizard yearlings utilize sperm competition, then the relative size of the testes in yearlings should be large compared with that of adults. For example, in fishes with sperm competition, relative testis size is much larger in males employing alternative mating tactics (reviewed by Petersen and Warner [1998]). We examined this possibility using data from an Arkansas population of C. collaris (Trauth 1979) to

Table 1-4   Mean (± 1 SE) frequency of intersexual encounters
initiated by adult (*n* = 15) and yearling (*n* = 13) males in
1991–92 and percentage of encounters when females rejected
males. Data from Baird, Acree, and Sloan (1996).

Male Age Class	Encounters/h	Rejection (%)
Adult	1.78 ± 0.12	5.2
Yearling	0.42 ± 0.22	35.0

determine the relationship between relative testis size and SVL. Because
Trauth (1979) did not record testis mass, we could not calculate gono-
somatic index, the most usual measure of relative gonad size. Instead,
we calculated a relative testis index by dividing the mean testis length
(millimeters) by SVL (millimeters) and multiplying the quotient by
100. Using this index, we found that the relative testis index of Arkansas
males the size of yearlings at AL ($\bar{X}$ index = 8.37, SE = 0.40) was similar (P = 0.82) to that of males the size of adult territorial males at AL
($\bar{X}$ = 8.52, SE = 0.50). The finding that relative testis index was not
larger in yearling males is not consistent with the hypothesis that yearlings in this species are competing by producing larger quantities of
sperm.

Because observations indicated that females preferred adult males
over yearlings, we used the results of experiments involving 4-d removal
of all adult males to examine the frequency with which yearlings initiated intersexual encounters when adult males were present and absent,
and the extent to which female responses to yearlings might change the
longer that adult males were absent. Yearlings were clearly disinhibited
from interacting with females during removals. The frequency of intersexual encounters initiated by yearlings increased significantly (2.5-
fold) when adults were removed and then declined again when adult
males were replaced (Baird and Timanus 1998). We used the duration
of courtship encounters as an index of female receptivity to yearlings,
because females terminated courtship by moving away from males. The
duration of courtship encounters initiated by yearlings during days 3
and 4 of removals was more than two times higher than during the first
two removal days, although this difference only approached statistical
significance (Baird and Timanus 1998). The tendency for longer
courtship encounters the longer that adult males were absent may indicate that females become more receptive to yearlings when they are not
courted by adult males, or that yearlings were more persistent when the

threat of aggression from adult males was removed. We suggest that the tendency for longer interactions is best explained by increased female receptivity because females appear capable of terminating the advances from even the most persistent males by fleeing.

Seasonal Variation in Agonistic and Courtship Behavior of Adult Males and Females

We examined the within-season temporal pattern with which adult males budget activities that are important for territorial defense and courtship by conducting focal observations on social and spatial behavior throughout the 1992 and 1997 seasons (Baird, Sloan, and Timanus 2001). We recorded five focal observations per male (1992, $n = 10$; 1997, $n = 14$) during six time periods (April 1–21 = APR; May 13–30 = MAY; June 1–14 = JN1; June 15–30 = JN2; July 1–15 = JL1; July 16–31 = JL2). Results revealed that males did not invest equally in territorial advertisement over the activity season. In 1992, for example, males patrolled and displayed significantly more frequently during June, followed by a decrease during July (Fig. 1-7A). There was no significant change in the frequency of intrasexual male encounters over these two seasons (Baird, Sloan, and Timanus 2001). However, males also courted more different females during June, and although not statistically significant, the frequency of courtship encounters was highest during this period as well (Fig. 1-7B).

These results indicate that collared lizard males established territory boundaries using relatively low rates of territorial activity at the beginning of the season. Instead, rates of territorial patrol and advertisement were highest long after territory boundaries had been established and matings for first and probably second clutches had already occurred. Interseasonal fidelity to the same territories by adult males was high at AL, and first time territory holders obtained areas that they at least partially overlapped as yearlings (Baird, Sloan, and Timanus 2001). Given the strong interseasonal site fidelity by males, the midseason pulse in patrol and advertisement may function to promote the establishment of territory boundaries in the following season with only a minimal early investment in display and patrol. Minimizing early season expenditures in territorial activity may allow males to invest more time courting early in the season. Why then were courtship interactions also most frequent during June, instead of the beginning of the season? Although first-year (yearling) females usually produced at least one clutch of eggs, they did not begin egg production until June; hence, only adult females (2 yr and

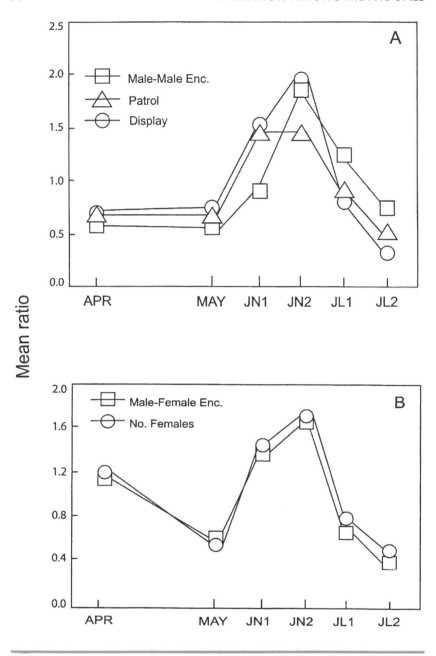

Figure 1-7 Seasonal changes in social and spatial behavior of adult male collared lizards in 1992. Data points are the ratio of the mean values for each time period (see text) to the season-long mean value for 1992. (*A*) Rate of patrol (m/h), displays/h, and male-male encounters/h; (*B*) male-female encounters/h and number of different females with which males interacted. Data from Baird, Sloan, and Timanus (2001).

older) became gravid in May. Males apparently concentrate their courtship early in the season with adult females to fertilize these earliest eggs. In June, yearling females produced their first clutches simultaneously with the second (and a little later sometimes a third) clutches produced by adult females. Therefore, more eggs were being carried by more females available for males to fertilize, probably promoting higher overall rates of courtship interactions (Baird, Sloan, and Timanus 2001).

Frequencies of agonistic behavior by females also fluctuated markedly during the season, depending on the timing of nest construction and oviposition (Sloan and Baird 1999). Daily focal observations on seven MS females for which we knew the locations of nests and dates of oviposition revealed that frequencies of intrasexual encounters, frequencies of aggressive acts during encounters, and travel rates were all significantly higher during the 5 d following oviposition relative to before egg laying. Moreover, the frequency with which these females initiated encounters was highest on the first day after emergence from egg laying and steadily declined with increasing time after oviposition. One possible explanation for the increase in female aggression following oviposition is that females were defending nesting sites from other females (Yedlin and Ferguson 1973). The nest defense hypothesis is not supported by our observations, however, because 85% of the aggressive encounters that we observed following oviposition occurred more than 13 m from nests (Sloan and Baird 1999). If increased aggression following oviposition functions in nest defense, females would be expected to locate nests in core areas where defense would require the least amount of energy to accomplish. Females probably move more efficiently when they are not carrying heavy eggs. Furthermore, egg-laying females were inhibited from repelling intruding consexuals for 1 to 2 d because nest construction and oviposition required them to burrow for this period of time. Therefore, increased aggression following egg laying may function to reestablish dominance over females' home ranges when it is necessary because they are emerging after a hiatus from home range occupancy and are no longer encumbered by heavy oviductal eggs (Sloan and Baird 1999).

Discussion

Our studies have revealed abundant evidence that in accordance with expectations of sexual selection theory, adult collared lizard males have been selected to compete intrasexually for access to multiple females through defense of all-purpose territories. Adult males defended territories from other adult males primarily by high rates of patrol punctuated by visually mediated displays from a distance. Although overt aggression between adult males occurred, its relative rarity suggests that the spatial relationships among males are established with only a few direct encounters (see similarly, Stamps and Krishnan [1994]), and that display and patrol are advantageous because they allow the maintenance of territory boundaries while avoiding the costs of escalated aggression. Furthermore, male collared lizards appear to reduce long-term territorial defense costs by budgeting their display and patrol such that they can reestablish themselves on proven high-quality territories without intense aggression early in the season, when it would detract from courting adult females. The importance of display and patrol in the control of territories is further exemplified by our findings that the relative frequency of these behavioral patterns appears stable from year to year, and that increases in display and patrol were the primary responses of yearlings when adult males were removed.

The social organization of collared lizards in our populations was characterized by territorial males consorting with multiple females (polygyny). However, at AL approximately 40% of females occupied home ranges that were partially overlapped by the territories of more than one adult male, even though these females primarily courted the male that overlapped the largest portion of their home ranges (Baird, Acree, and Sloan 1996). It is possible, therefore, that the mating system at AL is characterized by a significant number of females mating with more than one adult male, in which case the mating system would be more accurately described as polygynandrous, similar to some birds (Davies 1992). Paternity studies using DNA fingerprinting are necessary to determine whether males monopolize matings with females or the females mate with multiple males.

Large body size probably plays a role in the effectiveness of visual displays by males, and size is almost certainly important during escalated contests between males involving physical conflict (Le Boeuf 1974; Arak 1988; Howard 1988). In our population, however, variation in size among adult males was not pronounced, and larger adults did not achieve higher mating success than smaller adults (see similarly, [Abell 1998b]). Instead, adult males showed much greater variation in rates of patrol and particularly display, males with high rates of behavior that functions in the defense of territories had the highest estimated annual mating success, and there was statistically significant directional selection for high display rate. These results strongly suggest that although males must be large enough to acquire and control a territory, once they reach this qualifying size, behavioral parameters, particularly high rates of display, promote high mating success.

The ontogeny of male social behavior in collared lizards at AL can best be described as a system in which males are poised to obtain a breeding territory during their first year, but under the social conditions that usually prevail, yearlings defer reproductive efforts involving territorial defense. Acquisition of a territory is usually postponed until year 2, but from 1991 to 1998, 3 of 79 males did not establish their first territory until their third season. Yearlings (and these three 2-yr-old males) adopted subordinate tactics that allowed them to remain on adult territories, and occasionally to court females. For an ontogeny involving deferred territoriality to be favored by selection, individuals that adopt this pattern of behavior must be more successful than individuals that might adopt an alternative pattern, such as dispersal and attempting to establish a breeding territory as a yearling. Several factors appear to operate against dispersal as a feasible means of territorial acquisition by yearling males. Opportunities for yearling males to disperse to areas where they can acquire a territory appear to be very limited because the home ranges of all adult females on our study sites were overlapped by one or more adult males. Furthermore, yearling males have rarely been observed to change rock patches, probably because movement across grassy areas that lack refugia makes lizards vulnerable to snake and avian predators (Baird, Fox, and McCoy 1997). Even though yearlings disperse within habitat patches more frequently than among them, the areas to which they moved were usually already occupied by territorial adults (Baird and Timanus 1998). Together, these observations indicate that the potential for yearlings to acquire a territory through dispersal is low. Instead, male collared lizards appear to be making the best of a bad situation by channeling a large portion of their energetic budget into growth instead of territorial behavior. Although adult males are usually sufficiently abundant to inhibit establishment of territories by yearlings, there appears to be a limited potential for yearlings to acquire a territory when adult males die of natural causes and the nearest neighboring adult males are too far away to usurp the vacated area. Males have consistently established their territories in areas that at least partially overlapped their home ranges when they were yearlings (Baird, unpublished data). These observations suggest that yearlings may improve their chances for inheriting territories by using subordinate tactics to remain on habitat patches with proven high-quality areas rather than dispersing.

Consistent with the predictions of sexual selection theory, our results strongly suggest that female collared lizards are selected to acquire resources that are necessary for them to maximize the successful production and laying of eggs. The social behavior of female collared lizards was distinctly different from the territorial behavior of adult males. Females did not exhibit the conspicuous display and patrol that males used to advertise territories even on our site (MS) where intrasexual aggression among females was highest and the social status of females was dependent on age and size. Furthermore, even under environmental conditions in which high-ranking females aggressively monopolized core areas, the function of such behavior appeared to be priority of access to perches that are favorable for foraging while simultaneously maintaining access to a refuge. Fidelity by individual females to home ranges adjacent to those of several other females favors the evolution of polygynous or polygynandrous social systems (Emlen and Oring 1977). Therefore, variation in the level of female aggression at our two sites brings into question the costs to females of these social systems. At AL, where foraging perches were not limiting and intrasexual aggression was low, fidelity to a home range overlapped by one or more males clearly carried little cost (similarly, no-cost polygyny) (Davies 1991; Searcy and Yasukawa 1995; Jenssen and Nunez 1998). At MS, however, intrasexual female aggression was higher, and female growth rate and life span were lower than those of AL females. These findings show that there is a higher cost of site fidelity for females that settle on patches where perches favorable for foraging and refugia are scarce relative to habitats where these resources are more abundant (Baird and Sloan, unpublished data).

The social and spatial behavior of collared lizards in our populations share many similarities with that of other lizards, particularly other diurnal iguanians whose social behavior has been influenced by sexual selection. In several species, adult males utilize larger territories and have higher rates of display, patrol, and aggression than females (Jenssen 1970; Trillmich 1983; Jenssen and Nunez 1998; Abell 1999a), even in species in which females as well as males defend exclusive territories (Nunez, Jenssen, and Ersland 1997). Polygynous mating systems are common in lizards, particularly in species in which females exhibit a high level of fidelity to home ranges or exclusive territories (Ruby 1978, 1981; Stamps 1978; Ruby and Dunham 1987; M'Closkey, Deslippe, and Szpak 1990; Deslippe and M'Closkey 1991; Hews 1993), but males may also be able to monopolize mates that aggregate temporarily for other reasons (Trillmich 1983). Behavioral differences between classes of males have also been documented in other lizards. It is sometime the case that males of different age groups are spatially segregated (Stamps 1983b; M'Closkey, Deslippe, and Szpak 1990), but it appears more common that either subordinate adult or juvenile males coinhabit the same areas as in our populations of collared lizards. When different social classes overlap spatially, it is common for males to defer reproductive effort until they can effectively compete for territories and/or females (Berry 1974; Werner 1978; Cooper and Vitt 1987; Pratt et al. 1994). As a consequence, populations of these species are often characterized by an asymmetrical distribution of male mating success in which the largest individuals obtain several matings, and younger and/or smaller males obtain little if any success (e.g., Ruby 1981, 1984; Baird, Acree, and Sloan 1996). This pattern is common probably because many lizard species are restricted to patches of habitat with little possibility of dispersal to other suitable habitats (Baird, Acree, and Sloan 1996; Baird and Timanus 1998). Under these conditions, individuals have few alternatives but to make the best of conditions where they are, and then shift their behavior when opportunities arise.

Our data on *C. collaris* can be used to test a prediction of the hypothetical framework proposed by Zamudio and Sinervo (see chapter 3) to explain the evolution of lizard social systems. In this framework, both the spatial and temporal heterogeneity in the social landscape that individuals experience are important factors determining the evolution of lizard social systems. The coarseness of the social landscape is expected to depend on the probability that individuals will experience more than one set of social conditions during their lives, and the potential for exploiting different social environments is determined by species-specific variables such as dispersal ability and life span. In fine-grained social landscapes (*sensu* chapter 3), lizards can readily modify their social environment through dispersal or displacing a neighbor and usurping its territory. Selection for alternative mating tactics, especially those that are genetically fixed, is expected to be weak in species with fine-grained social landscapes because individuals can easily modify the social environment in which they find themselves. By contrast, when individuals have little chance of experiencing different social conditions during their reproductive lifetime due to reduced potential for dispersal and/or reduced social mobility, the social landscape may be considered coarse-grained (chapter 3).

The scale of the social landscape will also be influenced by temporal heterogeneity in the social conditions that individuals experience. The potential for individuals to experience different social environments should be low in short-lived species with limited dispersal. In these species especially, Zamudio and Sinervo's (chapter 3) framework predicts strong selection for the evolution of genetically fixed alternative social tactics, and the fixed behavioral and morphological polymorphisms among short-lived male *Uta stansburiana* (Sinervo and Lively 1996) and *Urosaurus ornatus* (Thompson and Moore 1991a, b; Carpenter 1995b) support this expectation. On the other hand, in longer-lived species, even when the social landscape is coarse-grained spatially, there may be greater potential for individuals to experience more varied social conditions (e.g., neighbors with different competitive abilities, different numbers of neighbors) because they survive for several breeding seasons. In longer-lived species, although the social environment is coarse over one season, multiple breeding seasons may make it possible for individuals to experience heterogeneity in selective regimes from year to year, resulting in a more fine-grained social environment over the lifetime of each individual. A coarse-grained landscape in the short term that is more fine-grained over time may not be expected to result in the evolution of genetically based alternative social tactics (chapter 3).

Collared lizards in central Oklahoma occupy spatially coarse social landscapes because they are restricted to isolated patches of rock outcrops and washes, and dispersal among patches is limited (Baird, Fox, and McCoy 1997; Hranitz and Baird 2000; Baird, unpublished data). Thus, once territory and home range boundaries are established during any one breeding season, it appears unlikely that individuals will experience more than one social context. However, collared lizards are also relatively long-lived. Males in our populations have survived for up to 6 yr, whereas females have a maximum life span of 7 yr. Zamudio and Sinervo (chapter 3) predict genetically based social tactics in species that occupy spatially coarse social landscapes like those of *C. collaris*. However, if the selective regime from the coarse-grained social environment in any one year is offset by finer-grained social conditions resulting from the long life span of collared lizards, then plastic social tactics may be expected to evolve (chapter 3). Yearling male collared lizards adopt subordinate social tactics that probably garner them only limited mating success, and then males shift to territorial tactics when they are older and larger. Similarly, our data indicate that in female *C. collaris* intense competition for limited resources can also result in age-related variation in social tactics. Older and larger females maintain access to preferred core areas, whereas younger females are displaced to adjacent home ranges that are apparently qualitatively inferior. Therefore, our results are fully consistent with the expectation of plastic alternative social tactics in a long-lived species, even though individuals occupy social landscapes that are coarse-grained in the short term.

Future Studies

There is a wealth of potential for future field studies using lizards of the genus *Crotaphytus* as study organisms. The application of contemporary molecular techniques to field studies of lizard mating systems is proving to be a powerful approach for examining the phenotypic attributes that influence the mating success of individual lizards (Abell 1999b; Zamudio and Sinervo 2000). Since it is possible to record the behavior of individual collared lizards over prolonged periods (perhaps the life spans of individuals), and to capture hatchlings, *C. collaris* is an outstanding candidate for studies that couple the use of molecular probes with field studies to determine the fitness of individuals displaying varied phenotypic attributes. Our studies to date also indicate that collared lizards are good subjects for continued studies of spatial behavior. Examination of the spatial behavior of male collared lizards prior to and following natural mortality of territorial males is a fruitful approach for testing predictions that are fundamental components of economic defensibility theory. Relatively little is known about how the spatial behavior of territorial individuals is influenced by factors such as age, the prior social experience of individuals relative to that of their neighbors, and changing local densities of competitors (Stamps 1994). Long-term field studies on individuals of known ages and social histories that address how these factors influence the territorial behavior of individuals are also possible because collared lizards are long-lived, and known individuals can be monitored over their entire life spans. Such studies within just a single population are challenging because they require long field observation periods. However, both our work at AL and MS and that of McCoy, Baird, and Fox (see chapter 5) show that interpopulation studies of this type offer opportunities for powerful tests of hypotheses central to the evolution of social behavior and sexual selection theory. Finally, even though the phylogeny of the genus *Crotaphytus* and the family Crotaphytidae has received thorough study using cladistic analyses of morphological, biochemical, and life-historical traits (McGuire 1996), information on the social behavior of other Crotaphytid lizards is limited. Therefore, there is great potential and need for detailed studies on the social behavior of other members of this family. Such studies will allow the application of contemporary phylogenetic methods (e.g., Brooks and McLennan 1991) to social behavior evolution in this family. It is our hope that the results reported here will prompt others to look toward collared lizards as model systems on which to conduct long-term behavioral ecological field studies of this type.

Acknowledgments

We wish to thank Bill Parkerson and Mary Sullivant of the U.S. Army Corps of Engineers and Terry Curly for access to study areas, Lt. Dan Davey of the U.S. Air Force for mapping study sites, and Stan Trauth for providing his unpublished data. We sincerely thank Tom Jenssen, Doug Ruby, and Kelly Zamudio for their constructive comments on an earlier draft. This work was funded in part by the Office of Graduate Studies and Research at the University of Central Oklahoma, Sigma Xi Grants-in-aid of Research (DKT and CLS), the American Museum of Natural History (DKT), and the Gage Fund of the American Society of Ichthyologists and Herpetologists (DKT).

This work is dedicated to the memory of John Albert Davis, a patient teacher, skilled outdoorsman, and keen observer of animal behavior. You will not be forgotten.

CHAPTER 2

Evolution and Maintenance

of Social Status-Signaling Badges

Experimental Manipulations in Lizards

Martin J. Whiting, Kenneth A. Nagy,
and Philip W. Bateman

Sexual selection theory aims to explain the evolution of extravagant traits that should seemingly impose a survival cost on the bearer (reviewed in Andersson [1994] and Andersson and Iwasa [1996]). This paradox is normally explained as a trade-off between reproductive success and survival. Two processes are thought to drive the evolution of extravagant traits: male contest competition and female choice (Andersson 1994). Traits that function in female choice are referred to as *ornaments*, while weapons and signals used in male contest competition are called *armaments* (Berglund, Bisazza, and Pilastro 1996). There is also increasing evidence that armaments frequently function in both aggressive interactions between males and directly in female choice (Kodric-Brown and Brown 1984; Møller 1988a; see Johnstone and Norris [1993] for a theoretical treatment; Morris, Mussel, and Ryan 1995; reviewed in Berglund, Bisazza, and Pilastro [1996]; Beani and Turillazzi 1999). Alternatively, successful males with large armaments may have a higher reproductive success as a result of competitive exclusion (e.g., *Lacerta agilis*, Olsson 1994a). Furthermore, animal signaling is complex and often consists of multiple signals (Møller and Pomiankowski 1993; Johnstone 1995a, 1996, 1997) and sometimes multiple ornaments that exploit different preferences, but with the same ultimate function (Brooks and Couldridge 1999). Signaling is further complicated by competition among signalers on information transfer during communication, motivational state, information content and reliability, the signaling environment, the receiver's sensory system, receiver psychology, predation risk, phylogenetic constraints, and a myriad of other factors (Fleishman 1986, 1988a, b, c, 1992; Ryan et al. 1990; Guilford and Dawkins 1991; Endler 1992, 1993; Arak and Enquist 1993; Dawkins

1993; McGregor 1993; Ryan and Rand 1993, 1995; Endler and Houde 1995; Godfray 1995; Greenfield 1997; Johnstone 1997; Leal and Ro-dríguez-Robles 1997; Zuk and Kolluru 1998; Leal 1999). Disentangling these various factors, searching for common patterns in diverse systems, and marrying empirical work with current signaling theory will be a challenge for some time to come. Furthermore, to properly understand signal design and evolution, we need to make use of many different biological disciplines (see Endler 1993) and even other scientific disciplines such as physics.

Animal signals have been the subject of intense study for the last two decades, increasingly so by modelers. As Grafen and Johnstone (1993) so succinctly point out, ESS (evolutionarily stable strategy) models of biological signaling are useful because they are explicit, more likely to be error-free, and allow more complicated exploration of the assumption of evolutionary stability. Recent work on signal terminology and concepts (Dawkins and Guilford 1991; Dawkins 1993; Guilford and Dawkins 1995; Maynard Smith and Harper 1995) and formal mathematical models (e.g., Grafen 1990; Getty 1998) have resulted in a framework in which signal content, reliability, and evolution can be examined. Animal communication biology has therefore entered an exciting new era in which theoretical models can be empirically tested. One branch of communication biology in which such interchange is potentially highly rewarding is the study of dominance disputes between males (status signaling or the badges-of-dominance game in particular) (Dawkins and Krebs 1978; Maynard Smith and Harper 1988; Kim 1995). Many animals possess markings (generally color patches) or features (e.g., lizard tails) that function as badges of status (Fox, Heger, and DeLay 1990; Krebs and Davies 1993). The term *badges of status* was coined for arbitrary structures (such as a color patch) that convey status signals and are uncostly to produce and potentially open to cheating (Dawkins and Krebs 1978; Roper 1986). Conversely, armaments such as antlers play a direct role in settling a contest, conferring a true advantage to the bearer (Senar 1999). When producing a badge does incur a cost, it may be argued that badges are no more than ordinary handicaps. It is also important to separate production from maintenance costs (e.g., Veiga and Puerta 1996).

According to the badges-of-status hypothesis, frequency-dependent selection maintains honest signaling of aggressiveness (Maynard Smith and Harper 1988). Alternatively, Rohwer (1982) proposed negative frequency-dependent selection to explain two badge-associated strategies in birds—dominants and subordinates—which are equally fit. The idea is that the two strategies are a form of resource exploitation that results either in cooperative exploitation (e.g., mutually beneficial foraging strategies in Harris' sparrows) or in alternative reproductive strategies.

Status signaling occurs in such diverse taxa as insects (Greenfield and Minckley 1993; Beani and Turillazzi 1999), fish (Wickler 1957 in Dawkins and Krebs 1978; de Boer 1980; Zimmerer and Kallman 1988; Morris, Mussel, and Ryan 1995), frogs (Davies and Halliday 1978; Arak 1983b), lizards (Fox, Heger, and DeLay 1990; Thompson and Moore 1991b; Olsson 1994a, b; Zucker 1994a, b; Carpenter 1995a, b), birds (e.g., Rohwer 1975, 1977, 1982; Rohwer and Rohwer 1978; Studd and Robertson 1985a; Møller 1987b; Whitfield 1987; Senar et al. 1993, 2000; Furlow, Kimball, and Marshall 1998; Senar and Camerino 1998; Senar 1999), and mammals (Clutton-Brock and Albon 1979) (see also Table 2-1). In the case of some mammals, birds, and frogs, status is frequently conveyed vocally (reviewed in Andersson [1994]; Leonard and Horn 1995). Also, in some species, individuals may have more than one armament that functions as a status-signaling badge (e.g., for tree lizards, see Zucker [1994a]; for birds, see Balph, Balph, and Romesburg [1979]). Status-signaling badges signal dominance and/or fighting ability and thereby prevent or reduce the costs of fighting (Rohwer 1975). However, badges may also signal aggressiveness (Rohwer 1982; Studd and Robertson 1985a, b; Maynard Smith and Harper 1988; Johnstone and Norris 1993). Aggressiveness may not be tightly linked to size or strength such that badges may be used to settle disputes over less valuable resources, whereas fighting ability is used in disputes over more valuable resources (Maynard Smith 1982; Maynard Smith and Harper 1988; Johnstone and Norris 1993). In addition, contests are more likely to escalate to fights between males of similar aggressiveness (Maynard Smith and Harper 1988). By using badges in asymmetric contests, individuals can avoid unnecessary energy expenditure and increased risk of injury or predation involved in fighting assessment (Rohwer 1982). Badges are most effective in quickly resolving conflicts when "large" asymmetries exist (Maynard Smith and Harper 1988); this is also predicted by the sequential assessment game (Enquist et al. 1990).

Status signaling is well known for its role in dominance disputes during the reproductive season in which the outcome determines access to mates. However, in nonbreeding, flocking birds, status signaling that determines access to food occurs in all age-sex classes (Rohwer 1975, 1977; partial review in Roper [1986]; Møller 1987b; reviewed in Whitfield

Table 2-1 Studies of status signaling that have in some way examined costs associated with signaling. Many studies did not explicitly test constraints to honest signaling, but the results provided correlative evidence or refuted specific handicap-based models. This table provides evidence for costly signaling (if any), how cheating is constrained, and any comments/ambiguities from the studies that may warrant further investigation; we included only those studies that referred specifically to status-signaling badges or those that dealt with an animal's status in the context of signaling/dominance costs since 1975. There are many other studies in which dominance and aggression were examined in a cost-benefit analysis, similar to status signaling. EA = experiment under artificial conditions (e.g., in aviaries); EF = experiment performed in the field; OF = observations in the field; OA = observed under artificial conditions. (If measurements were made [e.g., metabolic rate] but nothing [e.g., animal's phenotype, diet, or potential for interaction] was manipulated, a study was scored as observation based, whether in the laboratory or field.)

Species	Type of Study	Evidence for a Badge/Signal Cost	Constraints to Cheating	Comments/Ambiguities	Reference
Arthropods					
Tarbrush grasshopper (*Ligurotettix planum*)	OF, EF	Speculative	Likely morphological and physiological	Calling rate and number of shucks/call indicated fighting ability. High rate and call complexity should constrain cheating.	Greenfield and Minckley (1993)
Stenogastrine wasp (*Parischnogaster mellyi*)	EA	Yes	Social control	Males with manipulated badge (extra stripes) were challenged more frequently than controls.	Beani and Turillazzi (1999)

Table 2-1 *(Continued)*

Species	Type of Study	Evidence for a Badge/Signal Cost	Constraints to Cheating	Comments/Ambiguities	Reference
Amphibians					
Common toad (*Bufo bufo*)	OF, EA	No	Morphology	Call frequency was anatomically constrained: larger males had a larger larynx = deeper croak.	Davies and Halliday (1978)
Fish					
Siamese fighting fish (*Betta splendens*)	EA	Yes	Energetic cost	No badge as such; fighting ability was signaled during ritualized aggression.	Halperin et al. (1998)
Midas cichlid (*Cichlasoma citrinellum*)	EA	Yes	Energetic cost	No badge as such; aggression (not fighting ability) was signaled during rituals.	Barlow, Rogers, and Fraley (1986)
Firemouth cichlid (*Cichlasoma meeki*)	EA	Yes	Carotenoids	Red area on ventral surface displayed during aggressive interactions was dependent on amount of carotenoids in diet.	Evans and Norris (1996)
Lizards					
Sand lizard (*Lacerta agilis*)	EA, OF	Yes	Resource allocation social control?	A trade-off existed between somatic growth and badge size. Social control hypothesis requires testing.	Olsson (1994b); Olsson and Silverin (1997)

Table 2-1 *(Continued)*

Species	Type of Study	Evidence for a Badge/Signal Cost	Constraints to Cheating	Comments/Ambiguities	Reference
Tree lizard (*Urosaurus ornatus*)	(1) EA	No	Suggest that system is a mixed ESS	A good argument is provided to support this assertion.	Thompson and Moore (1991b)
	(2) EA	Speculative	Suggested costs: predation and/or increased thermal load	Signaling costs were not explicitly tested.	Zucker (1994a)
Mediterranean lizard (*Psammodromus algirus*)	EA	Yes	Social control	Small males with painted heads were attacked by larger males.	Martin and Forsman (1999
Augrabies flat lizard (*Platysaurus broadleyi*)	OF	Yes (correlative)	Metabolic cost	—	Whiting et al. (this study)
Birds					
Dark-eyed junco (*Junco hyemalis*)	EA	No	—	Failed to support incongruence hypothesis.	Holberton, Able, and Wingfield (1989)
	OA	No	Constraints discussed in general terms with reference to Rohwer's work	Weak support for status signaling; most likely via learned association (e.g., between plumage and body size).	

Table 2-1 (Continued)

Species	Type of Study	Evidence for a Badge/Signal Cost	Constraints to Cheating	Comments/Ambiguities	Reference
Collared flycatcher (*Ficedula albicollis*)	EF	Yes (correlative)	Social control and cost in parental care	—	Qvarnström (1997
Great tit (*Parus major*)	(1) OA	Yes (correlative)	Metabolic cost	Heart weight (relative to body weight) was greater in more dominant males.	Røskaft et al. (1986)
	(2) EF, EA	No	—	Speculate that system may be a mixed ESS. Criticized by Wilson (1992).	Järvi and Bakken (1984)
	(3) EA	Yes	Possibly incongruence hypothesis	Rejected social control hypothesis. Design/results were criticized by Wilson (1992) and Slotow, Alcock, and Rothstein (1993)	Järvi, Walsø, and Bakken (1987)
	(4) EF	Speculative	—	Weak evidence was found for status signaling in females and none in males. Suggests badge may reflect history of risk and energetic costs associated with successful fighting history.	Wilson (1992)

Table 2-1 (Continued)

Species	Type of Study	Evidence for a Badge/Signal Cost	Constraints to Cheating	Comments/Ambiguities	Reference
Pied flycatcher (*Ficedula hypoleuca*)	(1) OA	Yes (correlative)	Metabolic cost	—	Røskaft et al. (1986)
	(2) EF	No	—	Critical of above studies; little support for bright plumage functioning as badge.	Huhta and Alatalo (1993)
	(3) OF	Yes (correlative)	Predation	—	Slagsvold, Dale, and Kruszewicz (1995)
House sparrow (*Passer domesticus*)	(1) EA	Yes	Social control	Slotow, Alcock, and Rothstein (1993) questioned whether plumage variation in house sparrows even signals social status.	Møller (1987a)
	(2) EF	Speculative	Survival disadvantage	Simulated cheaters had lower survival; no benefit for cheaters in terms of offspring fledged.	Veiga (1993, 1995)
	(3) —	Yes	—	Badge size was negatively correlated to health; badge size was an honest signal of condition.	Møller, Kimball, and Eritzoe (1996)

Table 2-1 *(Continued)*

Species	Type of Study	Evidence for a Badge/Signal Cost	Constraints to Cheating	Comments/Ambiguities	Reference
	(4) OA, OF	Yes	Diet/access to food	Greater access to food during molt positively correlated with badge size; juveniles were forced to use more blood proteins during molt due to poorer body condition.	Veiga and Puerta (1996)
	(5) EA	No	—	No support for social control hypothesis; low sample size/statistical power.	Solberg and Ringsby (1997)
	(6) OF	Yes	Endocrine-immuno-suppressive trade-off	Breeding males with larger badges had lower immunocompetence.	Gonzalez, Sorci, and de Lope (1999)
	(7) EA	No	—	Birds fed on protein-rich and protein-poor diets. Diet had no significant effect on size and spectral qualities of badge.	Gonzalez et al. (1999)

Table 2-1 (Continued)

Species	Type of Study	Evidence for a Badge/Signal Cost	Constraints to Cheating	Comments/Ambiguities	Reference
	(8) EA	Yes	Endocrine-immuno-suppressive trade-off	Testosterone levels positively correlated with badge size; corticosterone reduced immunocompetence; relationship with testosterone uncertain.	Evans, Goldsmith, and Norris (2000); also see Poiani, Goldsmith, and Evans (2000)
Least auklets (Aethia pusilla)	OF, EF	Speculative	Social control and/or predation	Signaling costs were not explicitly tested.	Jones (1990)
Willow tit (Parus montanus)	OF	Yes (correlative)	Metabolic cost	Metabolic rate was related to dominance rank; badge size was not measured.	Hogstad (1987)
Yellow warbler (Dendroica petechia)	EF, OF	No	Different but equally successful strategies adopted by different morphs	Appears to be a mixed ESS.	Studd and Robertson (1985a)
White-crowned sparrow (Zonotrichia leucophrys)	(1) EF	No	As above	No support was found for social control hypothesis.	Keys and Rothstein (1991)

Table 2-1 *(Continued)*

Species	Type of Study	Evidence for a Badge/Signal Cost	Constraints to Cheating	Comments/Ambiguities	Reference
	(2) EA	No	Increased predation risk suggested as barrier to cheating	Rejected social control hypothesis.	Slotow, Alcock, and Rothstein (1993)
Harris' sparrow (*Zonotrichia quereula*)	(1) EF	Yes	Social control	Design was criticized by Shields (1977)	Rohwer (1975, 1977)
	(2) EF	Yes	Incongruence hypothesis	—	Rohwer and Rohwer (1978)
	(3) EF	N/A	Frequency-dependent selection (shepherds hypothesis)	Dominants and subordinates coexist in a mutualistic relationship; equal fitness.	Rohwer and Ewald (1981), also see Rohwer (1982)
Domestic rooster (*Gallus g. domesticus*)	(1) EA, OA	Yes (correlative)	Probably social control	—	Leonard and Horn (1995)
	(2) —	No	—	Hypothesis that crowing was energetically expensive (therefore preventing cheating) was rejected.	Horn, Leonard, and Weary (1995)
Red jungle fowl (*Gallus gallus*)	OA	No	—	Crowing had a minimal energetic cost.	Chappell et al. (1995)

Table 2-1 *(Continued)*

Species	Type of Study	Evidence for a Badge/Signal Cost	Constraints to Cheating	Comments/Ambiguities	Reference
Scarlet-tufted malachite sunbird (*Nectarinia johnstoni*)	OF	Yes	Uncertain	Traits were condition dependent; fluctuate with environmental conditions.	Evans (1991)
Ring-necked pheasant (*Phasianus colchicus*)	EA	Yes	Social control	Social control of cheating was not explicitly mentioned, but like-versus-like aggression suggests this is the case.	Mateos and Carranza (1997)
Siskin (*Carduelis spinus*)	OA	Yes	Stress	Subordinates had a higher metabolic rate due to stress-related encounters with dominants.	Senar et al. (2000)
Mammals					
Red deer (*Cervus elephas*)	OF, EF	Speculative	Physiological: energetic cost to roaring	—	Clutch-Brock and Albon (1979)

[1987]); although the results of some previous studies are increasingly being challenged (e.g., Whitfield 1987; Wilson 1992; Slotow, Alcock, and Rothstein 1993), while other studies have failed to find support for the status signaling hypothesis (SSH) as an explanation for plumage variability (reviewed in Whitfield [1987] and Maynard Smith and Harper [1988]). An increasingly recognized alternative to the SSH is the individual recognition hypothesis (Whitfield 1986, 1987, 1988), although both may occur simultaneously in the same population (Whitfield 1987).

In general, contests between males are settled through ritualized displays in which the dominance status of the signaler is transmitted to the receiver (Enquist and Leimar 1983; Kim 1995). The signal itself can take many forms (Johnstone 1996, 1997), but the intended purpose is common to all taxa: transmit information about dominance and/or fighting ability/aggressiveness, thereby avoiding or reducing the costs of fights with a predictable outcome (Maynard Smith 1982). The information content and honesty of the signal may, of course, vary greatly (Dawkins and Guilford 1991; Semple and McComb 1996).

There are two types of signals in agonistic communication: performance and strategic (Hurd 1997a). Performance signals are also referred to as unambiguous/assessment signals (Maynard Smith and Parker 1976; Maynard Smith 1982; Maynard Smith and Harper 1988) and condition-dependent handicaps (Grafen 1990). Performance signals are in essence, unbluffable. For example, in many cichlids, fighting ability is tightly linked to size. An early stage of sequential assessment is lateral displaying, in which cichlids pose side on with an opponent and smaller individuals will always appear smaller (Hurd 1997b). Conversely, all individuals are capable of making strategic signals (conventional signals, *sensu* Dawkins 1993), and these can all be bluffed or exaggerated to some degree (Hurd 1997a). (Badges of status are therefore conventional signals [Dawkins 1993].) Numerous game-theory models have attempted to formalize signaling strategies in contests (e.g., Enquist 1985; Maynard Smith and Harper 1988; Grafen 1990; Grafen and Johnstone 1993; Johnstone and Grafen 1993; Johnstone and Norris 1993; Kim 1995; Getty 1998; Számadó 2000). This has resulted in a framework in which to examine the evolution and maintenance of signaling systems and has set the stage for empirical testing of signaling theory in a variety of taxa.

Maintenance of Signaling System Stability

The last decade has seen a concerted attempt to understand signaler-receiver coevolution and the selective forces underlying signal evolution

(Ryan et al. 1990; Endler 1992, 1993; Arak and Enquist 1993; Dawkins 1993; Grafen and Johnstone 1993; Ryan and Rand 1993, 1995; Morris and Ryan 1996). Guilford and Dawkins (1991) proposed that signal design consists of two components: strategic design and efficacy. *Strategic design* refers to how a signal is designed by natural selection to ensure information transfer and a corresponding reaction from the receiver. *Efficacy* refers to the manner in which the information is transferred, such that it is easily measured by the receiver. Signals may be favored by natural selection if they are energetically cheap, effectively manipulate the receiver, and reduce the risk of predation (Johnstone 1997).

Signals may also be costly (reviewed in Zuk and Kolluru [1998]). For example, frog calls attract predators (Tuttle and Ryan 1981). The exact mechanisms driving the evolution and maintenance of badges are controversial. Possible mechanisms include natural selection (Rohwer 1975, 1982); sexual selection (Olsson 1994a); both acting simultaneously (Møller 1988a); and other less clear forms of selection, such as mimicry (reviewed in Andersson [1994]). However, it is now accepted that signal systems are complex and not always assignable to general selective pressures (Dawkins 1993; Greenfield 1997).

Occasional deception need not disrupt the signaling system provided that signals are on average honest (Johnstone and Grafen 1993). If signals were not honest, deception would ultimately result in an unresponsive receiver, and the system would become redundant (Johnstone 1997; but see Számadó [2000]). This is one of the most challenging areas in communication biology: the stability of signal systems and barriers to cheating (if any). Signals can vary greatly in their information content and honesty/reliability (Dawkins and Krebs 1978; Krebs and Dawkins 1984; Dawkins and Guilford 1991; Semple and McComb 1996; Viljugrein 1997). A major theoretical contribution to this area is the handicap principle, originally conceived by Zahavi (1975, 1977) and later formalized by Grafen (1990) (but also see Getty [1998]). There are two theoretical interpretations of this principle. The "strategic handicap" interpretation suggests that for maintenance of honesty, there must be a cost associated with the signal such that it only pays high-quality (or highly motivated) (Enquist 1985) individuals to make costly signals. In this "quality-dependent" interpretation (most commonly referred to as condition dependent), the benefits outweigh the costs only for superior individuals, such that selection favors quality-dependent expression of traits. (Both models are reviewed in Johnstone [1995b].) Another interpretation is that signals are not costly because they are "uncheatable"; instead, costly signals have evolved to enhance a trait's expression and these are termed *revealing handicaps* (Iwasa, Pomiankowski, and Nee 1991). The handicap principle, in the context of honest signaling, has received much theoretical attention (thoroughly reviewed by Johnstone [1995b]). However, much of this attention has been skewed toward female choice.

Here, we review recent theoretical contributions to status signaling; in particular, we review by major taxonomic group studies that in some way deal with constraints to cheating (Table 2-1). We also examine the factors necessary to maintain an ESS in light of honest signaling theory (Table 2-2). We have excluded studies that merely demonstrate status signaling without regard to signaling constraints and include only studies published since 1975 (when Zahavi published his first treatment of the handicap principle).

Signaling Costs

A central theme in discussions of status signaling is the issue of cheating (reviewed in Senar [1999]). Status-signaling badges (Dawkins and Krebs 1978) have been referred to as arbitrary indicators of social status (e.g., a color patch on a bird). In other words, badges are theoretically cheap to produce and they should therefore be easily mimicked. What, for example, prevents a low-ranking individual from assuming a dominant's badge? Clearly, cheating is minimal; otherwise, most signaling systems employing badges would be unstable and break down. Recent models (e.g., Johnstone and Grafen 1993; Johnstone and Norris 1993) support the notion that cheating occurs at a reduced level and have argued for contest-independent costs that exceed the costs of honest advertisement. For example, Veiga (1993) found that male house sparrows with experimentally enlarged badges acquired more nest sites but raised fewer offspring than control and badge-reduced males. Also, a follow-up study (Veiga 1995) showed reduced survival of yearling males with experimentally enlarged badges, suggesting a survival cost for cheaters. A number of costs have been proposed as constraints to cheating (reviewed in Table 2-1) and these fall into two categories: those that are independent of actual contests and those that are paid during contests. We deal with contest-independent costs first.

Elevated levels of androgens associated with aggression (e.g., testosterone) can be energetically costly (Marler et al. 1995) and also may compromise the immune system (Folstad and Karter 1992; Zuk, Johnsen, and Maclarty 1995; Møller, Kimball, and Erritzre 1996;

Table 2-2 Current theoretical models for maintenance of signaling systems, particularly as they apply to badges-of-status game. Signaling systems fall into two general categories: costly and cost-free (Hurd 1997a). We have focused on formal mathematical models but have included some references (indicated by an asterisk) whose models are inferred, based on empirical work.

Model	Costly or Cost-Free?	Maintenance of Signal Reliability and ESS
Zahavi (1975*, 1977*); Pomiankowski (1987); Grafen (1990); Johnstone and Grafen (1993); Maynard Smith (1991)	Costly	The handicap principle: a signal's cost enforces its reliability. (Companion models suggest that cheating is permitted, but incidence must be low.)
Rohwer (1977*); Maynard Smith and Harper (1988)	Costly	Socially enforced: the "social control hypothesis."
Rohwer and Rohwer (1978*).	Costly	Incongruence hypothesis
Owens and Hartley (1991)		
(1) "Cheat"	Costly	Social control.
(2) "Trojan sparrow"	Cost-free	In the absence of honest phenotypic limitations, the Trojan sparrow strategy can successfully invade honest populations. Result: mixed fighting strategies, badges-of-status model reduced to conventional hawk-dove model.
Dawkins and Guilford (1991)	Cost-free	Dependent on high cost of assessing signaler and low gain for extra information (to receiver). Occasional "probing" by receiver keeps cheating low.
Johnstone and Norris (1993)	Costly	A cost independent of a contest/aggression.
Maynard Smith (1994)	Cost-free	Participants place the possible outcomes of an interaction in the same rank order.
Krakauer and Pagel (1995)	Cost-free	Constraints on population spatial structure.
Adams and Mesterton-Gibbons (1995)	Costly	At ESS, only weak and strong individuals threaten, not intermediates. Reason: although threatening is very risky for weak individuals, the pay-off is high.

Table 2-2 *(Continued)*

Model	Costly or Cost-Free?	Maintenance of Signal Reliability and ESS
Hurd (1997a)		
(1) Conventional signals	Cost-free	ESS is maintained by signaler's state. Weak individuals avoid provoking stronger individuals by signaling their strength; stronger individuals avoid escalated contests.
(2) Conventional signal + cost	Costly	Use of a handicapped signal is more likely to influence weaker individuals, for which the cost will be greater. Stronger individuals of higher fighting ability gain less by avoiding conflict (escalated contests).
Keys and Rothstein (1991*); Slotow Alcock, and Rothstein (1993) and references therein*; Slagsvold, Dale, and Kruszewicz (1995*)	Costly	Increased predation is associated with more conspicuous coloration.
Studd and Robertson (1985a*); Thompson and Moore (1991b*)	Cost-free	Mixed ESS: different morphs adopt different but equally successful reproductive strategies.

Salvador et al. 1997; Qvarnström and Forsgren 1998; Evans, Goldsmith, and Norris 2000; Poiani, Goldsmith, and Evans 2000). If badge size signals aggressiveness, it is reasonable that variation in the ability of individuals to bear this cost will exist (Johnstone and Norris 1993). Another physiological cost of dominance is a higher metabolic rate. More aggressive individuals have more frequent encounters with rivals, and this is mediated proximally through elevated hormone levels (Hogstad 1987). Increased metabolic rates correlating with dominance have been confirmed for pied flycatchers and great tits (Røskaft et al. 1986) and willow tits (Hogstad 1987), while increased levels of testosterone resulted in an increase in social status in satin bowerbirds (Collis and Borgia 1992). Conversely, no significant relationship was found between social rank and androgens in dark-eyed juncos (Holberton, Able, and Wingfield 1989) and house finches (Belthoff, Dufty, and Gauthreaux 1994), and only a marginal relationship between metabolic rate and dominance was found for dippers (Bryant and Newton 1994).

Few costs that are directly linked to badge size have been identified. One risk associated with conspicuous plumage frequently mentioned in the sexual selection literature is increased conspicuousness to predators (Olsson 1993b; Andersson 1994). In support of this, Slagsvold, Dale, and Kruszewicz (1995) reported higher predation on bright male pied flycatchers relative to drab (female-like) males in the breeding season. Other studies have cited predation as a possible constraint to badge size, but without quantitative support. These include studies on tree lizards, least auklets, and white-crowned sparrows (references in Table 2-1). The idea is that males are differentially vulnerable to predation such that only superior males can bear the cost (increased risk of predation) of a large badge (the handicap principle). It has been argued that in birds, males with larger bibs are unlikely to be significantly more conspicuous to predators (Senar 1999). Instead, it is more likely that dominant males are more active or exposed than subordinates (with smaller badges), and therefore more susceptible to predation (Veiga 1993). Furthermore, a recent study on nest defense in house sparrows showed that males with larger badges, in the presence of a mounted mustelid predator, performed more risky behavior. This is explained by an improving cost/benefit ratio because males with larger badges likely have greater confidence in their paternity (Reyer et al. 1998).

Deception in many bird taxa is socially controlled during contests (Table 2-1). Under the social control hypothesis, a subordinate individual bearing a dishonestly large badge will pay the full cost of cheating if challenged by a dominant individual bearing an honest badge (Maynard Smith and Harper 1988). Social control assumes like-versus-like aggression such that individuals of the same age and sex challenge each other (Maynard Smith and Harper 1988). Slotow, Alcock, and Rothstein (1993) criticize previous studies invoking social control of cheating and argue that like-versus-like aggression may not be sufficient to prevent deception if the benefits outweigh the costs and may instead be randomly directed. Furthermore, they evaluated all studies of birds testing the social control hypothesis and concluded that the evidence thus far (1993) was "tenuous." Some studies subsequent to that of Slotow, Alcock, and Rothstein (1993) have suggested social control of cheating (Table 2-1).

Study Organism

The Augrabies flat lizard (Plate 1), *Platysaurus broadleyi* (formerly *Platysaurus capensis* [part]; Branch and Whiting [1997]), is a relatively small cordylid lizard (adult snout-vent length [SVL]: 64-84 mm) restricted to rocky terrain in the Gordonia-Kenhardt district of the Northern Cape Province, South Africa. At Augrabies Falls National Park, local populations are dense and large aggregations frequently occur along the Orange River in the presence of their primary prey, the black fly (*Simulium* spp.). Black flies occur in discrete plumes in the fast-flowing sections of the river, where conditions are most favorable for breeding. Consequently, the lizards have an extremely abundant food source and frequently occur in aggregations numbering in the hundreds. Males are territorial (Whiting 1999), although not all males defend territories, particularly in high-density areas where territories are at a premium (Whiting, unpublished data). As a result, the mating system appears to vary between a classical territorial system in which males offer the females no resources, to a resource defense polygyny in which males control areas with high prey abundance (Whiting, personal observation). In general, males approach many females but given a choice prefer larger females (Whiting and Bateman 1999).

In areas of high lizard abundance, the frequency of male-male interactions is high, but physical contests are generally avoided through the use of concealed badges of status that are flashed at rivals. Status-signaling badges in *P. broadleyi* are orange and/or yellow abdominal patches (Plate 2). The badge is bordered anteriorly by a black chest. The ventral surfaces of the thighs are also orange, as is a lateral stripe that generally extends about 30-40% of the distance between the front and hind limbs. The anterior edge of the lateral stripe matches the anterior edge of the ventral component of the badge. The orange ventral surface of the thighs and the lateral stripe may enhance the effect of the badge and serve as signal amplifiers (Hasson 1991; Plate 2). Males display their badge by raising one side of their body to an angle of 70-80°, such that a rival has a clear view of their abdomen. This behavior is termed a *ventral display* (Plate 3) and may be performed from either a stationary position or while approaching a rival. Males also sometimes chase a rival either before or after a ventral display. Occasional disputes do result in fights, although infrequently and generally when a high-quality resource is disputed (Whiting, unpublished data) (Plate 4). It is currently unknown whether abdominal pigment production is costly in flat lizards. Therefore, for the purposes of this chapter, we refer to the colored abdominal patch as a badge.

Many factors influence contest success in lizards and territorial lizards in particular. These include residency/resource-holding power, body size, motivational state, and androgen levels (Cooper and Vitt 1987; Olsson 1992, 1993a; Marler et al. 1995; Tokarz 1995a). Residency in *P. broadleyi* is important for contest success; however, in high-density areas males have smaller territories or may not defend territories at all (Whiting, unpublished data). Males also are readily visible to observers and sufficiently habituated to humans to allow relatively close observation. *Platysaurus broadleyi* at Augrabies is therefore well suited for experimental manipulation of status-signaling badges.

Hypotheses

We tested the following five hypotheses:

1. Badge size is positively correlated to contest success and therefore fighting ability.
2. The lateral stripe is a component of the badge and/or acts as a signal amplifier.
3. More contests initiated by males with reduced badges will result in draws due to signal confusion (an "inferior" signal perceived by the receiver).
4. Rohwer's incongruence hypothesis: enlarged badges are not matched by increased levels of aggression (contest initiation).
5. Badge size is maintained as an honest signal based on independent metabolic costs (i.e., males with larger badges have higher daily energetic costs during the breeding season).

Methods
Study Area

We conducted field work during September–October 1996 and 1997 at Augrabies Falls National Park (hereafter Augrabies) (28° 35′ S, 20° 20′ E), Northern Cape Province, South Africa. Detailed descriptions of the study area can be found in Branch and Whiting (1997) and Whiting and Greeff (1997). Briefly, the most favored habitat at Augrabies are the granitic banks of the Orange River. Granite at Augrabies varies between very smooth and coarse, and lizards use crevices and exfoliating flakes for refuge. The numerous Namaqua fig trees also serve as gathering points. Lizards appear to use these trees for shade and also feed on ripe figs and insects attracted to the trees (Whiting and Greeff 1997, 1999; Greeff and Whiting 2000). Lizards act as dispersal agents for figs (Greeff and Whiting 1999) as well. Rainfall is erratic and occurs during the summer months (<4 to about 39 cm). Augrabies experiences cold winters (as low as $-2.9°C$) and hot summers (as high as $42.9°C$) (Weather Bureau 1996 in Branch and Whiting 1997).

Badge–Body Size Relationships

The same individual (MJW) took the following measurements on all males: SVL (nearest mm), head length (± 0.01 mm), head width (± 0.01 mm), and body mass (± 0.1 g). We assumed that any measurement error would be equal for all treatment groups and therefore did not assess measurement error through repeatability of measurements. Prior to release, we uniquely marked each male on the dorsum, using white enamel paint. We quantified the allometric relationship between badge and body size (SVL and mass) by first tracing badge area onto Plexiglas and then onto paper (brand and quality standard for all lizards). Then we cut out and weighed the tracing on a digital balance (± 0.0001 g). We traced badges from 383 lizards used in a separate study but from the same location. We also traced badges for males used in the testosterone study. To convert from mass to area (mm^2), we weighed 10 pieces of paper of known area and generated the following equation: badge area $= 0.2015 + 126.82$ (badge mass) ($r^2 = 0.9996$, $F_{1,8} = 19{,}443.2$, $P < 0.0001$). To ensure comparison of dimension-free variables, we cube root transformed mass, square root transformed badge area, and natural log transformed SVL. We also calculated Olsson's (1994a) condition index using the cube root of mass and SVL ($mass^{1/3}/SVL$). We investigated allometric relationships using simple linear regression and estimated the relationship between badge area and body size and condition using Pearson correlation coefficients.

Badge Manipulation

HYPOTHESES 1 AND 2 (EFFECTS OF BADGE SIZE AND LATERAL STRIPE) AND HYPOTHESIS 3 (SMALL BADGE AND SIGNAL CONFUSION)

We manipulated the status-signaling badges of free-ranging males using orange vermilion acrylic paint (Chromacryl®) to increase badge size and black to reduce badge size. Both colors closely resemble natural orange and black on lizards to a human observer. We did not measure spectral properties of badges, and, therefore, the reflectance values of the paint likely did not match that of the badges. This is not ideal and may introduce some bias but should be ameliorated by the controls (discussed next). Further, many lizard taxa have been shown to respond to paint-manipulated individuals in what has been perceived as normal behavior (reviewed in Cooper and Greenberg [1992]; Martín and Forsman 1999).

By using orange we controlled for color polymorphism; also, we painted the chest area on all males black (natural color) to control for possible extraneous effects resulting from variable black coloring. We randomly allocated lizards to the following treatments:

1. Control 1: sham painted with a dry brush ($n = 13$)
2. Control 2: badge painted with no adjustment to badge size ($n = 12$)
3. Badge increased by 30% ($n = 14$; the 30% increase in badge size was visually estimated and may have marginally exceeded natural badge size of the largest individuals)
4. Badge reduced by 30% ($n = 17$)
5. Lateral stripe erased ($n = 24$)
6. Badge erased ($n = 19$)

Of the 119 males we captured and manipulated, 99 (83%) were resighted for behavioural observations. The unpredictability of resighting individual males resulted in differences in sample size among treatment groups.

HYPOTHESIS 4 (ROHWER'S INCONGRUENCE HYPOTHESIS: BADGE SIZE AND AGGRESSION)

Rohwer's incongruence hypothesis suggests that individuals that do not behave in accordance with the status that their badge conveys will be persecuted by dominants. In Harris' sparrows, only individuals that received an enlarged badge in addition to a testosterone boost rose in social status (Rohwer and Rohwer 1978). We tested whether male aggression in flat lizards was congruent with badge size. We predicted that males with enlarged badges would be more aggressive and therefore initiate more contests. (Contest initiation was used as an index of aggression.) For all contests, we scored which individual was the initiator and quantified the relationship between aggression (contests initiated) and badge size (see Statistical Analyses).

Testosterone Manipulation

We further explored the relationship between aggression and badge size by manipulating the testosterone levels of free-ranging males. We evaluated the influence of testosterone on aggression and contest success by using two testosterone regimens. The first experimental group of males ($n = 17$) was injected with 600 ng of testosterone (200 µl of a 3 mg testosterone/ml saline solution), and the second group ($n = 16$) was similarly injected with 1,000 ng of testosterone (100 µl of a 10 mg testosterone/ml saline solution). A control group ($n = 17$) was handled similarly and pierced with a syringe needle, but not injected. Released lizards quickly resumed normal behavior, and we collected behavioral data 1.5 or more days after injection, depending on how quickly lizards were resighted in the field.

Costs of Aggression

HYPOTHESIS 5 (BADGE SIZE AND FIELD METABOLIC RATE)

Field Protocol. We measured field metabolic rates (FMRs) on free-ranging individuals using doubly labeled water (DLW) (Lifson and McClintock 1966; Nagy 1980, 1989). We captured 30 males during morning hours, and each received an intraperitoneal injection of sterile water containing 300 mg of 10 atom percent $H_2^{18}O$ and 2.5 mg of 99.9 atom percent ^2H_2O. (DLW-injected males were not used in any other aspect of this study.) We then drew blood from the suborbital sinus using 75-µl microhematocrit capillary tubes 2 to 3 h later. We also drew blood from a further four males that were not injected, to be used for measurement of background isotope concentrations. Depending on when they were captured, we released lizards either on the same day, during the late afternoon, or the following morning. Most lizards immediately resumed normal behavior and some males began displaying shortly after release. We flame sealed all blood samples in glass capillary tubes and refrigerated them. We weighed all lizards (± 0.1 g), measured them (SVL, ± 1 mm), and had their badges traced within a few hours of capture. We began recapturing lizards after 9 d, and after 13 d had recaptured 14 of the 30 injected lizards for a second blood sample.

Laboratory Protocol. We sent blood samples to the University of California, Los Angeles (UCLA), for isotopic analyses. Blood was distilled under vacuum to obtain pure water, which was analyzed, along with diluted injection solutions, for ^{18}O concentration by proton activation analysis (at UCLA) (Wood et al. 1975), and for deuterium concentration by gas-isotope-ratio mass spectrometry (at the Boston University isotope laboratory). We estimated body water volumes, necessary for the calculations of FMR, for the times of injection from the dilution volumes of injected oxygen-18 (Nagy 1983). We estimated body water volumes at recapture from body mass, assuming individuals maintained the same fractional body water content during the study. We calculated FMRs using equation 2 in Nagy (1980), as modified from Lifson and

McClintock (1966). We converted rates of CO_2 production to units of energy (joules) using the relationship 25.7 J/ml of CO_2 produced for an insectivorous diet (Nagy 1983).

Field Observations

We worked in an area in which lizards are habituated to tourists and respond to thrown food. When a marked male was resighted in the field, we altered natural spatial patterns by throwing bread to other nearby males. By reducing neighbor distances, we increased the likelihood of male-male contests between the manipulated male and nearby conspecifics. An independent groups design was employed in which each male was observed once. We recorded the following data for all contests during timed, 10-min trials: (1) initiator, (2) agonistic display (ventral display, ventral display–chase, or chase), and (3) outcome. A ventral display occurs when a male raises the side of his body and flashes his badge at an intruder. Ventral displays may or may not precede a chase and sometimes occur after a short chase. There were no significant (Mann-Whitney tests, $P > 0.3$; two-tailed) differences in contest outcome for ventral displays and ventral display–chases for any treatment group. Therefore, we pooled ventral displays and ventral display–chases for statistical analysis. We determined contest outcomes when one individual left the immediate area in response to agonistic behavior by its rival. If both lizards displayed agonistic behavior and neither individual withdrew, we considered the contest a draw. We timed (seconds) the duration of agonistic interactions only for the testosterone and DLW experiments.

Statistical Analyses

We assessed differences in body size among treatment groups using the χ^2 approximation of one-way Kruskal-Wallis analysis of variance (ANOVA) (two-tailed). Males that were not resighted in the field (and therefore lacked behavioral observations) were excluded from this analysis. Contests involving status signaling were analyzed separately from those involving a chase. We made the distinction because a chase always resulted in a win for the initiator and precluded status signaling. We therefore tested for significant differences in contest success as the ratio of ventral display–chase wins to the total number of contests excluding chases. Significant differences among treatment groups in the number of draws were analyzed as the ratio of total draws to the total number of contests excluding chases. Because of the high number of zeros in many of the behavioral categories, we used nonparametric Mann-Whitney tests and Kruskal-Wallis ANOVA for comparison of treatment means. For significant differences, we performed a comparison of mean ranks in which subsets of homogeneous means were grouped together using $\alpha = 0.05$ (Analytical Software 1996). We assessed behavioral correlations using Spearman rank correlation coefficients. For badge measurements, we established normality using rankit plots and Wilkinson-Shapiro tests. We used Spearman rank tests to assess the relationship between badge size and FMR and between FMR and body size (mass). Unless otherwise stated, tests were one-tailed for directional hypotheses. Means are reported ± 1 SE. All differences were considered significant at $\alpha < 0.05$.

Results

Badge–Body Size Relationships

Badge area correlated significantly with mass ($r_s = 0.523, n = 379, P < 0.0001$), SVL ($r_s = 0.493, n = 383, P < 0.0001$) and body condition (mass$^{1/3}$/SVL) ($r_s = 0.506, n = 379, P < 0.0001$). Badge area scaled allometrically to SVL by the equation: badge area $= -13.720 + 4.2886 \log_e$ (SVL) ($F_{1,381} = 122.21, P < 0.0001$), while it scaled allometrically to body mass by the equation: badge area $= 0.5511 + 2.0179 \log_e$ (mass) ($F_{1,377} = 142.25, P < 0.0001$) (Fig. 2-1).

Badge Manipulation

The treatment groups were not significantly different in SVL ($H = 5.27$, $P = 0.38$), head length ($H = 5.21, P = 0.52$), head width ($H = 5.34, P = 0.38$), and mass ($H = 6.66, P = 0.25$) (Table 2-3).

The behavioral categories for the two control groups showed no significant differences; therefore, the two groups were combined for all statistical comparisons (Mann-Whitney, $n = 25, P > 0.4$ in each case; two-tailed).

HYPOTHESES 1 AND 2 (EFFECTS OF BADGE SIZE AND LATERAL STRIPE)

Contest success (proportion of ventral display and ventral display–chase wins) was significantly different among treatment groups ($H = 19.03, P = 0.0008$). Males with enlarged badges won significantly ($P < 0.05$) more contests using ventral displays and ventral display–chases than lizards in the other treatment groups in the order 30% reduction > lateral stripe removed > badge erased. However, contest success between males with enlarged badges and the control group was not

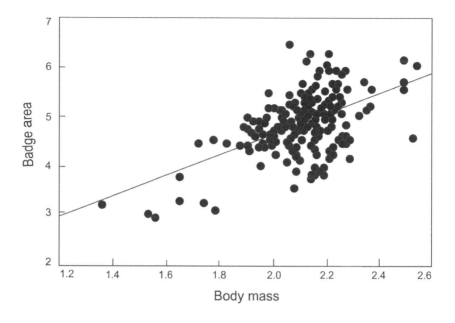

Figure 2-1   Relationship between badge area (square root transformed) and body mass (cube root transformed). The relationship is given by the equation: badge area = 0.5511 + 2.0179 $\log_e$ (mass).

Table 2-3   Descriptive statistics of body measurements for six treatment groups. Means are given ± 1 SE. Only lizards for which behavioral interactions were recorded are included (n = 99).

	n	SVL (mm)	Head Length (mm)	Head Width (mm)	Mass (g)
Sham painted	13	76.08 ± 0.51	16.45 ± 0.10	13.10 ± 0.09	10.75 ± 0.20
Control painted	12	76.50 ± 0.60	18.92 ± 2.50	13.12 ± 0.16	10.16 ± 0.33
30% Increased	14	76.29 ± 0.67	16.25 ± 0.11	13.00 ± 0.11	10.64 ± 0.22
30% Reduced	17	76.18 ± 0.40	16.22 ± 0.12	13.19 ± 0.13	10.54 ± 0.27
Lateral stripe erased	24	77.17 ± 0.29	16.59 ± 0.10	13.25 ± 0.11	11.04 ± 0.21
Badge erased	19	76.58 ± 0.49	16.33 ± 0.09	13.22 ± 0.11	10.40 ± 0.28

significantly different ($P > 0.05$), nor were there differences among the remaining three treatment groups with reduced badges and the control group ($P > 0.05$).

HYPOTHESIS 3 (SMALL BADGE AND SIGNAL CONFUSION)

The proportion of contests with no outcome (draw) was significantly different among treatment groups ($H = 11.94$, $P = 0.018$), with individuals from reduced badge treatment groups involved in more contests with no outcome (30%, lateral stripe erased) (Table 2-4). However, although an overall treatment effect was detected, the experimentwise error rate was sufficiently high to preclude significant differences ($P > 0.05$) during pairwise comparisons.

HYPOTHESIS 4 (ROHWER'S INCONGRUENCE HYPOTHESIS: BADGE SIZE AND AGGRESSION)

The mean frequency of all behavior by treatment group is reported in Table 2-4. The total number of contests initiated by all treatment groups was not significantly different ($H = 6.28$, $P = 0.09$), although the number of contests initiated with a ventral display was marginally significant ($H = 7.88$, $P = 0.05$). The number of ventral displays performed was significantly different ($H = 8.91$, $P = 0.03$); males with enlarged badges initiated significantly more contests than males with their badges erased ($P < 0.05$).

Initiation of contests was significantly correlated with contest outcome (wins) (control group: $r_s = 0.878$, $P < 0.0001$; 30% enlarged: $r_s = 0.975$, $P < 0.0001$; 30% reduced: $r_s = 0.725$, $P < 0.002$; lateral stripe removed: $r_s = 0.802$, $P < 0.0001$; badge erased: $r_s = 0.97$, $P < 0.0001$). A contest sometimes ended in a draw if lizards were equally matched or if the initiator's opponent failed to see the challenger (Table 2-4).

Manipulation of Testosterone

The two treatment groups and the control were not significantly different in SVL ($H = 2.275$, $P = 0.321$), head length ($H = 3.284$, $P = 0.194$), head width ($H = 0.331$, $P = 0.847$), mass ($H = 3.785$, $P = 0.151$), and badge area ($H = 0.105$, $P = 0.949$) (Table 2-5).

Of the 51 marked lizards, 41 were resighted for behavioral observations (Table 2-5). The number of contests initiated among the three groups was significantly different ($H = 6.202$, $P = 0.023$). Specifically, the second testosterone group initiated significantly more contests than

Table 2-4 Descriptive statistics of behavioral frequencies performed by males with manipulated badges during 10-min observation periods. Means are given $\pm$ 1 SE. See text and Table 2-3 for treatment group sample sizes.

	Control	30% Enlarged	30% Reduced	Lateral Stripe Removed	Badge Erased
Initiated with					
ventral display	1.28 ± 0.26	2.31 ± 0.49	1.24 ± 0.47	1.24 ± 0.24	0.88 ± 0.33
Total initiated	1.88 ± 0.40	3.15 ± 0.71	1.71 ± 0.53	1.62 ± 0.30	1.24 ± 0.40
Ventral display–					
chase wins	0.88 ± 0.21	2.31 ± 0.52	0.35 ± 0.19	0.33 ± 0.14	0.29 ± 0.14
Ventral display–					
chase: no					
result	0.44 ± 0.14	0.15 ± 0.10	0.76 ± 0.34	1.00 ± 0.21	0.47 ± 0.19
Total ventral					
displays	1.32 ± 0.25	2.46 ± 0.50	1.29 ± 0.48	1.33 ± 0.23	0.88 ± 0.33
Total won	1.52 ± 0.35	3.08 ± 0.75	0.76 ± 0.29	0.71 ± 0.18	0.88 ± 0.30

the control group ($P < 0.05$), but neither differed from the first testosterone group ($P > 0.05$) (Table 2-6). The amount of time engaged in agonistic interactions was significantly different among the three groups ($H = 7.413, P = 0.012$). Males in the second testosterone group also spent significantly more time ($P < 0.05$) engaged in agonistic interactions than control males, but neither differed from the first testosterone group ($P > 0.05$) (Table 2-6).

Initiation of contests was significantly positively correlated with contest outcome (wins) (control group: $r_s = 1.0, P < 0.0001$; testosterone group 1: $r_s = 0.926, P < 0.0001$; testosterone group 2: $r_s = 0.988, P < 0.0001$). In a few instances contests resulted in draws ($n = 2$), the opponent ignored the challenge ($n = 3$), or the opponent did not see the challenger ($n = 5$). Contest outcome (wins) was also significantly positively correlated with the amount of time engaged in agonistic interactions (control group: $r_s = 0.986, P < 0.0001$; testosterone group 1: $r_s = 0.858, P < 0.0001$; testosterone group 2: $r_s = 0.936, P < 0.0001$).

Costs of Aggression

HYPOTHESIS 5 (BADGE SIZE AND FMR)

Mean measurements for the 14 recaptured lizards relating to FMR are reported in Table 2-7. There was a significant positive correlation between FMR and badge size ($r_s = 0.31, P = 0.03$) (Fig. 2-2), but no significant relationship between FMR and body mass ($r_s = -0.08, P > 0.5$).

Table 2-5 Descriptive statistics of body measurements for testosterone and control groups. Means are given ± 1 SE. Only lizards for which behavioral interactions were recorded are included ($n = 41$).

	n	SVL (mm)	Head Length (mm)	Head Width (mm)	Mass (g)	Badge Area (mm²)
Testosterone group 1	14	76.73 ± 0.46	16.27 ± 0.11	13.22 ± 0.13	10.59 ± 0.27	2.76 ± 0.15
Testosterone group 2	12	77.50 ± 0.36	16.51 ± 0.11	13.25 ± 0.12	11.34 ± 0.20	2.80 ± 0.16
Control	15	77.80 ± 0.59	16.53 ± 0.13	13.56 ± 0.30	10.89 ± 0.26	2.58 ± 0.22

Table 2-6　Descriptive statistics of behavioral observations from testosterone experiment (10-min trials). Means are given ± 1 SE.

	Testosterone Group 1 ($n = 14$)	Testosterone Group 2 ($n = 12$)	Control ($n = 15$)
Contests initiated	3.64 ± 1.08	4.17 ± 1.02	1.60 ± 0.70
Duration of contests (s)	8.86 ± 3.31	14.08 ± 6.31	2.27 ± 1.01
Total contests won	3.21 ± 1.05	4.00 ± 0.96	1.60 ± 0.70

Table 2-7　Body mass and field metabolic rates (FMRs) of male flat lizards ($n = 14$) in relation to badge size.

	Mean	SE	Range
Body mass (g)	9.14	0.19	7.90–10.30
Measurement period (days)	12.20	0.43	9.00–15.00
Badge area (mm²)	2.26	0.05	2.10–2.82
FMR (l of CO_2/kg/d)	6.91	0.45	3.04–9.18
FMR (kJ/d)	1.61	0.11	0.75–2.20

We were able to collect limited behavioral data on 12 of the 14 lizards for which FMR was calculated. There was no significant relationship ($P > 0.5$) between aggression (contests initiated/min, type of aggressive behavior, contest duration) and badge size, but this was largely confounded by alternative reproductive strategies (Whiting, unpublished data). The sample split equally into territorial and floater males.

Discussion

Badge Area and Body Condition in Flat Lizards

In an analysis independent of the badge manipulation experiment, body size (mass and SVL) and body condition correlated significantly with badge area. The badge manipulation experiment controlled for differences in body size among treatment groups and thus demonstrated a role for badge area in contest success independently of body size/condition. This suggests that larger and/or older males are therefore superior fighters and badge area is a reliable index of fighting ability. However, the correlation coefficients (range: 0.493–0.523) between body size variables and badge area also show a high degree of scatter. This variation indicates differing abilities of males to settle contests and bear the costs of dominance and suggests that variation in fighting ability is not simply age related. Body condition may be the best indicator of fighting

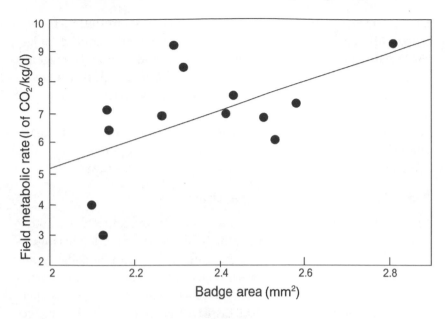

Figure 2-2   Relationship between FMR and badge area. There was a significant ($P = 0.03$) correlation between FMR and badge area (see text). The relationship is given by the equation: FMR $= -4.0334 + 4.6087 \times$ badge area.

ability because, unlike SVL, which may be more tightly linked to age in an organism with indeterminate growth, body condition could reflect resistance to parasites (Hamilton and Zuk 1982; Møller, Dufva, and Erritzøe 1998) and/or foraging ability (Møller and de Lope 1994), both of which ultimately may affect fitness.

Badge Area and Contest Success in Flat Lizards

Males with enlarged badges won more contests than other treatment groups. This finding confirms the presence of a status-signaling badge in *P. broadleyi* and demonstrates the importance of badge area in contest success. Status-signaling badges in *P. broadleyi* are polychromatic, which may influence contest success. However, we controlled for color polymorphism by painting the entire badge and chest area using the same paint for all lizards.

Males without a lateral stripe lost significantly more contests than badge-enlarged lizards, but not significantly more than control males, although a trend was apparent. Because lateral stripes are always visible and correlate with ventral badge area, they may passively convey information about the bearer's fighting ability. Further, during signaling, the lateral stripe may function as a signal amplifier (Hasson 1991). The negligible differences among the treatment groups with reduced badges (30% reduction, lateral stripe removed, and badge erased) suggest a threshold below which further badge reduction makes little difference to contest success. Contests resulting in a draw were most frequent among the treatment groups with reduced badges. These lizards were conveying signals that underestimated their fighting ability. This signal confusion may have increased the likelihood of an opponent returning the challenge and resulted in a greater proportion of draws.

Badge Area and Aggression in Flat Lizards

Males with enlarged badges did not initiate more contests than other treatment groups. However, when this analysis was restricted to contests initiated using a ventral display, the difference was marginally significant. Males with enlarged badges initiated the most contests, and males with their badges erased, the least. Males with enlarged badges also performed significantly more ventral displays than males with their badges erased but did not perform more ventral displays than the other groups. The absence of a strong relationship between contest initiation and badge size among all the treatment groups (excluding enlarged versus erased) could be due to the ventral placement of the badge. Most males were observed the day following badge manipulation, which may not have been sufficient time for the lizard to become aware of its new phenotype. Olsson (1994a) tested the relationship between badge area and contest success in the lizard *Lacerta agilis*. Although *L. agilis* with enlarged badges initiated more contests, this could be because the lateral placement of badges in *L. agilis* in contrast to the ventrally placed badges in *P. broadleyi* allowed immediate recognition of a manipulated badge. In addition, recognition of this new phenotype is likely to be proximally mediated by testosterone levels. In Harris' sparrows, an increase in badge size failed to influence status because their behavior was incongruent with their new badge size (incongruence hypothesis of Rohwer and Rohwer [1978]). However, males treated with testosterone implants in addition to enlarged badges demonstrated behavior congruent with their new badge size and attained a higher social rank. Male flat lizards with enlarged badges won more contests than males with reduced badges but only initiated more contests after testosterone boosts. This result therefore supports the incongruence hypothesis and highlights the importance of testosterone in agonistic behavior, as has been found for other lizards (e.g., Moore and Lindzey 1992; Tokarz 1995a).

Honest Signaling and Maintenance of Signaling Reliability in Flat Lizards

ESS models suggest that frequency-dependent selection should maintain badges of status as an honest signal, but only if there is a cost independent of fighting (Johnstone and Norris 1993). Badge (or pigment) production was previously thought to be energetically cheap (Kodric-Brown and Brown 1984; Krebs and Dawkins 1984), and this has been shown for melanin production in house finches (*Carpodacus mexicanus*) (Hill and Brawner 1998; but see Veiga and Puerta [1996]). However, carotenoid-based pigments are derived from food and are condition-dependent indicators of quality and are considered costly (Hill and Montgomerie 1994; Olson and Owens 1998). A growing body of literature supports the view that carotenoids are valuable because they are scarce (alternatives are reviewed by Olson and Owens [1998]). Therefore, brightness in males may reflect superior foraging ability and, as a consequence, quality. In general, herbivores ingest more carotenoids than carnivores and omnivores fall somewhere between the two (Olson and Owens 1998). Dietary requirements for birds are obviously very different from lizards (endo- versus ectotherms), especially lizards that are largely carnivorous and have fewer carotenoids available to them. However, *P. broadleyi* do occasionally ingest Namaqua figs and are thus considered omnivorous (Whiting and Greeff 1997). Figs are not readily available and are considered an unpredictable resource (Whiting and Greeff 1997, 1999). Given that *P. broadleyi*'s badge is orange and/or yellow, the role of carotenoids in badge development and expression cannot yet be excluded. However, both the presence and levels of carotenoids in Namaqua figs must first be determined.

Another hypothesis for constraints to cheating is that frequent aggressive encounters mediated through hormones could translate to a higher metabolic rate, providing a proximate cost to badge size (Marler et al. 1995). (In birds, there is also evidence of a three-way interaction among androgen levels, immunocompetence and badge size/dominance [Møller, Kimball, and Eritzøe 1996; Evans, Goldsmith, and Norris 2000; Poiani, Goldsmith, and Evans 2000] [see Table 2-1].) A positive correlation between badge size and metabolic rate has been demonstrated for great tits, pied flycatchers (Røskaft et al. 1986), and willow tits (Hogstad 1987). We found a significant correlation between badge size and FMR, independent of body size. This suggests that a physiological cost may constrain badge size, such that it honestly signals fighting ability. Unfortunately, we had insufficient data to explore properly the relationship between FMR and levels of aggression (also confounded by males adopting alternative reproductive strategies.)

A further cost of bright coloration is increased risk of predation (e.g., Darwin 1871; Endler 1980; Zuk and Kolluru 1998). Thus, under the handicap principle, ornaments are honest signals because the bearer is able to avoid predation under conditions of increased conspicuousness (either through possessing a larger badge or through behavior such as increased activity that correlates with badge size) (Zahavi 1975, 1977). It is difficult to exclude this possibility in the case of flat lizards; however, their badges are ventrally concealed and in most cases are only briefly exposed during contests.

For a signal to be effective, it must be easily detectable by the receiver in a manner that complements its physical habitat (Fleishman 1988a, 1992; Johnstone 1997). In a rocky landscape devoid of vegetation, individuals of *P. broadleyi* are readily visible. In addition, the orange and/or yellow badge is conspicuous against a black chest. Consequently, their display is simple and devoid of rapid, repeated motion, such as displays used by vegetation-dwelling lizards (e.g., *Anolis auratus*, Fleishman 1988a). In a highly dense population such as at Augrabies, in which male-male contests occur frequently, a readily visible badge of status effectively reduces conflict and risk of injury.

Future Directions

Of all the work on badges to date, very few studies have dealt with badge ontogeny (see Møller and Erritzøe [1992] and Senar, Copete, and Martin [1998] for birds, and Carpenter [1995b] for lizards). This is a neglected component of badge studies that could shed new light on the influence of badge developmental plasticity on future dominance and, ultimately, on reproductive success. An important issue is whether badges are fixed at birth, or a function of later interactions (continuous). A related issue to this is badge size heritability, for which estimates are reported for great tits (*Parus major*) and house sparrows (*Passer domesticus*) (Pomiankowski and Møller 1995). One way to explore the influence of badge ontogeny is to raise juvenile males in the laboratory under varying social conditions. Some treatment groups could be subjected to controlled interactions in neutral arenas to examine the effects of prior experience on badge development. This would also allow a proper analysis of badge development in relation to body size. The tracing of badges in the field may mask subtle, important differences in badge size. Digital analysis of badge size using pixel counts may be more reliable and should allow fine-grained analysis of variation in badge size in relation to body size and fighting ability. Finally, female mate choice (or preference) could be evaluated in individuals with a known history and in relation to known badge traits (chroma and area).

To develop further theory on the evolution of badges and ritualized encounters between males, more field studies on a variety of taxa are needed. The majority of studies have been on birds, and results of these have sometimes been unequivocal (see Slotow, Alcock, and Rothstein 1993; Senar 1999). In addition, much of the work has focused on the role badges have played in access to food rather than mates. Studies of status signaling in lizards number less than a handful (Table 2-1) but suggest potentially rewarding systems for future work. Compared to birds, many lizards have polychromatic badges (tree lizards, Zucker [1994a, b] and Carpenter [1995a, b]; flat lizards, this study) and even dual status-signaling badges (Zucker 1994a). Furthermore, unlike most birds (except red-winged blackbirds, Metz and Weatherhead [1992]), flat lizards have concealed badges that are flashed when using distinct behavior, while tree lizards use throat color to signal status. Different selective forces are likely to be at play here.

Unfortunately, there is a poor understanding of what may constrain cheating in lizards. Our study suggests a physiological cost independent of contests. Studies on the relationship between badge size and hormone levels are greatly needed (Whiting and Hews, unpublished data). A recent study by Olsson and Silverin (1997) showed that in sand lizards (*Lacerta agilis*), testosterone plays a trigger function during badge development; phenotypic expression of badge traits (area and chroma) are constrained by resource allocation. The possible link between badge size and resource holding potential (Maynard Smith and Harper 1988), possibly mediated through hormones, requires further investigation. We hope to evaluate more fully badge size and aggression in light of trade-offs among metabolic rate, androgen levels, and immunocompetence, relative to honest signaling and constraints to cheating. Additionally, the potential role of carotenoids in badge development in lizards with yellow/orange coloration requires investigation. Finally, numerous models now exist showing that deception is likely to occur to some degree (e.g., Johnstone and Grafen 1993; Johnstone and Norris 1993; Semple and McComb 1996). Although many studies have manipulated phenotypes to introduce "cheaters" into a population, almost nothing is known of the incidence of deception in natural populations (but see Adams and

Caldwell [1990] and references therein). This is likely to be an extremely challenging line of research but could be key to a better understanding of social evolution.

Acknowledgments

Permission for this work was granted by the National Parks Board of South Africa. The van der Walts, Mombergs, and staff at Augrabies provided logistical support and encouragement. We thank Hayley Komen for painstakingly cutting and weighing badges. MJW is extremely grateful to Debbie Bellars for 2 yr of expert field assistance on several different lizard projects and to Hayley Komen for assistance in 1997. We also thank Francis Thackeray and Bill Cooper for statistical advice. Comments from Troy Baird, Dirk Bauwens, Bill Cooper, Rufus Johnstone, Johnathan Losos, Anders Møller, Mats Olsson, Trevor Richter, and Hannes van Wyk on earlier versions of the manuscript greatly improved its quality. We acknowledge that the reviewers were not always in agreement with our conclusions and acknowledge full responsibility for the viewpoints published here. Juan Carlos Senar and some of the reviewers listed above kindly provided reprints and key references. Graham Alexander provided advice and equipment for the DLW component of the study; Duncan Mitchell and Helen Laburn did the same and generously provided the deuterium at no charge (all University of the Witwatersrand). Robert Michener (Boston University) measured deuterium in samples. MJW thanks Stanley Fox, Troy Baird, and Kelly McCoy for inviting him to take part in the symposium and for generous financial assistance via the National Science Foundation and the Herpetologists' League. Financial assistance from Hannes van Wyk (University of Stellenbosch) and the Department of Biology at the University of California at Los Angeles made the DLW study possible and is especially appreciated. Additional support was provided to MJW by the National Research Foundation, South Africa.

CHAPTER 3

Ecological and Social Contexts for the Evolution of Alternative Mating Strategies

Kelly R. Zamudio and Barry Sinervo

Alternative mating strategies have long fascinated evolutionary biologists. A system in which all males are not equal in physical attributes, and are caught up in the most basic competition for reproductive success, is certain to capture the attention of any natural historian. Fueling this fascination even further is the anthropomorphic and colorful terminology (for better or worse) that is often applied to different mating strategies. We hear of sneaker and bourgeois males, dominants and satellites, reproductive parasites, egg thieves, and pirates. Our curiosity about alternative mating strategies started with very careful observations of mating systems and behavior long before our understanding of the common patterns that exist among widely divergent species. Only recently has it become clear that alternative mating strategies occur in almost all animal lineages (Le Boeuf 1974; Gross 1991; Shuster and Wade 1992; Erbelding et al. 1994; Lank et al. 1995; Sinervo and Lively 1996; Thomaz, Beall, and Burke 1997; Jones, Ostlund-Nilsson, and Avise 1998; Lanctot et al. 1998; Vinnedge and Verrell 1998; Yamane 1998). The challenge for future studies will be to document more fully the diversity of strategies that exists in nature, as well as the conditions favoring the evolution and maintenance of these systems. Such an endeavor should focus initially at the level of inter- and intrapopulation variation in alternative strategies, rather than on comparison of mating strategies among species. Focusing our attention at the intraspecific level of analysis allows us to explore mechanisms such as natural and sexual selection that shape variation among and within populations. The generalities among species will thereby become clearer.

Each decade the study of biology benefits from theoretical and technological advances, and this has been true for the study of mating strategies as well. First, Maynard Smith's application of economic game theory to alternative mating strategies and other evolutionary games provided the theoretical and analytical framework for the newly developing field (von Neumann and Morgenstern 1953; Maynard Smith and Price 1973; Maynard Smith 1982). Second, the relatively young field of environmental endocrinology (Wingfield and Farner 1975) documented the mechanistic underpinnings of many vertebrate alternative mating systems in terms of endocrine regulation of behavior and life history (Wingfield and Moore 1987; Moore and Thompson 1990, Cardwell and Liley 1991; Moore 1991; Ketterson and Nolan 1994; Sinervo et al. 2000). Finally, the application of molecular techniques has answered explicit questions about parentage and reproductive success and offered us a window through which we can quantify the selective outcome of alternative reproductive strategies (Schartl et al. 1993; Jones and Avise 1997; DeWoody et al. 1998).

Here we focus on the first and third of these three areas, because they reflect the causal selective mechanisms that promote the evolution of alternative strategies. The second area, examining proximate endocrine mechanisms, is an important component of the analysis of how alternative mating strategies operate and how they develop, but it is still unclear how they might impact the outcome of selection. Proximate mechanisms may constrain the evolution of certain alternative strategies, as has been shown for endocrine regulation of certain life history traits (Marler and Moore 1988; Ketterson et al. 1991, Sinervo and Licht 1991a, b; Sinervo et al. 1992; Sinervo and DeNardo 1996; Sinervo and Doughty 1996; Sinervo and Svensson 1998; Sinervo 1999, 2000). In these cases, selective mechanisms may be channeled by constraints of the endocrine system into some evolutionary avenues for the diversification of alternative strategies but not others. Furthermore, basic endocrine mechanisms are broadly conserved across a diversity of vertebrate taxa (Licht et al. 1977), and this may limit reproductive strategies in a wide diversity of taxa to a small number of options. The fact that we can classify most mating tactics in a handful of categories reflects this limited set of options. Alternatively, this conservatism may be the outcome of frequency-dependent selection, which results in a limited set of behavioral alternatives regardless of underlying mechanism. This remains a major unresolved issue underlying the evolution of mating systems, but one that will necessarily require additional data on precise selective regimes underlying this diversification (Sinervo 2000).

Alternative mating strategies can be divided into two distinct groups, depending on their degree of genetic determination: those in which mating strategy polymorphisms are labile (condition dependent, facultative, or phenotypically plastic) and those in which alternative strategies are genetically determined (or at least partially so) (Austad 1984; Caro and Bateson 1986). Although these two groups usually exhibit interesting similarities, it is important to distinguish between them because the conditions necessary for their evolution are different (De-Woody et al. 1998), as are the implications of competitive outcomes between or among morphological types (morphs) adopting different alternative strategies. In facultative or plastic alternative strategies, morphs will evolve in cases in which the benefits of a specific strategy outweigh the cost. Obviously, the context in which these morphs coexist will influence the cost-benefit ratio; however, advantageous contexts for each morph need only exist some of the time for these behavioral polymorphisms to be maintained in a population. On the other hand, genetically based strategies are not labile; individuals exhibiting a specific strategy usually do so for life and have no opportunity to change strategies midstream. In this situation, very precise conditions must exist for the maintenance of polymorphisms to occur over evolutionary time. If one morph routinely obtains higher fitness than other morphs in a genetically based system, the less fit morphs will be removed by the action of selection. Thus, one of two conditions must exist: the morphs must have exactly the same fitness in every reproductive bout, or the morphs may vary in fitness but on a frequency-dependent basis, such that the fitness of morphs over the long term averages to equal levels. In particular, negative frequency-dependent selection refers to a system in which morphs are maintained because each morph has a strong fitness advantage when rare (Maynard Smith 1982, Sinervo and Lively 1996). Even in situations in which the morphs have exactly equal fitness, it is likely that frequency-dependent selection maintains this equilibrium. Thus, researchers should focus on measuring the frequency-dependent selection, and hence the selective processes by which morphs are maintained in a population.

Many theoretical studies have examined the exact requirements for the evolution of alternative mating strategies; however, most empirical evaluations in natural populations have focused on the success rates of morphs in facultative systems. In addition, although many alternative mating systems have been described, few have actually probed the social context of the competition between morphs, and the mechanisms that promote the evolution and maintenance of mating polymorphisms in nature. Here, we describe a well-studied case of alternative mating systems found in some populations of the side-blotched lizard (*Uta stansburiana*) and report on the genetic assessment of paternity in this population. In the first part of the chapter, we focus on intrapopulation social contexts that we believe promote positive selection for maintenance of the morphs. In the second part, we extend these ideas to comparisons among populations and examine the social and ecological contexts that may explain why alternative systems such as these have evolved in some populations and species but not in others.

The Study System

The side-blotched lizard (*U. stansburiana*) is a small and common territorial lizard that is widely distributed in North America. Because this species reaches fairly high densities in most parts of its range, it has been a model for studies of population dynamics, demography, and behavior. Like most other phrynosomatid lizards (Martins 1994), males of this species are highly territorial, and the mating system has been generally described as resource-defense polygyny, with large territory holders gaining access to many females whose home ranges are included within their territories.

Monographic studies of this species (Tinkle, McGregor, and Dana 1962; Tinkle 1967a, b; Fox 1983) have focused on specific aspects of territoriality such as size, degree of overlap, and individual interactions between competing males. Some populations of this species in the coastal range of California exhibit a curious combination of alternative male mating strategies that has been described previously as a rock-paper-scissors game (Sinervo and Lively 1996). As in the children's game, the three alternative strategies interact in a system that has no single winner. Instead, each morph has strengths that allow it to outcompete one other strategy, but weaknesses that leave it vulnerable to the tactics of the third. Males adopting these three strategies have complex behavior and specific morphologies (Plate 5). Orange-throated males are aggressive, have higher levels of testosterone (Sinervo et al. 2000), and vigorously defend large territories (Calsbeek and Sinervo, unpublished data), which presumably affords them access to a larger number of females. Blue-throated males are also territorial but are mate guarders and stay with their females after copulating. Mate guarding may prevent their females from copulating with other males; however, this behavior may also interfere with territorial defense, potentially limiting access by blue males to additional mates. Finally, the yellow-throated males are nonterritorial; these morphs are sneakers that make opportunistic forays into territories of other males and copulate with their females. To do so, sneakers not only behave surreptitiously (thus avoiding detection by the territorial male), but also rely on female mimicry—their throat and dorsal coloration are most similar to patterns found in females.

The social and behavioral interactions among these morphs set the stage for the competitive mechanisms that underlie the rock-paper-scissors game. Each morph has specific behavioral attributes that allow it to outcompete only one of the other morphs. Orange-throated males are able to outcompete the blue-throated mate guarders through sheer aggression (Calsbeek and Sinervo, unpublished data). On the other hand, the mate-guarding behavior of the blue-throated males allows them to outcompete the yellow-throated sneakers, because they effectively deter sneakers from copulating with their females. Finally, the yellow-throated sneakers are especially effective in sneaking copulations from females within the large territories of the orange-throated males.

The rock-paper-scissors game is a genetically based system (Zamudio and Sinervo 2000), and thus requires very specific evolutionary conditions. Previous behavioral estimates of fitness (Sinervo and Lively 1996) found that the three morphs exist in an evolutionarily stable state. Negative frequency-dependent selection maintains each phenotype in the population and all three morphs may have equal fitness in the long term (Zamudio and Sinervo 2000). Morph frequency cycles in this population accordingly, and the rare morph always has high fitness, until it becomes more common in the population (Fig. 3-1). For example, the fitness of sneakers will be highest when orange males are present in large numbers, because this should offer sneakers ample opportunity to sneak copulations from within the territories of these males (Sinervo and Lively 1996).

Lizard Territories: Focal Points for Male Competition

An analysis of the long-term fitness of the three cycling strategies must average the frequency-dependent fitness of each morph at all stages of the cycle and across the entire population. An analysis of this sort requires not only long-term demographic data but also long-term paternity data sets. By contrast, an analysis of the local frequency-dependent processes that underlie these patterns provides a more rapid assessment

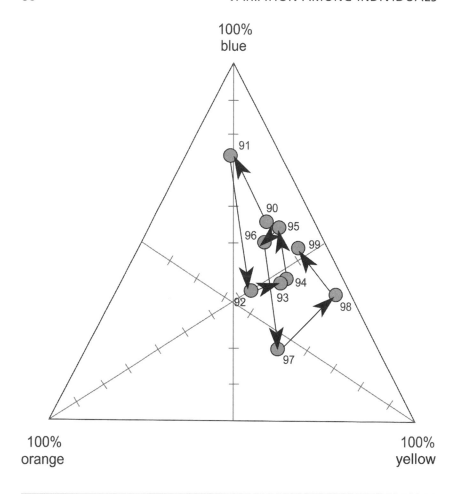

Figure 3-1 The proportion of males adopting each of three alternative strategies varies in a cyclical fashion due to frequency-dependent sexual selection. From 1990 through 1999, the cycle was completed twice in the Los Baños population.

of the forces maintaining alternative strategies in this population. Demonstrating frequency-dependent selection provides a better estimate of the global stability of the system, particularly if the mating system is maintained by negative frequency dependence in which rare morphs have a fitness advantage. The scale at which these behavioral interactions occur is the level of competing groups of neighboring males, in that the exact composition of males within a neighborhood is expected to determine the fitness of all males within that group. For example, the success of any one focal orange male should depend on the number of males of the other two morphs that come in direct competition with him. Therefore, an orange male in a neighborhood consisting of only blue males should have high success if orange males are in fact successful at aggressively outcompeting blue males, whereas we would predict that an orange male in a field of yellow males would be the target of many sneakers, and thus his fitness would be substantially lower (Fig. 3-2). It is neighbor-neighbor interactions that determine the fitness outcome for any particular male (Sinervo and Lively 1996; Zamudio and Sinervo 2000).

Given the importance of these interactions at the level of neighborhoods, we investigate how reproductive success is partitioned at a microgeographic spatial scale. Using a paternity analysis based on microsatellite loci (Zamudio and Sinervo 2000), we examined two spatial parameters relevant to male reproductive success. First, we examined the degree of proximity among females and males of each reproductive morph to determine whether mating opportunities vary for the different male types. Second, we extended our analysis beyond competition among males and again used spatial associations and paternity to explore whether female choice may play a role in male success.

During the 1992 mating season, we mapped the exact location of all male territories and female home ranges in one population of side-blotched lizards near Los Baños, California (see Sinervo and DeNardo [1996] for details on the field site). Territories were mapped by making multiple passes over the field site early in the reproductive season (during the first reproductive bout). At this time, male densities were highest and individual interactions within neighborhoods most intense (Sinervo et al. 2000). We reconstructed territory sizes by minimum convex polygons using the program MacTurf (Sinervo 1996). We estimated territory sizes for all males with at least three sightings, and on average we had 8.4 sightings/male (range = 3–16). Estimates for males with low numbers of sightings may underestimate true territory sizes (Rose 1982). However, this does not seem to be a significant source of error in this study because our territorial estimates are comparable with those previously described in a more detailed analysis of males on a neighboring outcrop at the same location (DeNardo and Sinervo 1994). The sizes of male territories were not equal among morphs, and their distributions on the outcrop were not random. Orange-throated territorial morphs had significantly larger territory sizes than the home ranges of the sneaker yellow-throated males (mean size ± 1 SE: orange = 39.5 ± 8.8 m^2, $n = 29$; blue = 22.8 ± 4.7 m^2, $n = 50$; yellow = 20.5 ± 4.8 m^2, $n = 46$; Fisher's LSD post-hoc comparisons: orange versus yellow, $P =$

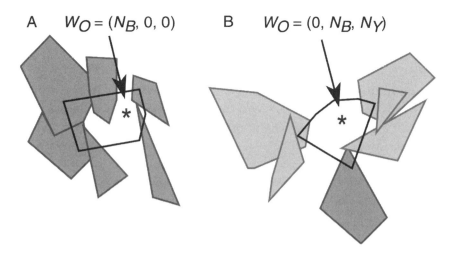

A $W_O = (N_B, 0, 0)$ **B** $W_O = (0, N_B, N_Y)$

Figure 3-2 The social context (neighborhood) of any one individual male has important consequences for his fitness. In this example, the polygons represent the territories of different male morphs; male fitness (*W*) is directly related to the number of males of each morph that are in the immediate vicinity. An orange-throated male in an exclusively blue neighborhood (*A*) will have higher fitness than an orange-throated male in a neighborhood with many yellow-throated sneakers (*B*).

0.03; orange versus blue, $P = 0.05$; blue versus yellow, $P > 0.75$). In addition, yellow-throated males associated with blue-throated males less often than would be expected by chance, suggesting that the mate-guarding strategy of the blue males is effective against sneakers (Zamudio and Sinervo 2000).

It is often assumed that territory size is an important determinant of mating opportunity. In territorial species, whether based on resource defense or mate defense, males with larger territories will possibly have access to more females available for mating. We used our field and laboratory data on the side-blotched lizards to ask two simple questions: (1) Does territory size correlate with mating opportunity in males in this population? and (2) Does overlap with female home ranges in fact translate into reproductive success for the males? The answer to the first question is positive: males with larger territories overlap with home ranges of more females (Fig. 3-3; $n = 126$, $r^2 = 0.591$, $F_{1,124} = 179.3$, $P < 0.001$); thus, at least theoretically, they should have higher mating opportunities. On the other hand, female home range size does not correlate directly with the number of overlapping males; female home ranges are generally smaller than male territories, and the number of

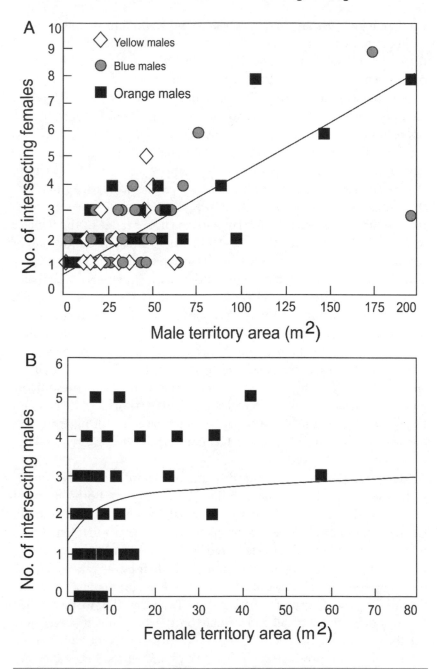

Figure 3-3   (*A*) Male territory size was positively correlated with access to females (judged by the degree of overlap between male territories and female home ranges). (*B*) Female home range sizes were smaller overall and not positively correlated with the number of overlapping males.

overlapping males is best described by a nonlinear relationship (number of intersecting males = 1.339 + 0.385 × ln(female area); n = 66; r^2 = 0.137; $F_{1,64}$ = 10.166, P < 0.05) with an average of 1.9 males on each female's home range (SD = 1.3; range = 0–5).

Of course, in a system with alternative mating strategies, degree of territory overlap may very well not be a good indicator of reproductive opportunity. Furthermore, if we consider that female choice may also have an effect (Alonzo and Sinervo 2001), it is quite possible that overlapping a female's territory does not guarantee successful reproduction. If we examine the spatial associations of females with males that sire their offspring in this population, we see that most sire territories (62.8%) do not necessarily overlap with the home range of the females with which they reproduce (Fig. 3-4). In other words, spatial overlap of territory alone is not sufficient to guarantee reproductive success every year (Baird, Acree, and Sloan 1996), and in some years, territorial males do not monopolize females. We would predict this pattern from the cyclical nature of the rock-paper-scissors game: in years during which orange territorial males are the most common, defending territories may in fact not result in higher fitness, because in these years male sneakers are the most effective. On the other hand, in years of high orange fitness (when they are rare), their despotic behavior presumably will result in monopoly of females on their territories (Sinervo and Lively 1996). It is this context-dependent fitness (resulting from frequencies of male morphs) that determines the payoff for territorial defense and maintenance.

It has traditionally been quite difficult to separate the effects of male-male competition from female choice (Tokarz 1995b). Both aspects of sexual selection can have, under some situations, the same outcome in terms of male reproductive success. Nonetheless, the fact that females do not necessarily have offspring fathered by the resident male suggests that females may be selecting mates from outside their home ranges. We explored whether females go outside their home ranges to mate and, if so, with which males they preferentially mate. To address this, we performed a neighbor analysis of all individuals on the field site to determine whether the linear distances between territories or home ranges were correlated with reproductive success. We estimated the linear distance between the centroid of a male's territory and the centroid of the home range of the female with which he sired offspring. Distances between females and the sires of their offspring were not even for all male morphs. The average distance between yellow males and the females with which

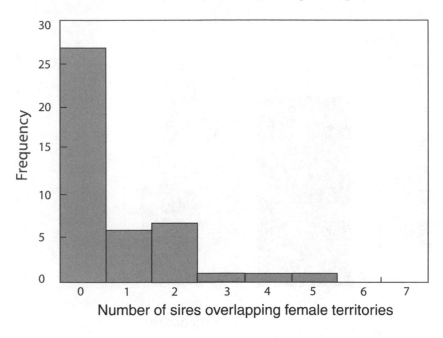

Number of sires overlapping female territories

Figure 3-4 The DNA-based paternity analysis was used to estimate the frequency with which males actually overlapped the territories of females with whom they sired off-spring. A large proportion of offspring produced during the 1992 reproductive season was sired by males that were not present on the female's territory. Thus, monopoly of females through territorial defense does not occur in this system.

they successfully reproduced was significantly smaller than that for the other two morphs (Fig. 3-5). In other words, if females are choosing yellow males, they do not have to go very far to do so.

The exact nature of sexual selection that is operating in this system is not yet clear, but our spatial analyses suggest that more than just male-male competition is determining the fitness of males in this population. Theory suggests that in any given year of the cycle, females should choose the male morph that maximizes reproductive success of their progeny (Alonzo and Sinervo 2001). Estimates of male fitness within all neighborhoods on the outcropping (derived from the 1992 DNA paternity data) (Fig. 3-6) indicate that all male morphs have an advantage when rare in their neighborhood (Zamudio and Sinervo 2000). Thus, a female should mate with a male that produces sons with the rare male strategy (Alonzo and Sinervo 2001) in order to maximize a son's fitness. In 1992, females should have preferred the rare yellow male, because yellow sons in 1993 would have enjoyed high fitness; the close association

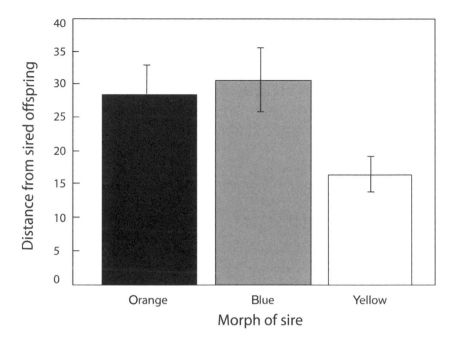

Figure 3-5   The distance between a male's territory and the home range of the female with which he sired offspring was not the same for all morphs in this population. In particular, females that mated with yellow-throated sneaker males did so with individuals that were significantly closer than fathers of the other two morphs (mean distance of sire ± 1 SE: orange = 28.7 ± 4.2 m, n = 53; blue = 30.6 ± 4.8 m, n = 49; yellow = 16.8 ± 2.6 m, n = 50; Fisher's LSD post-hoc comparison of orange versus blue, P = 0.75; orange versus yellow, P = 0.04; blue versus yellow, P = 0.02).

of females with yellow fathers may be indicative of active mate choice. Paternity data from other years (when other morphs should be favored) will help clarify the relative contribution of female choice and male competition to individual fitness in this population.

The Social Landscape: Dispersal, Competition, and the Scale of Environmental Heterogeneity

Social interactions of territorial species are spatially local and are thus restricted primarily to interactions that occur among nearest neighbors. Because of this spatial restriction, the "grain," or scale, of environmental heterogeneity can have important consequences for evolutionary dynamics. Many theoretical studies have focused on the role spatial variation in selection plays in maintaining polymorphism (Levene 1953; Gillespie 1974, 1975). Theory predicts that heterogeneity alone is

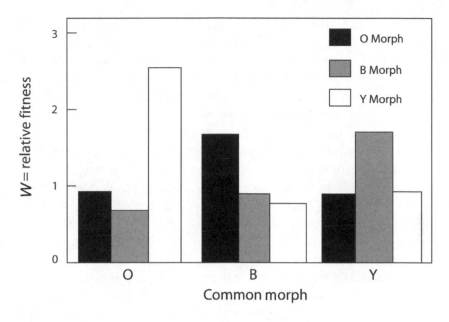

Figure 3-6   Male morphs have higher fitness when rare and when competing with a common morph. The payoff matrix of relative fitness when morphs are present at different frequencies was estimated by examining male frequencies and paternity rates in all neighborhoods on the outcrop. The fitness estimates confirm the predictions of the rock-paper-scissors model previously proposed for this system and underscore the importance of social context in determining an individual's success.

not sufficient to promote polymorphism or specialization, but the selection coefficients must result in disruptive selection in which each "morph" is strongly favored in a particular situation (Maynard Smith and Hoekstra 1980). Environmental heterogeneity at the appropriate scales obviously will also have effects on selection regimes (Hedrick, Ginevan, and Ewing 1976; Hedrick 1986; Brown and Pavlovic 1992), although the resulting magnitude of fitness variation is not always proportional (Bell and Lechowicz 1991; Stratton 1994).

The analysis of paternity as a function of mate proximity described earlier provides an index of the grain of the social environments or local neighborhoods in this species. In general, female and male side-blotched lizards overlap with only a few neighbors (0–5), and neighborhood size does not increase dramatically if we consider the area in which they sire progeny (1–10). Here we review the theory that predicts the evolutionary outcome of selection in fine and coarse-grained environments and apply that theory to social environments.

In general, spatial variation in performance promotes specialization and local adaptation, whereas homogeneity promotes general strategies (Levins 1968). Empirical and theoretical studies have demonstrated that the degree to which heterogeneous environments promote differentiation depends also on the scale of dispersal and competition (Mitchell-Olds 1992). For example, reduced migration makes the establishment of polymorphism more likely (Felsenstein 1976); therefore, spatial variation is clearly important when it occurs at scales large enough to allow differentiation of subgroups into local variants. Of particular importance is the scale of the environmental variation (or selective environment) environment relative to the scale of dispersal and competitive neighborhoods (Stratton and Bennington 1996, 1998). The influence that selective environments have in promoting differentiation is dependent on the total proportion of the selective environment to which each individual is exposed (or experiences). Thus, if the selective environment is very fine scale, one population or individual will likely experience many different selective regimes within the confines of its dispersal and competition realms. On the other hand, if the selective environment is coarse grained, an individual will likely experience only one selective regime (and one genotype by environment interaction) in its lifetime, resulting in the promotion of specialization or a different morph for each selective environment found in a population. Many mechanisms can act to reduce the total proportion of available heterogeneity experienced by any one individual, and thus promote diversity. A simple example is one of spatial segregation due to microhabitat specialization. Specialization for particular microhabitats results in spatial genetic structuring of the population, which, in turn, reduces between-genotype competition and dispersal among habitats. Genotypes thus tend to remain in or disperse to habitats where they have high fitness, further reinforcing the maintenance of specialized types.

These theoretical considerations are important in describing the relationships among dispersal, competition, and scale of heterogeneity in the selective environment. Likewise, understanding how these variables interact in particular systems will help us understand the total range of possible evolutionary trajectories for the evolution of specialization. We apply these ideas to the consideration of the evolution of alternative mating strategies in lizards and ask whether these three variables may help explain the array of systems observed in populations of the side-blotched lizard and perhaps even among species of lizards. Lizards often show finite levels of dispersal (Doughty and Sinervo 1994), competitive interactions over varying distances, and complex patterns of spatial variation in fitness (Sinervo and DeNardo 1996; Sinervo and Lively 1996). Thus, it would appear that these three variables may help us predict the evolutionary dynamics for species with certain characteristics. We outline a heuristic selection landscape for a breeding lizard and examine the conditions that should promote the evolution of specialist and generalist breeding tactics.

Evolution of Genetically Fixed Alternative Strategies

Ideas about the role of environmental heterogeneity in maintaining diversity can be extended to explore the context for the evolution of various lizard mating systems. Many lizards are territorial, and this has several important implications for the social interactions an individual will experience in its lifetime: (1) dispersal is somewhat limited because of spatial monopoly, (2) competitive neighborhoods are relatively stable, and (3) "social mobility" (the probability of changing neighborhoods) is low. Because, as we have argued previously, the social context of a lizard can have important consequences for an animal's fitness, we can think of a "social landscape" analogous to the fitness landscape already described. In other words, because social context and the outcome of dyadic interactions among the morphs are so closely linked to fitness, the social landscape can be viewed as an additional layer that ultimately affects fitness (Table 3-1). To this landscape we bring the idea of scale and "grain": if the social landscape is fine grained (if lizards experience many different social contexts, each affording different fitness outcomes), then we would predict the evolution of a generalized social strategy. On the other hand, if the landscape is coarse grained, an individual lizard will not experience many different social contexts. Because of either limited dispersal or reduced "social mobility," a lizard will find itself in a competitive neighborhood, where its fitness is fixed. This coarse-grained social landscape should promote the evolution of diversity, in this case, alternative mating strategies. Social environments can vary temporally as well, which also will affect the grain of the social landscape. In short-lived organisms, social environments are usually much coarser than for long-lived organisms. Long-lived organisms will experience a diversity of neighbors, especially as they grow and age and their dominance status changes.

We propose that all mating strategies can be categorized somewhere along this social landscape axis (Table 3-1), ranging from fine to coarse grained. Of course, dispersal ability and competition will determine the

Table 3-1   Environmental, life history, and social characteristics that promote either generalized or specialized mating systems. We hypothesize that the social context of lizards, together with environmental conditions, and life history characteristics may be important determinants in the evolution of alternative mating strategies. Characteristics of the environment interact with the temporal and spatial scale at which competition and dispersal events happen, favoring the evolution of either generalized or specialized reproductive strategies.

	Mating System	
Environment/Species Characteristic	Generalized	Specialized
Environment/distribution of resources	Homogeneous	Heterogeneous
Fitness variation among environments	Low	High
Potential for monopoly	Low	High
Social landscape grain	Fine	Coarse
Opportunity for "social mobility"	High	Low
Life span	Longer	Shorter
Breeding season	Longer	Shorter
Dispersal	High	Low

relative grain of the same landscape for different species, but the interaction between these factors will result in either generalized or specialized mating systems. One obvious question is, How does the structure of the environment itself influence the grain of the social landscape? The answer is that it is probably very important. The role of the environment, such as the patchiness of resources, the degree to which they can be monopolized, and total resource availability, have long been appreciated as important factors in determining the cost-benefit trade-off involved in resource defense and reproduction (Emlen and Oring 1977), and certain classes of habitat may be associated with the evolution of specific mating systems or characteristics (Wiens 1999). We believe that the environment plays a similar role (and through many of the same mechanisms) in determining the grain of the social landscape (Table 3-1). Territorial lizards will have a greater chance of monopolizing resources, and thus reducing dispersal of conspecifics, if resources are limited and those resources are concentrated and defendable to the exclusion of other individuals (Emlen and Oring 1977).

The situation we observe in *Uta* can be used as an example of how distributions of resources might promote despotic behavior in nature. The *Uta* at Los Baños live on large sandstone outcroppings in grasslands. The outcroppings are composed of a jumble of boulders of different sizes and in different spatial arrangement, resulting in large microhabitat variability. Because lizards are ectotherms, one important aspect of their territory is the diversity of thermal resources that a male controls. A male that monopolizes various microhabitats, with different sun exposures, that maintain different temperatures during the day may be active much longer (Adolph and Porter 1993; Grant and Dunham 1988, 1990) and may be able to control his territory more effectively (Ruby 1981). Likewise, a juvenile settling in a thermally heterogeneous territory may grow larger because this affords longer daily growing seasons (Sinervo and Adolph 1989, 1994; Sinervo 1990; Adolph and Porter 1993).

As we have shown, side-blotched lizard male territories vary in size and can occupy one small boulder or monopolize many boulders with heterogeneous microhabitats. The exact structure of the environment has three important consequences for lizards and their social landscape. First, this kind of structured environment lends itself well to monopoly by despotic individuals (orange-throated males). Boulders and rock piles have natural borders, and the differences in height allow for effective patrolling by individual males. Second, because of this monopoly, movement of other lizards across the outcropping is unlikely and, in fact, quite unusual (DeNardo and Sinervo 1994). Territory positions, although not fixed, tend to be relatively stable, limiting the opportunity for lizards to change their position in the social landscape, as well as limiting large dispersal events. Thus, a coarse scale of grain in the social landscape and the limits imposed on movement and competition mean that a side-blotched lizard male that reaches maturity and establishes a territory on such a rock outcropping will very likely experience only one set of neighbors during his reproductive life. The probability of changing neighbors is extremely limited, and most males and females die after their first breeding system (Sinervo and DeNardo 1996; Sinervo and Lively 1996; Sinervo 1999); this situation should promote specialized breeding tactics. Third, the complexity of the environment per se allows a sneaker to move along the margins of territories or surreptitiously along the bottom edges of boulders (yellow-throated males). The only defense against the sneaky strategy would be a mate-guarding male (blue-throated male) that attempts to keep females within viewing distance. By contrast, the wide-ranging dominant strategy of orange males is particularly vulnerable to sneakers (Zamudio and Sinervo 2000).

Studies of other species inhabiting different environments will serve as an important contrast in examining the role of environment in the evolution of alternative mating strategies. Strategies such as the rock-paper-scissors game are not found in all populations of side-blotched lizards. In fact, despite many careful and long-term studies focusing specifically on population dynamics and territoriality in this species (e.g., Tinkle, McGregor, and Dana 1962; Tinkle 1967a, b; Fox 1975, 1978, 1983; Ferguson and Fox 1984; Wilson 1991), alternative strategies have been reported only in populations in the coastal range. It is possible that what varies among these populations are the basic components favoring the evolution of diversity (and thus alternative tactics). In other words, it may prove fruitful to examine the same demographic and environmental parameters (the grain of the social landscape) across populations in various parts of this species' range.

For example, some of the previous detailed demographic studies of *Uta* populations took place in Texas, in a habitat quite different from the situation described in the Los Baños population. Adult side-blotched lizards at these sites maintain territories, but they inhabit sandy expanses in primarily flat and scrubby areas. This habitat is more continuous and does not have the "hard edges" that so clearly form territorial boundaries on specific rock boulders in the Los Baños population. Less distinct territorial boundaries may result in a larger degree of movement. Alternatively, the marginal gain in fitness of large movement and defense of mates in widely separated areas may favor mate guarding and a monogamous pattern in which males defend a single female (Orians 1969; Emlen and Oring 1977). Ultimately, environmental structure contributes to a change in the grain of the social environment experienced by adult lizards. In the case of the side-blotched lizards in Texas, there is no opportunity for despots to control resource. Thus, the social landscape is relatively fine grained relative to the dispersal capability of individual lizards; lizards may experience more than one social context in the course of their reproductive season, and their fitness may vary in those various contexts. This will promote a generalized breeding strategy, perhaps one that works relatively well over the whole spectrum of contexts in that social landscape.

Variation in fitness promotes the evolution of alternative mating systems, particularly if frequency dependence yields a disruptive pattern of selection that favors particular morphs at different times. Lizard species afford us with a wealth of opportunities to examine intraspecific variation in mating systems and the underlying patterns of selection. In some groups of lizards, variation in genetically based alternative mating strategies seem to be coupled to the evolution of conspicuous throat colors (Thompson and Moore 1991a, b; Carpenter 1995b; Sinervo and

Lively 1996), head bobs, and push-up displays (Carpenter and Ferguson 1977; Martins 1993). In particular, *Urosaurus ornatus* exhibits tremendous population variation in throat color. A population in one geographic locale may possess one set of throat color types (e.g., orange, green, yellow; Carpenter 1995b), whereas an entirely different set is displayed in another population (e.g., orange and orange-blue; Thompson and Moore 1991a, b). Other phrynosomatid lizards in the genus *Uta* (Sinervo and Lively 1996) and, to a lesser extent, in the genus *Sceloporus* (Rand 1990, 1992) exhibit similar variation. A recent comparative analysis by Wiens (2000) demonstrated that the evolution of display morphology is uncoupled with the evolution of display behavior in phrynosomatid lizards, and that the display morphology he examined (presence/absence of belly patches) was more labile than changes in behavior. It is possible that the variation in throat color that we observed among populations also reflects this evolutionary lability (Wiens 2000).

This variation among populations in throat color "ensembles" in phrynosomatids needs to be distinguished from variation found in the family Polychrotidae, a large neotropical clade. In polychrotids, males typically exhibit a single color, which can vary dramatically among species and populations, but less so among individuals within a population (Duellman and Schwartz 1958; Christman 1980; Losos 1985b; Cooper and Greenberg 1992). The role of throat color variation in this lineage is thought to play an important role in species recognition (Williams and Rand 1977; Losos 1985a). On the other hand, variation in ensembles of throat color variation found in many short-lived phrynosomatid lizards reflects an underlying diversity of mating systems.

We hypothesize that this dichotomy in the action of sexual selection, in which ensembles of throat colors arise in some groups but not others, is driven by an underlying difference in life history that impacts the grain of a lizard's social environment. In short-lived species such as *Uta* or *Urosaurus,* in which breeding is "explosive" and tied to either spring or summer monsoon rains, sexual selection is extremely intense and individuals are likely to experience only a small subset of territorial neighbors. Such short time scales and coarse-grained social environments promote the evolution of distinct genetically based polymorphisms. By contrast, in longer-lived temperate sceloporine lizards or species with year-round opportunity for breeding as in Neotropical anoline lizards, an individual is likely to experience a diversity of social environments during its lifetime. Even short-lived *Anolis* lizards have a very long breeding season with numerous single-egg clutches (Rand 1967c;

Andrews and Rand 1974); their social environment is thus relatively fine grained because of the potential for temporal heterogeneity. When temporal heterogeneity exists within an individual's lifetime, the coarseness of the social environment will not promote genetically based polymorphisms. If alternative mating strategies evolve, they are likely to be plastic or facultative. Thus, both temporal and spatial heterogeneity or grain can influence an animal's experience in the social landscape.

We make a specific prediction that absence of genetically based mating strategies within populations is causally related to environmental conditions (e.g., short breeding season, low environmental potential for monopoly) and certain life history traits (such as life span) that limit a species' position on the social landscape. Whether or not the reproductive tactics, if present at all, are phenotypically plastic will be secondarily determined by the degree to which the physical environment enhances the concentration of resources, and thus a despotic distribution of male success. By contrast, the presence of genetically based diversity in mating strategies within a population is causally related to environmental and demographic conditions favoring short life span or nearly annual life history. Under such conditions, temporal heterogeneity in social environment will be coarse grained in that individuals are likely to experience only one social neighborhood. In addition, the degree to which the physical environment results in concentration of resources and favors despotic males is still a crucial factor because this makes the social environment spatially coarse grained. Thus, we propose that temporally and spatially coarse-grained social environments are necessary (but not sufficient) conditions for the evolution of genetically based strategies. We have explored these ideas with specific reference to the side-blotched lizard, *U. stansburiana,* in which a complex alternative mating system, the rock-paper-scissors game, has evolved in one population while variation in other populations seems quite unspectacular. Such a post-hoc analysis of one system will be useful only if the information we have presented herein inspires others to collect comparative data in other populations and other species of lizards in which such polymorphisms are present or absent.

Alternative Strategies in Other Lizard Species

Our own natural history observations on other phrynosomatid lizards provide additional qualitative support for our hypothesis of spatial and temporal social grain. In addition they suggest ecological settings where future research might be focused to test the role of spatial and temporal grain in promoting plastic versus genetically fixed alternative male strategies. During a decade of work on the sagebrush lizard, *Sceloporus graciosus* (Sinervo and Adolph 1989, 1994), we realized that in some populations this species adopts a mating system that appears plastic in that males adopt alternative mating strategies with age. In particular, in populations on Throope Peak (2,000–3,000 m) in the San Gabriel Mountains of California, we observed satellite and despotic mating strategies. Very old and large males (10–11 g) defend isolated rock piles and fallen logs in the relatively open habitat. Females aggregate around these rock piles and logs, resulting in a polygynous system in which large males are apparently capable of defending a group of females. The rock piles and logs provide the best oviposition sites in the dry sandy ridge inhabited by these populations. However, each large male also has a rosette of younger satellite males that presumably attempt to sneak copulations from the females. The large dominant males have belly patches comprising at least three distinct colors of blue ringed by black. By contrast, the coloration of the smaller males is greatly reduced and more closely resembles the pale blue patches of nonbreeding females.

At lower elevations (<2,000 m) in the San Gabriel range, *S. graciosus* populations are sympatric with their larger congener, the western fence lizard, *Sceloporus occidentalis.* In areas where the two species co-occur, even the largest male sagebrush lizards have a markedly reduced intensity of blue coloration, with only one or two blue colors on the belly patches. Moreover, *S. graciosus* males are significantly smaller (6–8 g) and appear to defend only one or two females. No other obvious alternative strategies are present in the sympatric populations. In these populations, large western fence lizard males typically monopolize the best thermal environments and tend to exclude the sagebrush lizards, *S. graciosus,* to lower-quality thermal territories. In the case of the high-elevation sagebrush lizards, if strategies are present at all, they appear to have an age component and are thus presumably plastic and not genetically fixed. Sagebrush lizards are relatively long-lived (Ruth 1977), and, thus, we would predict the evolution of plastic alternative strategies, provided that the environmental grain in the physical or resource environment favors the evolution of despotic male strategies. It would be fruitful to examine intraspecific variation in this species to confirm whether the demographic and social environment favor plastic alternative strategies in some areas and lack of alternative strategies in others.

A similar situation is also found in subspecies of the large Sierra fence lizard, *S. occidentalis taylori,* which is the only species occurring at high elevation in the Sierra Nevada of California. Males of this sub-species exhibit an ontogenetic transformation of belly patches (E. L. Bell, personal communication; Stebbins 1985; Bell and Price 1996). Pre-sumably, *S. occidentalis taylori* in the high Sierras has a long life span, al-though detailed demographic studies have yet to be completed for this subspecies. The striking ontogenetic transformation of males may also reflect a plastic set of alternative mating strategies that afford modest opportunities early in life and tremendous mating opportunities to the oldest and largest males.

An analogous situation is found in one subspecies of the eastern fence lizard, *Sceloporus undulatus erythrocheilus,* in which bright orange and yellow male types are found during the breeding season at higher elevation (Rand 1990, 1992). By contrast, low-elevation subspecies of *S. undulatus* tend to be relatively monomorphic and have blue on their throats and bellies. One male morph in *S. u. erythrocheilus* defends a despotic distribution of resources and females, while the other morph appears to adopt an alternative nonterritorial strategy (M. S. Rand, per-sonal communication). Whether or not the trait is phenotypically plas-tic or genetically determined has yet to determined, but the endocrine basis of the strategies has already been documented (Rand 1992).

A detailed comparison of the demography and reproductive success in populations at different elevations and the frequency-dependent re-productive schedule of male lizards (e.g., parentage as a function of ter-ritorial neighbors and age) will be required to test our hypothesis of temporal and spatial grain of the social environment in shaping alter-native strategies in lizards. Reproduction in montane environments is explosive on a yearly basis as the breeding season is abbreviated to 2 or 3 wk. Thus, temporal coarse grain could promote alternative tactics (despotic and satellite males), but the fine grain of the social environ-ment over the lifetime of a lizard may favor phenotypic plasticity (males transform from satellite to dominant) and the ability to change strate-gies. DNA paternity analysis in these mating systems is clearly war-ranted. The occurrence of throat and belly pattern and behavioral poly-morphisms in high-elevation, allopatric populations of three species of *Sceloporus* suggests that some common environmental factor may un-derlie the evolution of alternative strategies in these environments. In addition to the proposed role of environmental and spatial grain in these montane environments, an obvious biotic effect that deserves attention is the role of sympatric congeners in reducing the occurrence of alternative tactics at lower elevations. Reproductive character displacement and species competition for ecological resources may restrict the diversity of male types found within a single species, thereby allowing the two species to partition the habitat. For example, the lack of bright blue coloration in sagebrush lizards at middle elevations in the San Gabriel Mountains, while males are more splendid at high elevation, may be driven by selection arising from the presence of the large brightly colored western fence lizard at middle elevations. Such interspecific competition may thus limit the evolution of alternative mating and ecological polymorphism within a single species.

A detailed understanding of the exact relationship among the environment, the social landscape, and the consequences for individual male lizards requires a level of knowledge about one particular system that is rarely available. Testing our hypotheses about the environmental heterogeneity and its consequences for the evolution of alternative mating strategies will necessarily involve an understanding of how male fitness varies in those environments, and to what degree individual males experience different parts of the social landscape over their lifetime. Answering these questions is difficult enough for one population, let alone many; nonetheless, it may be useful to think of the relationship among these variables as a framework for studies of alternative mating strategies and their differences among populations and species.

A second potential avenue for research would be a comparative study of phrynosomatid lizards. Using the most comprehensive phylogeny for this group (Reeder and Wiens 1996; Wiens and Reeder 1997), one could examine the phylogenetic distribution of mating systems and their concordance with specific life history and demographic parameters. A phylogenetic analysis such as this will require detailed natural history observations on the mating systems of many more taxa along with corroborative estimates of reproductive success. The estimates of parentage serve two vital functions: first, the frequency dependence of the strategies, and thus selection, can be determined from paternity data (Sinervo and Lively 1996; Zamudio and Sinervo 2000); second, long-term parentage studies in natural populations also provide information on whether the strategies have a Mendelian or quantitative genetic basis (Sinervo and Doughty 1996; Sinervo and Zamudio 2001) or whether they are plastic strategies.

Finally, we have only begun to determine the role of female choice in the origin and maintenance of alternative mating strategies. Theory predicts that context-dependent mate choice will evolve in mating systems in which male and/or female morphs cycle in frequency (Sinervo, Svensson, and Comendant 2000; Alonzo and Sinervo 2001). Carefully controlled mate choice studies of the sort commonly performed in fish (Houde and Endler 1990) would be a valuable complement to studies of male-male competition and parentage (Tokarz 1995b).

CHAPTER 4

Social Behavior and Antipredatory Defense in Lizards

William E. Cooper, Jr.

Because social behavior is complex and crucially linked to reproductive success, its various forms, factors affecting them, and relationships to reproductive success constitute a major focus of the behavioral literature. Until fairly recently, the relationships between reproduction and other key autecological elements such as foraging and avoidance of predation have received very little emphasis. Although antipredatory adaptations are better studied due to their crucial role in survival, most of the research has been conducted without regard to social behavior. Yet, because reproductive success demands survival as well as appropriate social behavior, strong interactions may be expected between antipredatory adaptations and social behavior.

What are the nature, extent, and importance of these interactions? This question has not been answered fully but has been addressed increasingly by behavioral ecologists in the past 20 yr. Optimality theory, which attempts to predict behavioral decisions from a balance between costs and benefits (Krebs and Davies 1993), is especially useful in predicting trade-offs between opportunities to engage in social activities and needs to escape from approaching predators. Its use has led to the discovery of such trade-offs in many taxa. In this chapter, I review connections between social and escape behavior in lizards in light of predictions of optimal escape theory regarding the effects of social encounters on escape decisions.

Predators affect social behavior primarily at early stages of predator-prey interactions, i.e., the detection, identification, and approach phases of encounters, rather than during the subjugation and consumption phases. The temporary presence of predators in the area is likely to suppress or alter movement by potential prey, and thus conspicuous behavior such as fighting and courtship, until after the predator has left the area. For example, the intensity of courtship by male threespine sticklebacks (*Gasterosteus aculeatus*) decreases greatly in the presence of a predator (Candolin 1997). Dominant males of many taxa temporarily inhibited by predators from display or fighting might lose matings to subordinates, especially those not in the predator's field of view and those employing stealthy mating tactics. Males that stop courting in the presence of a predator may lose the attention of females to other males that continue courting, suggesting that males must balance the benefits of mating opportunities against avoidance of predators (e.g., Candolin 1997). Approach by or the presence of predators may also affect decisions regarding social antipredatory behavior such as mobbing, but such behavior patterns are unknown in lizards.

When a predator approaches, prey must decide whether and when to flee. Optimal escape theory predicts that prey should flee when the predator approaches to the point at which the costs of remaining exceed the costs of fleeing (Ydenberg and Dill 1986). A major factor affecting the costs of remaining is the detectability of the prey. Prey that remain immobile are less likely to be detected, especially if cryptically colored, but the probability of being captured if detected increases as the predator nears. Prey that are immobile when they first detect a predator must weigh the joint probability of being detected and of being captured if they are detected while deciding whether and when to flee. Prey engaged in searching for mates, signaling to them, or active sexual or aggressive behavior have greater risk of detection and capture than immobile individuals (Magnhagen 1991; Jakobsson, Brick, and Kullberg 1995). Risk of detection also is affected by the directness of a predator's approach (Burger and Gochfeld 1981, 1990; Cooper 1997b) and eye contact (Burger, Gochfeld, and Murray 1992; Cooper 1997c), which are unlikely to be strongly affected by the prey's social behavior, and by the distance of the prey from refuge, which may be related to social activities. For example, copulation in some species may occur in or close to refuges to reduce the probability of detection by predators and conspecifics. Other prey living in open habitats where obstructions to view might conceal predators may conduct their social activities far removed from cover.

The distance between the prey and the approaching predator at which escape is initiated is called the *flight initiation distance* or *approach distance*. For prey at a fixed distance from refuge, the most obvious factor affecting flight initiation distance is the escape speed of the prey relative to the pursuit speed of the predator (Ydenberg and Dill 1986). Escape speed may be altered by consequences of social behavior. For example, gravid females may be slowed as a consequence of previous sexual behavior (Shine 1980), and individuals of either sex may be slowed by injuries incurred during intraspecific fighting. Individuals thus slowed may be expected either to flee when a predator is farther away to maintain a margin of safety or to remain nearer to a refuge.

Many factors affect the cost of fleeing, and therefore the optimal flight initiation distance. For species that flee to refuge, the energetic cost of fleeing is likely to be a minor consideration (Ydenberg and Dill 1986) relative to risk of predation, but for species that must flee for long distances to escape, energetic expenditure may importantly affect escape decisions. Animals in poor physical or energetic condition due to social behavior might lack the energetic resources needed for sustained flight or pursuit-deterrent display (FitzGibbon and Fanshawe 1988). For example, debilitated males recently deposed from dominance over mammalian harems might be forced to defend themselves in situ because they cannot flee effectively.

The most thoroughly studied costs of fleeing are those of lost opportunity (Ydenberg and Dill 1986). Time spent fleeing and hiding may result in lost social opportunities to fight, mate, or guard mates against conspecific rivals. These losses can have severe consequences for fitness. Reluctance to flee should increase as the probability that fertilizations may be lost increases and as residual reproductive value declines. Lost opportunities to feed due to fleeing and hiding may have indirect social consequences because time lost for feeding may have to be made up later at some cost to time allocated to social activities.

In addition to its current effects on relationships between social behavior and defense, the focus of this review, predation has a strong influence on the evolution of social structure, mating systems, colonial breeding, flocking, roosting, and other aspects of sociality (Lima and Dill 1990). If existing social structures have been determined in part by historical predation pressure, there may be correlations between social structure and types of antipredatory adaptations that are effective under the selective regime that produced both. Methods for reconstruction of behavioral states on phylogenies and for correlating traits on the same phylogeny (Brooks and McLennan 1991; Harvey and Pagel 1991) offer excellent opportunities for studying evolutionary relationships between lizard social behavior and antipredatory adaptations.

Social Behavior and Antipredator Behavior in Lizards

Although there have not been a great many studies of social-antipredatory links in lizards, enough information is available to show

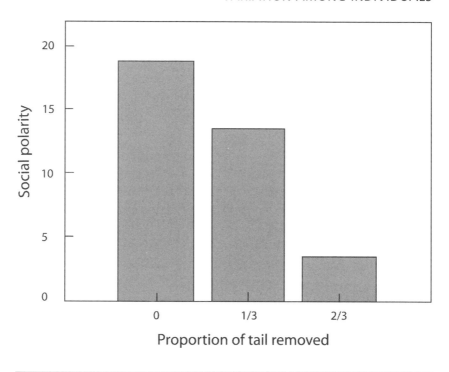

Figure 4-1 Effect of tail loss on dominance in juvenile *Uta stansburiana*. Loss of two-thirds of the tail caused a significant reduction in social polarity, which is the difference in dominance scores between dominant and subordinate individuals in paired aggressive encounters. Based on data from Fox and Rostker (1982).

that they are important in several contexts. Social behavior may be suppressed or strongly modified by antipredatory behavior. This is best documented for trade-offs between escape and several social activities, including courtship, mate guarding, and intraspecific agonistic behavior (Cooper 1999b; Díaz-Uriarte 1999; Cooper and Vitt, unpublished data). Tail autotomy, an antipredatory adaptation occurring widely among lizards, has profound effects on social behavior (Figs. 4-1 and 4-2) (Fox and Rostker 1982; Fox, Heger, and DeLay 1990; Salvador, Martín, and López 1995). As a consequence of previous reproductive behavior, gravid lizards may become sufficiently slowed that they must alter their escape behavior, especially by remaining close to refuges (e.g., Cooper et al. 1990; Braña 1993).

This review emphasizes relationships between social and escape behavior and between tail autotomy and social behavior, but there are

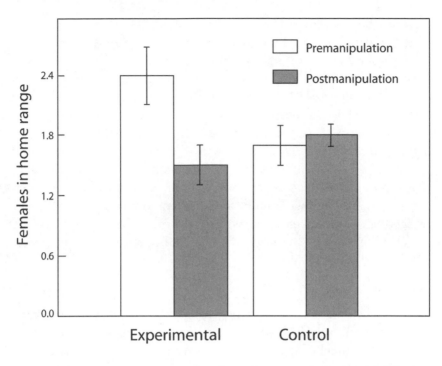

Figure 4-2   In *Psammodromus algirus*, males from which tails were removed subsequently had home ranges that overlapped with those of significantly fewer females, but no such change occurred in control males having intact tails. Based on data from Salvador, Martín, and López (1995).

other important interfaces between predation and social behavior that are noted only in passing. It has been realized recently that lizards direct pursuit-deterrent displays to predators, informing predators that they have been detected and, therefore, have greatly reduced probability of capturing the prey (Dial 1986; Hasson, Hibbard, and Ceballos 1989; Leal and Rodríguez-Robles 1997; Cooper, in press).

Parental care following birth or eclosion is absent or minimal in lizards, but females that remain in nests with their eggs may greatly reduce predation on the eggs by their social antipredatory behavior. Females of several species in the scincid genus *Eumeces* and anguids of the genus *Ophisaurus* stay in the nest with their eggs during incubation (Noble and Mason 1933; Vitt and Cooper 1985a, 1989). In *E. septentrionalis* (Somma 1985) and *E. obsoletus* (Evans 1959), females may aid the hatching process, but experimental tests of *E. laticeps* and *E. fasciatus* revealed no association between mothers and their hatchlings in the first few days after hatching (Vitt and Cooper 1989). Any defense of young against predators is presumably limited to the incubation period. The presence of female *E. okadae* in the nest greatly enhances survival of the eggs (Hasegawa 1985), but it is uncertain whether or to what extent maternal defense against egg predators contributes to egg survival. Brooding female *E. fasciatus* vigorously defend themselves and/or their eggs against a variety of small predators including mice, lizards, and small snakes (Noble and Mason 1933). Brooding female *E. fasciatus*, *E. inexpectatus*, and *E. laticeps* in the laboratory all consume insects that might eat eggs (Cooper, unpublished observations). That females may weigh the costs of defense against the survival of the eggs is suggested by abandonment of the eggs when faced by much larger predators (Noble and Mason 1933).

Escape and Social Behavior

Surprisingly little is known about effects of social behavior on escape decisions by lizards. This section focuses on details of the only four studies of optimal escape by lizards in relation to social costs (Tables 4-1 and 4-2).

COURTSHIP

A male lizard that is actively courting a female has a chance to increase his fitness materially by fertilizing the female's eggs. Disruption of the mating opportunity by a predator could be very costly, especially if the opportunity might not occur again, which is especially likely for males courting unfamiliar females that may be transients unlikely to be encountered again. Thus, when a predator approaches during courtship, escape may result in a very costly lost mating opportunity. According to optimal escape theory, the optimal flight initiation distance is the point at which the costs of escape are exactly balanced by the risk of remaining (Ydenberg and Dill 1986).

For courting males the costs of escape are greater than for noncourting males in the same setting. Assuming that the male has detected the approaching predator, its risk of being captured should the predator attack is the same as that of a noncourting male at the same distance from the predator. Optimal escape theory predicts that a courting male should allow a predator to approach closer before attempting to flee than a noncourting male. In effect, the large cost of a lost mating opportunity makes greater risk acceptable.

Table 4-1   Effects of courtship and mate guarding on escape behavior by lizards. All effects are for *Eumeces laticeps* with the exception of that reported for *Psammodromus algirus* by Martín and López (1999b). FID, flight initiation distance.

Courtship	
Males given the opportunity to court had shorter FID than when females were absent.	Cooper (1999b)
Isolated males given the opportunity to court had shorter FID than guards.	Cooper (1999b)
Many males that fled from the predator when a female was present often returned to the female, drawing closer to the predator; control males did not.	Cooper (1999b)
Latency to approach a female was greater when a predator was closer to the female.	Cooper (1999b)
Mate guarding	
Mate-guarding males had shorter FID than isolated males.	Martín and López (1999b)
Females had longer FID than guards, leaving guards exposed after the females fled.	Cooper and Vitt (unpublished data)
After females fled from a predator, guards followed them more slowly, prolonging risk.	Cooper and Vitt (unpublished data)

Table 4-2   Effects of agonistic behavior on initial escape and reemergence from refuge following escape by lizards. Data in Cooper (1999a) are for the skink *Eumeces laticeps*. Those in Díaz-Uriarte (1999) are for the iguanian *Tropidurus hispidus*. FID, Flight initiation distance.

Approached by predator during agonistic encounter	
Given the opportunity to fight, males had shorter FID than when other males were absent.	Cooper (1999b)
Given the opportunity to fight, mate guards had shorter FID than isolated males.	Cooper (1999b)
Males in ongoing encounters had shorter FID than males that had finished agonistic encounters 5 min before being approached.	Díz-Uriarte (1999)
Males in ongoing encounters and males that had finished agonistic encounters 5 min before being approached had similar times to reemergence.	Díaz-Uriarte (1999)
Approached by predator 5 min after encounter	
Males that fought and control males had similar FID.	Díaz-Uriarte (1999)
Males that fought took longer to reemerge from refuge than control males.	Díaz-Uriarte (1999)

In the broad-headed skink, *E. laticeps,* the breeding season is brief. Females are nearly synchronously sexually receptive during a brief span of a few weeks in late spring (Vitt and Cooper 1985a; Cooper and Vitt 1997). Males fight for opportunities to mate with females (Vitt and Cooper 1985a; Cooper and Vitt 1987), and large males guard females, typically remaining within 1 to 2 m from the females while following them for a week or longer (Vitt and Cooper 1985a; Cooper and Vitt 1997). The short breeding season and length of time spent guarding females greatly limit opportunities for males to mate with multiple females, but males do attempt extrapair matings when the opportunity arises (Cooper and Vitt 1997). Given these reproductive features, opportunities to court unfamiliar females may be very valuable for males, offering the possibility of greatly increasing fitness.

The broad-headed skink is thus a good species for testing the prediction of optimal escape theory regarding the trade-off between escape and courtship. When on the ground, this semiarboreal species escapes by fleeing toward or into refuges, which may be trees, bushes, or surface litter large enough to conceal these skinks (Cooper 1997a–c, 1998a). Several aspects of the broad-headed skink's escape behavior are consistent with the predictions of optimal escape theory. Flight initiation distance increases with distance from refuge and the angle of the line of flight to refuge with respect to the predator's path of approach (Cooper 1997a). In both cases, the increase in flight initiation distance is predicted by the increased risk of being captured. Broad-headed skinks flee at a greater distance from approaching human experimenters simulating predators when they approach more rapidly or more directly, when they change direction to move toward the lizard, and when they have approached previously (Cooper 1997b, c, 1998b). These effects, too, represent increases in optimal flight initiation distance compensating for higher risk due to greater predator speed, higher probability of having been detected and of being under attack, and predator persistence indicated by recurring attacks.

In contrast to these studies on the effects of increased risk on escape behavior, studies of social effects on escape focus on the predicted effects of altering costs of escaping. The effect of courtship on escape by male broad-headed skinks was studied by introducing tethered females to males in the field (Cooper 1999b). I introduced a tethered female to a male by positioning it 0.5 m from the male by means of a fishing rod, and then I stood motionless 2.7 m from the male. There were two groups: isolated (solitary) males and mate-guarding males. I stood motionless until courtship began, as indicated by the male approaching the female, tongue flicking her, and attempting to grasp her neck skin in his jaws, which is typically the final step of courtship before attempting intromission (Cooper, Garstka, and Vitt 1986). I then slowly approached the male being tested (about 35 m/min). When it fled, I stopped moving and recorded flight initiation distance and whether or not the male returned to the introduced stimulus when I stopped approaching.

Both groups permitted very close approach before fleeing, consistent with the high cost of losing a mating opportunity, but isolated males allowed closer approach than did mate-guarding males (Fig. 4-3). The prediction that introduction of tethered females would decrease flight initiation distance owing to the cost of lost courtship opportunity was robustly confirmed. Isolated males detected the predator but gave little sign of incipient flight until the predator approached very closely, typically allowing the experimenter to approach within 0.1 m before retreating. The flight initiation distances of both groups to which females were introduced were much shorter than for control males in a related experiment on agonistic behavior (Cooper 1999a).

This finding that isolated males take much greater risks than mate-guarding males (Cooper 1999b) is predicted by optimal escape theory because isolated males have a greater expected payoff for successful courtship, implying a greater lost opportunity cost of escape. Both isolated and guarding males have an opportunity to possibly mate with the introduced female. Guarding males that leave their mates briefly run the risk of losing fertilizations to competitors; those that abandon the previously guarded females to guard the introduced female would run a greater risk of losing fertilizations of the abandoned female. For isolated males, the potential mating represents a much higher proportion of expected seasonal reproductive success. Isolated males may also guard the new female without risk of losing fitness due to loss of fertilizations after abandoning a previous female. Given the greater potential payoff for isolated males, the cost of escaping is greater for them. The results confirm the prediction by optimal escape theory that the lizards adjust their escape decisions by accepting greater risks of being captured when the social costs of fleeing are greater. Risk behavior is traded off against the opportunity to court.

As initially formulated (Ydenberg and Dill 1986), optimal escape theory did not directly consider return to a specific activity after escaping. However, because lost opportunity costs increase over time, the probability that the activity will be resumed increases, especially if the

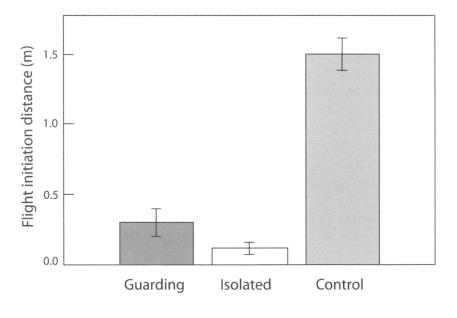

Figure 4-3 Effect of opportunity to court an introduced female on flight initiation distances of isolated and mate-guarding male *Eumeces laticeps*. No females were introduced to controls. Based on data from Cooper (1999b).

threat of predation decreases (e.g., Fig. 4-4; Sih 1992; Cooper 1998b; Martín and López 1999a, c). When the simulated predator stopped approaching as soon as an *E. laticeps* fled and then remained motionless, the lizards presumably assessed a lessened risk of being attacked. Under these circumstances, optimal escape theory predicts a higher probability of resumption of the former activity with the greater lost opportunity cost.

Males that fled after introduction of females returned to the females significantly more frequently than did control males, which did not return at all (Fig. 4-5) (Cooper 1999b). This finding is consistent with the optimality prediction. Contrary to optimal escape theory, there was no difference between the proportions of isolated and mate-guarding males that quickly returned to the tethered females (Cooper 1999b). However, it is possible that the qualitative prediction was too crude to detect real differences between isolated and mate-guarding males.

At least three factors might have obscured any such differences in tendency to return to the introduced females. Return was recorded only within 1 min after escape and cessation of approach by the predator, and latency to return was not recorded. Because the female may leave the

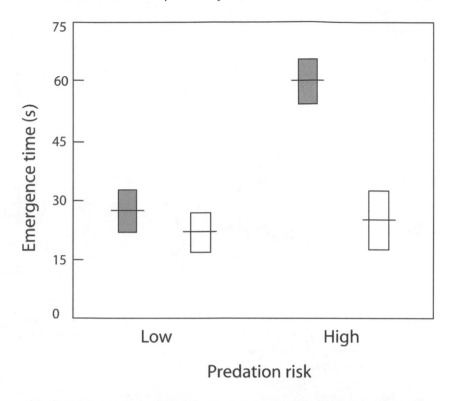

Figure 4-4 Time to emerge from refuge after escape in the lizard *Lacerta monticola* is greater when the threat of predation is greater. In the low-risk situation, an experimenter approached on a path bypassing the lizard. In the high-risk situation, the experimenter approached the lizard directly. Shaded bars represent lizards that emerged at the same site where they entered refuge, and open bars represent those that emerged away from the site at which they entered refuge. Reprinted with permission from Martín and López (1999c).

area or mate with another male, the probability that the mating opportunity will be lost increases with the length of time spent in refuge or otherwise engaged in antipredatory behavior. Optimal escape theory thus predicts that when it is possible to resume the interrupted activity, latency to resumption should decrease as the cost of losing the opportunity increases. Tests of this prediction would be valuable. Another factor clouding interpretation of the similar probabilities of return to tethered females by isolated and mate-guarding males is that mate-guarding males often returned close to the guarded females when they returned to the tethered female. Predictions of optimal escape theory are not clear in this case.

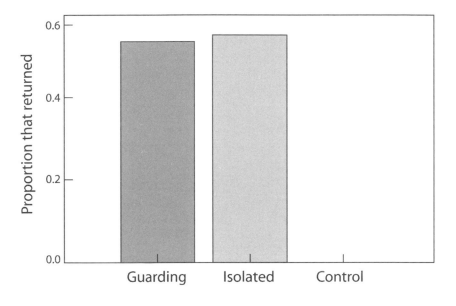

Figure 4-5 Proportions of isolated, mate-guarding, and control male *Eumeces laticeps* that returned after fleeing to site where females or control stimuli were introduced. Based on data from Cooper (1999b).

There is a possible interpretive pitfall for this experiment (Cooper 1999b). For both isolated and mate-guarding males, the risk of predation may have been reduced by the presence of the female being courted. The presence of the guarded female may have further diluted the risk of being captured for mate-guarding males. However, if dilution of risk is important, it would be predicted that the mate-guarding males would permit closer approach due to lesser risk than would isolated males, the opposite of the actual results. The shorter flight initiation distance for isolated males suggests that the greater cost of possibly losing a mating opportunity for isolated males was sufficient to overcome any dilution of risk by the presence of a second female.

In a second experiment, I introduced tethered females to isolated males to study how predator proximity affects latency to approach the females (Cooper 1999b). Females were introduced as in the previous experiment while I stood 1.2 or 2.7 m from the male. The latency to approach females was recorded as the time in seconds from placement of the female until the male first contacted the female by tongue flicking. When the male and female were closer to the predator, the mean latency to approach the female was significantly greater than when the predator was farther away. In addition, within 1 min after the female was introduced, 11 of 12 males approached when the predator was 2.7 m away and only 5 of 12 approached when the predator was 1.2 m away. The greater risk of moving into close proximity with the predator clearly affected the probability and latency of approaching the female.

MATE GUARDING

During the breeding season, males of some species of lizards guard females to prevent other males from mating with them. This behavior has been studied in few lizards and is best documented in the scincid *E. laticeps* (Vitt and Cooper 1985a; Cooper and Vitt 1997; Cooper and Vitt, unpublished data), the teiid *Ameiva plei* (Censky 1995, 1997), and the lacertid *Psammodromus algirus* (Salvador, Martín, and López 1995; Martín and López 1999b). In *E. laticeps,* a guarding male follows the female closely, usually staying within 1 m even while foraging, and often resting in physical contact with her (Vitt and Cooper 1985a; Cooper and Vitt 1997). He chases other males that attempt to approach the female (Cooper and Vitt 1986b, 1997). If a guarding male becomes separated from the female, which often occurs while chasing or fighting rivals, he attempts to relocate her either visually or by scent trailing (Cooper and Vitt 1986b). In *A. plei,* guarding males also follow females, staying with them throughout their receptive period (Censky 1995). In *P. algirus,* large, old males follow and court females for a prolonged interval prior to mating and then guard them after copulating, but younger males neither court nor guard (Salvador, Martín, and López 1995). Although mate guarding has not been studied in many taxa, it may occur much more widely than is currently appreciated; for example, it occurs in several anguid lizards (Bowker 1988).

Males that guard females against matings by other males have a greater lost opportunity cost when escaping than do nonguarding males because disruption of guarding entails the possibility that fertilizations may be lost. Optimal escape theory predicts that guarding males have shorter flight initiation distances. This effect has been studied directly only in *P. algirus* by Martín and López (1999b). These investigators approached large isolated and mate-guarding males and single and guarded females by walking directly toward them at a medium walking speed. They recorded the flight initiation distance for both sexes and distance fled for single and guarding males. As predicted by optimal escape theory, mate-guarding *P. algirus* had significantly shorter approach distances than males not near females. Distances fled, which are not predicted by optimal escape theory, did not differ significantly for guarding and nonguarding males. Although the effect of mate guarding on flight initiation distance has not been measured in *E. laticeps,* my observations extending over many breeding seasons strongly suggest that, as in *P. algirus,* guarding males permit closer approach by a simulated predator than do isolated males.

Flight initiation distances did not differ between guarded and isolated *P. algirus* females. This finding is of interest, but its interpretation is not entirely clear. The most likely explanation is that females have little or nothing to lose if separated from the male because other mates are readily available. On the other hand, it is possible that they might gain opportunities for extrapair copulations while separated from a guarding male and/or suffer harassment by other males. If females prefer large males as mates, as do female *E. laticeps* (Cooper and Vitt 1993), they are unlikely to find better mates quickly because only large males are guards. Harassment is unlikely to be a large cost because either the original male would soon relocate the female or another would replace him. Yet another possible factor is that the presence of a guard decreases the risk of being captured for the guarded female by diluting risk or being more conspicuous than the female. If so, a lack of difference in flight initiation distance between guarded and single females could be the result of exactly offsetting tendencies of females to flee to escape guarding males and of lower predation risk at any fixed distance for guarded than single females. Such offsetting factors seem highly unlikely. It seems much more likely that lost opportunity costs of escape do not differ between guarded and single females and that the presence of the guarding male does not dilute the risk of capture enough for females to permit closer approach.

Mate guarding may carry reproductive costs because guards may lose opportunities to mate with other females and risk being injured in fights with other males, with subsequent loss of status or ability to compete for matings. Guarding is also likely to have energetic costs of following the female and fighting and chasing other males, and possibly of reduced feeding. However, such costs have not been studied. It also seems likely that mate guards may incur costs in terms of predation risk. Such costs remain poorly known, but one study strongly suggests that mate guarding may increase the risk of predation (Cooper and Vitt, unpublished data).

Adult male broad-headed skinks are larger and more strikingly colored than females and have bright orange heads during the breeding season (Vitt and Cooper 1985a; Cooper, Mendonca, and Vitt 1987; Cooper and Vitt 1988). Their large size and bright head coloration make the males more conspicuous than the females. Male consorts are more conspicuous than their females even when they are not engaged in chasing or fighting. Among 25 pairs observed of which both lizards were on similar backgrounds, males were detected by a human experimenter first in 20 cases, females in 2 cases, and both were detected simultaneously in 3 pairs (Cooper and Vitt, unpublished data).

The sex difference in detectability due to size and color differences is not a cost of mate guarding for males, but the female's presence may increase the probability that a predator will detect the male. This effect on detectability is presumably smaller than that of the male's presence for the female, but detection of two consort males after my attention was first attracted by the females that they guarded suggests that remaining with the female may entail some risk of predation for males. This increase in risk is presumably greatest when female movement attracts the attention of a predator and the male is detected while moving to follow her.

Although the greater detectability of males is not a cost of mate guarding for males, it may be a cost for females if detection of the male increases the likelihood that a predator would detect the female. Because the sex ratio of lizards observed in the field is much lower than the 10:1 ratio of males:females detected first in pairs, male presence is very likely to increase the detectability of guarded females. Of course, some of the females would have been detected later in the absence of the males, but for almost all pairs, the observer detected the second member within 2 s after having detected the first. From many years of field work with this species, I have the impression that single females are much harder to detect than single males, and may often go undetected. Although movements were not a factor in the study (Cooper and Vitt, unpublished data), male movements are likely to attract the attention of predators, thus increasing the probability of detection for females. The presence of a guarding male may increase the chances that a predator will detect the female, but it is not clear whether mate guarding is beneficial or costly to females.

Females may tolerate mate guarding because it reduces harassment by other males that might carry costs outweighing any increase in predation risk. Female pheromones released at the onset of sexual receptivity may attract several males simultaneously (Cooper and Vitt 1986a; Cooper, Garstka, and Vitt 1986; Cooper and Garstka 1987), suggesting that sexual harassment might be a substantial cost. The presence of a large consort male greatly reduces exposure of females to courtship by smaller males, which they would reject as mates by aggressive behavior if necessary (Cooper and Vitt 1993). Alternatively, the net effect of guarding may be costly to females, but they may not be able to avoid consorts that can follow their scent trails (Cooper and Vitt 1986b).

Additional costs of mate guarding may be associated with differences in escape behavior between guards and the guarded females. Because males are easier to detect, risk at a particular distance from an approaching predator is greater for the guard than for his mate. On the other hand, a male that escapes pays a lost opportunity cost not suffered by the female. Without knowing the magnitudes of these effects, one cannot predict which sex among pairs should have the shorter approach distance or even whether there should be a difference. Because guarding males must follow females to maintain contact to prevent extrapair copulations, their attempts to relocate females after either or both have escaped might expose males to additional risk.

A first assessment of these possible costs was made in a field study of broad-headed skinks conducted in the breeding seasons of 1994–96 in Charleston County, South Carolina (Cooper and Vitt, unpublished data). In that study a human experimenter approached 30 pairs consisting of a female and her consort until one or both fled and then stopped to record which lizards fled and which fled first. For 21 of the 30 pairs, the experimenter remained immobile for 2 min to record subsequent behavior. Although the experimenter attempted to select a path of approach along which the two lizards were equidistant from the experimenter, there were a few trials in which one was slightly closer than the other.

Although males were more readily detectable, females were significantly more likely to flee first than were males (Table 4-3) (Cooper and Vitt, unpublished data). The female usually fled first even in a few trials in which the male was closer to the experimenter. When the female fled, the male was left exposed to the predator until it fled, increasing risk for the male. That the female fled first even though less detectable suggests that the lost opportunity cost for males may have outweighed the increased risk of predation to males due to greater detectability. An alternative explanation that seems less likely, but cannot be ruled out a priori, is that by remaining until the females fled, males might have been protecting their reproductive investments in the females by deflecting attacks to themselves. These explanations are compatible.

Table 4-3   Detectability, escape, and subsequent responses by male and female *Eumeces laticeps* in pairs with a male mate guard when approached by an experimenter. Data from Cooper and Vitt (unpublished data).

	Male	Female	Tie
Detected first by experimenter	20	2	3
Fled first when approached	6	20	4
Proportion that followed partner	0.76	0.00	—

Whatever its basis may be, delayed male escape places the male at risk. First, males are likely to have permitted closer approach due to the female's presence than they would have in the absence of a female. Second, although the motion of the female's flight might trigger attack by the predator, a predator that has detected both is extremely likely to attempt to capture the nearer prey that is still or has begun its escape attempt later. When females fled first, they typically entered refuges or went under surface litter, disappearing from view while the male was plainly visible. In many cases, the female disappeared before the male began to flee (Cooper and Vitt, unpublished data). Rather than entering refuges themselves, males typically remained exposed on the ground.

The behavior of males and females also differed after the predator stopped approaching (Cooper and Vitt, unpublished data). A large majority of males followed females, but no females followed males (Table 4-3). While following females, males moved much more slowly than while fleeing, prolonging their exposure to the predator. Following males sometimes entered refuges with females, but only after a delay. Both sexes usually fled in the same direction, but in two cases in which they fled in different directions, the males exposed themselves to greater risk by returning to the initial site and rapidly tongue flicking (Cooper and Vitt, unpublished data), presumably to locate the female's scent trail (Cooper and Vitt 1986b).

It seems clear that mate-guarding male *E. laticeps* incur risks of predation greater than those of isolated males due to shorter flight initiation distances, increased probability of being detected because of female presence and activity, females fleeing first, and following and attempting to relocate females by the guarding males. However, neither the overall magnitude of the increase in risk nor the relative contributions of the factors contributing to it are known. These are important foci for future research.

AGONISTIC BEHAVIOR

Although intraspecific fighting may be injurious and energetically costly to male lizards, these potentially high costs can be borne because successful agonistic behavior can permit some dominant individuals to mate with multiple females and decrease the likelihood that those females will mate with additional males. Males actively engaged in intraspecific aggressive encounters may be directly contesting immediate mating opportunities or indirectly contesting future mating opportunities with females residing nearby. A mate-guarding male forced to escape from a predator may be in especially great danger of losing contact with the female long enough for another male to copulate with her. Even in the immediate absence of a female, a male forced to flee by a predator may lose a valuable opportunity to exclude a rival from his territory or to establish dominance permitting rapid, safe resolution of future encounters with the same male.

Intraspecific aggressive behavior is a major means by which male lizards exclude conspecific males from the immediate vicinity of females or from territories that may overlap with those of females. Species that exhibit mate guarding, such as *E. laticeps* and *A. plei,* tend to be nonterritorial active foragers, guarding females rather than territories (Stamps 1977b; Vitt and Cooper 1985a; Censky 1995; Cooper and Vitt 1997). Among active foragers that do not exhibit mate guarding, males typically fight with other males when encountered during the breeding season but do not attempt to exclude them from the home range (Stamps 1977b). Typically, ambush foragers are territorial and attempt to aggressively exclude adult conspecific males from the territory (Stamps 1977b). Effects of intraspecific aggressive behavior on antipredatory behavior have been studied in only one nonterritorial, actively foraging lizard, *E. laticeps* (Cooper 1999b), and one territorial, ambush-foraging species, *Tropidurus hispidus* (Díaz-Uriarte 1999).

In male broad-headed skinks, *E. laticeps,* the short mating season and mate guarding limit mating opportunities for mate guards and nonguards alike. Establishment of dominance by fighting is crucial to reproductive success of males in fairly dense populations on the barrier islands of coastal Georgia and South Carolina (Cooper and Vitt 1987, 1997). The opportunity to fight or chase a substantially smaller male should be beneficial to males because the larger male is almost certain to win the encounter and is unlikely to be injured. Because small adult males often follow females, it may be especially important for larger mate-guarding males to prevent them from gaining access to the females. The greater immediate potential for loss of fitness due to a rival near a guarded female predicts that mate-guarding males should permit closer approach by a predator than isolated males. Because of the lost opportunity to fight or chase a smaller male, isolated males and mate guards near smaller males are predicted to permit closer approach before fleeing than are males in the absence of any other lizards.

To test these predictions, I observed responses of three groups of male *E. laticeps* to tethered males introduced as in the experiment in which females were introduced (Cooper 1999b). In addition to groups of isolated and mate-guarding males, control males were isolated males treated like the other groups except that there was no male on the tether. After placing the introduced male 0.5 m from an experimental male, I remained immobile until a fight started in trials with the isolated male and mate-guarding groups and for 20 s with the control group (a latency by which most males approach tethered females). Starting from 2.7 m away from the experimental male, I slowly approached at approximately 35 m/min and recorded the flight initiation distance.

Mate-guarding males allowed closer approach than did isolated males, which, in turn, permitted much closer approach than control males (Fig. 4-6); the differences between all pairs of conditions were significant (Cooper 1999b). These results confirmed both predictions of optimal escape theory, indicating that increasing costs of lost opportunities to engage in intraspecific agonistic behavior cause decreasing flight initiation distances. Lost opportunity costs associated with courtship, mate guarding, and agonistic behavior all affect the escape decisions of broad-headed skinks. Furthermore, the relative strengths of the effects of introducing sexual rivals and females on flight initiation distance are opposite for isolated males and mate-guarding males; guards permit closer approach than do isolated males when a rival male is present, but isolated males permit closer approach when a female is introduced. This difference mirrors the relative magnitude of the lost opportunity costs in the two situations.

Males that are fighting or courting may permit closer approach than at other times due to impaired ability to detect the predator while focusing on a conspecific lizard. However, in the present experiments on *E. laticeps*, the lizards observed the simulated predator approach and introduce the tethered lizard. Any risks of being captured during subsequent social behavior could be factored into decisions to initiate the social activity. When females were introduced to male *E. laticeps*, impaired ability to detect the predator approaching due to courtship should have

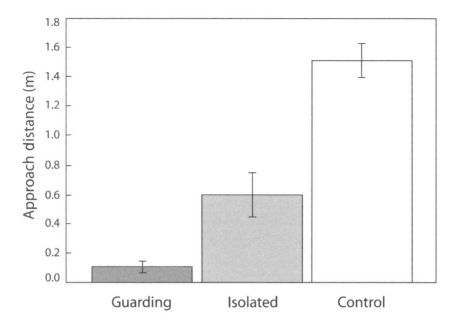

Figure 4-6 Effect of opportunity to fight an introduced male on approach distances (i.e., flight initiation distances) of isolated and mate-guarding male *Eumeces laticeps*. No females were introduced to controls. Based on data from Cooper (1999b).

been equal for the two groups of males or greater for the mate-guarding groups if such males attempted to monitor the guarded females while courting the tethered ones. Thus, the shorter flight initiation distance for the isolated males contradicts the impairment hypothesis. In the experiment with tethered males, any additional impaired ability to detect the predator while fighting due to the presence of the guarded female would predict that mate-guarding males would permit closer approach. However, because guards had to move away from the females to fight, it is unlikely that guarded females impaired the ability to detect the much larger predator. There were no obvious differences in intensity of fights in trials with isolated and mate-guarding males. Therefore, although minor differences in impairment between groups are possible, they are unlikely to account for the large differences in approach distance between groups, which is better explained by the difference in lost opportunity cost.

As in the experiment on the effects of mating opportunity on escape behavior, introduction of the tethered male *E. laticeps* may have diluted the risk of predation for both isolated and mate-guarding males, and risk may have been further diluted for guards by the presence of the guarded female. By contrast, no other lizards were present to dilute the risk of predation for control males. If dilution of risk is an important effect, it might be an alternative explanation for the shorter flight initiation distances by males in the two groups to which males were introduced than by control males. It might also be a contributory factor to differences based in part on loss of opportunity to fight in the two experimental groups. Control males would have both lesser lost opportunity cost and greater risk at a given distance from the predator. For the mate-guarding group, the magnitude of the lost opportunity cost of escape as well as the dilution of risk are greater than for isolated males that court tethered females. Thus, the effects of cost of escape and risk of predation are confounded in the experiment on agonistic behavior.

Although the importance of lost opportunity was clear in the experiment on courtship, it was not clear whether there was any effect of dilution of risk. The finding that mate guarding does not affect flight initiation distance in female *P. algirus* (Martín and López 1999b) suggests that dilution of risk is not an important factor for them. At the low density of lizards in social situations, dilution of risk may be generally unimportant. The results of the experiment on fighting in *E. laticeps* show strong differences among groups that very likely reflect a lost opportunity to fight due to escape. Nevertheless, further study using experimental designs that eliminate possible differences based on dilution of risk are desirable to demonstrate this effect unequivocally.

In territorial species, a male must defend his territory against conspecific males to minimize access by nonresidents to females whose home ranges or territories may partially or entirely overlap his territory. If a male actively engaged in territorial defense flees from a predator, he loses the opportunity to exclude a rival that may become more difficult to exclude if the initial attempt is unsuccessful and might mate with a female while the resident is elsewhere, especially if he enters a refuge. The longer the resident remains in refuge, the greater is the cost of lost opportunity to defend the territory. Furthermore, monitoring the territory may be particularly important for a territorial male that has recently defended it because the probability of further intrusion is high. Competitors in some species acquire or expand territories by persistent intrusion (Stamps and Krishnan 1995). The expected costs of interrupting an ongoing aggressive encounter are greater than those for a completed encounter.

Díaz-Uriarte (1999) predicted that resident males in the territorial *T. hispidus* should permit closer approach by a predator and should emerge from refuge sooner when a predator approaches during an ongoing agonistic bout than shortly after its completion. Decisions to emerge from refuge are essentially a decision to terminate the escape and resume other activities. It has recently been shown that decisions to emerge are predictable from both lost opportunity costs and risk factors in the skink *E. laticeps* (Cooper 1998b, 1999a) and the lacertids *L. monticola* (Martín and López 1999c) and *Podarcis muralis* (Martín and López 1999a).

Díaz-Uriarte (1999) tested hypotheses about the effects of ongoing and recently concluded intraspecific agonistic behavior on escape and the closely related decision to emerge after entering a refuge in *T. hispidus*. Experiments were conducted in outdoor enclosures, each of which contained a resident male and at least one female. Intruders were introduced and removed by means of suspended fishing lines manipulated from a blind. This method has the advantage of permitting the agonistic encounter to begin or be completed with the resident unable to detect the experimenter, who eventually simulated a predator by approaching the resident.

In an experiment on the effects of a recently completed agonistic encounter (Díaz-Uriarte 1999), each resident male was tested in two conditions in counterbalanced sequence. In the experimental condition, a male intruder was introduced for a maximum of 15 min, for at most 3 min after he was attacked, or until he was attacked three times. In the control condition, a stick of roughly the same size and color as a male was introduced and removed after 3.75 min (the median time that intruders remained in the enclosure in pilot tests). Five minutes after the intruder or the stick was removed, the experimenter approached the resident and recorded his flight initiation distance, minimum distance (closest flight initiation distance if the resident fled more than once before entering refuge), and time to reemerge from refuge.

As predicted, male *T. hispidus* reemerged significantly sooner after fighting than in control trials (Table 4-4). Minimum distance did not differ between conditions. The results for flight initiation distance were complicated by variation in enclosure size. There was a significant interaction between enclosure size and treatment: flight initiation distance increased with enclosure area in the control condition, but not in the experimental condition. Fighting 5 min earlier did not significantly affect flight initiation distance.

Table 4-4 **Effects of agonistic behavior on median flight initiation distance (FID) minimum distance, and time to emerge from refuge in *Tropidurus hispidus* approached by an investigator simulating a predator.** Based on data from fig. 2 in Díaz-Uriarte (1999) and personal communication.

Experiment and Conditions	Median FID (m)	Minimum Distance (m)	Time to Emerge (s)
Experiment one: 5 min after encounter			
Agonistic encounter	6.9	3.6	127
Control introduction	6.2	4.3	384
Experiment two: ongoing versus completed encounter			
Ongoing encounter	4.0	3.2	108
Encounter completed 5 min earlier	5.2	4.1	140

A second experiment (Díaz-Uriarte 1999) was conducted to test the effects of ongoing agonistic behavior on flight initiation distance and time to reemergence in *T. hispidus*. This experiment was similar to the first, but the two conditions were approach by the experimenter 5 min after removal of the intruder or immediately, with the intruder still present. Other differences were that enclosure area was held constant, the intruder was left in the enclosure for 2 min after being attacked or until six attacks occurred (whichever occurred first), and the intruder was removed as soon as the resident hid in the immediate approach condition.

Flight initiation distance and minimum distance were both significantly shorter during an ongoing agonistic encounter than 5 min after the end of an encounter (Table 4-4), confirming the prediction based on greater expected cost of abandoning the fight (Díaz-Uriarte 1999). As in the experiments with *E. laticeps*, dilution of risk must be considered an alternative explanation for the behavior of *T. hispidus*. In the immediate approach condition, the intruder's presence may have reduced the risk below that experienced by residents 5 min after removal of an intruder. Dilution and cost of escape both predict that residents should permit closer approach in the immediate approach condition. Díaz-Uriarte (1999) argued that dilution is unlikely to have affected flight initiation distance because flight initiation distance was the same when females were in or outside refuges. He also recognized that the shorter flight initiation distance of males while fighting might be a consequence of limited ability to detect the predator.

In contrast to the results of the previous experiment, time to reemerge did not differ between residents approached 5 min after fighting or during an ongoing encounter (Table 4-4). Díaz-Uriarte (1999) interpreted the differing findings of his two experiments in relation to the temporal distribution of risk of predation and the cost of avoiding it. Because a predator may attack and then rapidly depart if unsuccessful, risk may have returned to baseline levels quickly, which would account for the absence of any observed difference between the ongoing and completed fighting conditions. By contrast, the cost of remaining in refuge remains high when a rival is known to be nearby and likely to return, which accounts for the shorter time to reemerge following approach 5 min after fighting than in the control condition.

That flight initiation distance for male *T. hispidus* is shorter during an encounter than 5 min after an encounter is consistent with the findings for male *E. laticeps* (Cooper 1999b). In the latter species, males allowed closer approach when interacting agonistically with an introduced male than in a control condition with no other male present. In light of the rapid decrease in risk following attack hypothesized by Díaz-Uriarte (1999), the seemingly different control conditions in the two studies may have had roughly equivalent risk relative to baseline for each species.

Tail Loss, Social Behavior, and Antipredatory Behavior

Tail autotomy has long attracted the interest of herpetologists as a specialized escape mechanism of lizards (reviewed by Arnold [1984, 1988]). Loss of most or all of the tail by autotomy has several costs that vary in severity among species, including loss of energy stored in the tail (Congdon, Vitt, and King 1974), the need to divert energy from growth and reproduction to regenerate the lost portion of the tail (e.g., Congdon, Vitt, and King 1974; Dial and Fitzpatrick 1981; Niewiarowski et al. 1997), decreased running speed (e.g., Ballinger, Nietfeldt, and Krupa 1979; Punzo 1982; Martín and Avery 1998), loss of full capacity for using autotomy as a defense until regeneration is complete (e.g., Congdon, Vitt, and King 1974; Dial and Fitzpatrick 1984; Daniels, Flaherty, and Simbotwe 1986), and decreased abilities to fight and obtain mates (e.g., Fox, Heger, and DeLay 1990; Salvador, Martín, and López 1995). Autotomy might also affect social signaling in species using tail displays and signaling to predators in species that use the tail for pursuit-deterrent signals. Social behavior can also affect the ability to use autotomy as an antipredatory mechanism if lizards break the tails of conspecifics during fighting. The focus here is on the effects of autotomy on dominance and reproductive success. Table 4-5 summarizes the known relationships between social behavior and caudal autotomy.

Table 4-5 Effects of tail autotomy on social behavior in lizards.

Behavior	Reference
Dominance	
Adult male *Sauromalus ater* lose dominance after autotomy.	Berry (1974)
Juvenile *Uta stansburiana* lose dominance after autotomy.	Fox and Rostker (1982)
Male *Lacerta monticola* lose dominance after autotomy,	Martín and Salvador (1993a)
Male *Anolis sagrei* do not lose dominance after autotomy.	Kaiser and Mushinsky (1994)
Male *Agama agama* develop a clublike structure at the tip of the tail during regeneration that might be a weapon.	Schall et al. (1989)
Postautotomic loss of dominance in *U. stansburiana* can be prevented in females, but not males, by reattachment of the tail after 2 wk	Fox, Heger, and DeLay (1990)
More force is required to break male than female tails in *U. stansburiana,* but second breaks require equal force (tailless males may be unable to mate but have little or no further immediate mating cost from second breaks).	Fox, Conder, and Smith (1998)
Courtship and mating success	
Tail loss decreases courtship and copulation by male *L. monticola* and decreased frequency of copulation by females.	Martín and Salvador (1993a)
Tailless male *L. monticola* are less active and stay closer to refuge, which may reduce mating opportunities.	Martín and Salvador (1997)
After tail loss, large male *Psammodromus algirus* decrease activity, stay closer to cover, and have smaller home ranges that overlap with home ranges of fewer females.	Martín and Salvador (1993b): Salvador, Martin, and López (1995); Salvador et al. (1996)

EFFECTS OF AUTOTOMY ON DOMINANCE

After autotomy a lizard might suffer loss of social dominance. Fighting ability is impaired owing to decreased mass, lack of a tail if it is used in fighting (e.g., *Agama agama,* Harris [1964]; *Dipsosaurus dorsalis,* Carpenter [1961]), and possibly altered balance. To reduce risk of predation while running ability is diminished and capacity for autotomy is curtailed, lizards may make adjustments to antipredatory behavior incompatible with defense of a territory or a female. While energetic allocations for tail regeneration are high, they might avoid fighting to conserve energy.

Field observations show that tail loss sometimes is followed by loss of dominance in adult male *Sauromalus ater* (Berry 1974), but there have been no studies demonstrating that tail loss causes a reduction in dominance in agonistic contests between adult territorial lizards. However, this effect has been shown clearly for juveniles of the phrynosomatid lizard *Uta stansburiana* (Fox and Rostker 1982; Fox, Heger, and DeLay 1990). Fox and Rostker (1982) staged encounters between juvenile *U. stansburiana* matched for size and sex but differing in the degree of tail loss. Dominance relationships were determined initially for 30 pairs of juveniles having intact tails. Immediately after the encounter, the distal one-third of the tail was removed from the dominant individual. After being housed for 2 wk in their solitary home cages, the same pairs were again tested for dominance. The effect of losing one-third of the tail on dominance was not significant (Fig. 4-1). Following the second agonistic contest, half of the remaining tail was removed from individuals that had been dominant in both encounters. Two weeks later, third encounters were staged between former subordinates with intact tails and former dominants with two-thirds of their tails removed. The difference in agonistic scores between dominants and subordinates decreased significantly in the final trials compared with each of the earlier trials, indicating that lizards having lost two-thirds of their tails suffered a decrease in dominance (Fig. 4-1).

Because large body size confers a great advantage in intraspecific agonistic encounters between lizards (e.g., Ruby 1981; Tokarz 1985; Cooper and Vitt 1987), Fox, Heger, and DeLay (1990) conducted a study to separate the effects of decreased body size from other effects of autotomy such as energetic loss, impaired balance, and mechanical injury. Encounters were staged as in the study by Fox and Rostker (1982). After an initial trial in which both lizards had intact tails, two-thirds of the tails of dominant lizards were removed and the autotomized portions were frozen. For another test 2 wk later, thawed tails were reattached by adhesive to some formerly dominant individuals. Other formerly dominant juveniles serving as a control group were handled similarly, including placement of adhesive on the tail stump, but their tails were not reattached. Controls of both sexes lacking two-thirds of their tails exhibited the expected decreased dominance (Fox, Heger, and DeLay 1990). Dominant females having restored tails retained their initial degree of dominance (Fig. 4-7). This finding was interpreted as indicating that perceived body size in females determined the dominance relationships, so that the loss of dominance in females following autotomy is due to decreased body size. Most of the change in dominance was attributable to the difference in responses of subordinate females to former dominants having or lacking reattached tails.

In contrast to females, dominant males suffered a significant decline in dominance despite having reattached tails (Fig. 4-7), indicating that tail size alone cannot account for the loss of dominance. Fox, Heger, and DeLay (1990) suggested that female tails are status-signaling badges (in the same sense that size and intensity of color patches signal dominance in some birds [Rohwer and Ewald 1981; Rohwer 1982, 1985] and lizards [see chapter 2]). Because neither sex of *U. stansburiana* uses the tail as a weapon or for defense in fighting, differential tail use cannot account for the sex difference.

There may be an alternative explanation for the sex difference in the effects of reattachment. The disparity in dominance between pair members was much greater for males than for females in the tests prior to autotomy. In subsequent tests, the aggressiveness of formerly dominant females declined, as did that of subordinates faced with an apparently unchanged dominant. Unless there is some indication that ability to dominate has changed, such decreases in dominance are to be expected in all groups on the basis of previous experience in which dominance relationships have been established. Alternatively, the sex difference might be a consequence of a difference in aggressiveness. In the 1-h trials, subordinate females might not have detected any changes in the ability of females with reattached tails to dominate. The greater decline in dominance for males than females may have provided a more obvious cue to subordinate males. In addition, the more aggressive males may have been more likely to test potential changes in ability to dominate.

The force required to experimentally autotomize intact tails of *U. stansburiana* is greater for males than females having equal tail thickness, and the sex difference increases with body length (Fox, Conder, and Smith 1998). Although both sexes suffer loss of dominance and ability occupy the best habitats after autotomy, reproductive costs may be lower for females than males because females are likely to mate, whereas the ability of males is likely to be impaired by the inability to hold a territory (Fox, Heger, and DeLay 1990; Fox, Conder, and Smith 1998). Fox and colleagues also discuss the possibility that females lacking tails may adopt a subordinate strategy in which they occupy poorer home ranges and avoid

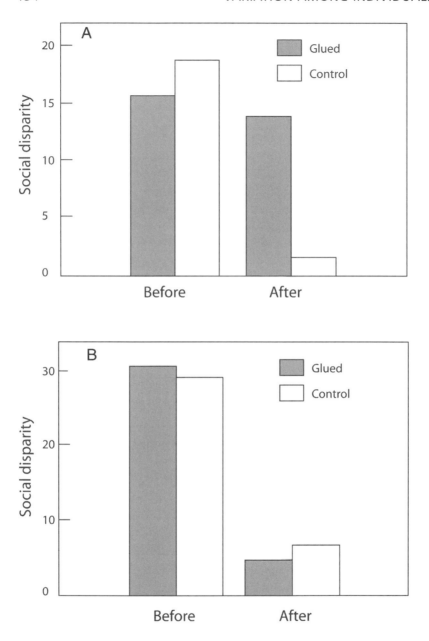

Fig. 4-7. After having the tail reattached 2 wk after removal from the formerly dominant member of pairs of juvenile *Uta stansburiana*, females (*A*), but not males (*B*), were able to retain their former dominance status. Social disparity is the difference in dominance scores between dominant and subordinate individuals in paired aggressive encounters. Based on data from Fox, Heger, and DeLay (1990).

agonistic encounters, thereby saving energy for reproduction. Because males lacking tails may be forced out of the best habitats and avoid agonistic contests with dominant males, the crucial sex difference appears to be the greater ability of autotomized females to obtain fertilizations.

First and second tail breaks were equally difficult to induce in females, but second breaks were significantly easier to induce than first breaks in males (Fox, Conder, and Smith 1998). Because the tails of females were equally difficult to break the first and second times, females do not appear to require a decrease in force for autotomy in compensation for reduced ability of the incomplete tail as an antipredatory device. Fox, Conder, and Smith (1998) argued that females have evolved ease of autotomy because they can be reproductively successful even after autotomy. Although males may require greater force for the initial tail autotomy owing to their greatly reduced social status and ability to mate, Fox, Conder, and Smith (1998) suggested that males that have already lost much of their tails autotomize more readily because they have much less to lose from further autotomy. However, because no test was presented to determine whether the sexes differed in the degree of change in force needed for first and second autotomies (as would be shown by a significant trial by sex interaction), it cannot be concluded that there was a true sex difference.

Although there have been no studies of the proximal mechanisms underlying differences in ease of autotomy, the sex difference in ease of first autotomy and the difference in males between first and second autotomies might be the result of voluntary decisions based on costs of autotomy. This hypothesis could be tested by comparing forces required to autotomize the tail in alert lizards with those in anesthetized or freshly killed lizards. If the effects disappear in death or under anesthesia, the hypothesis that assessment of costs affects voluntary decisions to autotomize would be supported. If they persist, current differences in mechanical resistance to breakage, perhaps reflecting the results of past selection based on differential costs, would be implicated.

In the lacertid *Lacerta monticola*, experimental removal of the tail 17 mm distal to the cloaca was followed by loss of social status in a study conducted in a seminatural enclosure (Martín and Salvador 1993a). Despite similar handling of all lizards except for tail removal, males with intact tails won 92.3% of encounters with tailless males. Each of the five males with intact tails was dominant over all five males lacking tails. No direct statistical test was made for a difference in dominance between groups, but a binomial test shows that tailed males dominated tailless males at above chance levels. The probability that one tailed male would dominate all five tailless males if dominance were equal for the two groups is 0.03. The probability that all five tailed males would dominate all five tailless males is only 3.0×10^{-8}.

The possibility that the effects of autotomy on dominance, ability to maintain territories, and aggressiveness may vary among species with aspects of natural history suggests some avenues for research. Potentially important factors include the degree to which tail loss impairs escape ability, predation pressure, and adjustments to microhabitat use to reduce exposure to predators that affect ability to occupy or defend all of the area occupied prior to autotomy. Autotomy may have smaller effects on dominance and territory size in species that are very cryptic and slow moving than in species that are more readily detectable. Autotomy may be predicted to have strong effects on active foragers because their motion makes them readily detectable by predators that rely on visual cues. This is especially so for species that suffer large decreases in running speed due to autotomy (Ballinger, Nietfeldt, and Krupa 1979; Punzo 1982; Arnold 1988). In species that use the tail as a weapon in intraspecific fighting, autotomy might greatly reduce the ability to dominate.

In contrast to the detrimental effects of autotomy just discussed, a possible advantage of autotomy in fighting has been proposed in the agamine *Agama agama* (Schall et al. 1989). Following autotomy, 65.5% of the regenerated tails of males and 36.4% of those of females are expanded distally into a rounded blunt end having a clublike appearance. Because male *A. agama* use the tail as a weapon to whip adversaries in intraspecific agonistic encounters, Schall et al. (1989) hypothesized that the hard, swollen tip of the regenerated tail may increase the effectiveness of the tail as a whip. This intriguing hypothesis has not been tested.

In other species, autotomy may not strongly affect social status. For example, there were no detectable changes in dominance in adult males of the polychrotid *Anolis sagrei* (Kaiser and Mushinsky 1994). However, it is premature to conclude that autotomy does not affect dominance in *A. sagrei*. Significant effects might have been detected using an experimental design comparing the differences in aggressiveness between dominant and subordinate individuals before and after autotomy and comparing these data for experimental and control pairs.

EFFECTS OF AUTOTOMY ON COURTSHIP AND MATING SUCCESS

A major consequence of high dominance status for males is greater access to mates, which translates into greater reproductive success and higher fitness than for males having lower social status (Dewsbury 1982). Decreased ability to dominate owing to autotomy may be inferred to have potentially severe consequences for male reproductive success. Female preference for mates with intact tails is another potential source of a fitness cost of autotomy for males.

The best evidence for a reproductive cost of autotomy for male lizards has been reported for lacertids. In *L. monticola*, males having intact tails had a distinct reproductive advantage over males whose tails had been artificially removed (Martín and Salvador 1993a). Males with tails had a significantly higher mean number of courtships and courted significantly more female individuals than tailless males (Table 4-6). Males with tails copulated significantly more times than tailless males (Table 4-6). There was a similar, but nonsignificant, trend for males with tails to copulate with more females. Tailless females copulated less frequently than tailed females, suggesting that males might prefer females with tails as mates, possibly due to decreased energy available to tailless females for reproduction (Martín and Salvador 1993a). An alternative interpretation might be that the sexual behavior of both sexes was altered by changes in activity and/or distance from refuge in compensation for increased risk of predation following autotomy.

Males having tails traversed a wider diversity of habitats (Martín and Salvador 1993a), were more active (Martín and Salvador 1995), had larger home ranges, and spent less time close to refuges (Martín and Salvador 1997) than tailless males, which may account for the differences in courtship and copulation frequencies. These findings suggest that adjustments in antipredatory behavior to reduce vulnerability of tailless males to predation rather than female choice may be responsible for the reduced frequency of courtship and copulation by tailless males. Martín and Salvador (1993a) suggested the intriguing hypothesis that lack of a tail or possession of a regenerated tail reveals a lower ability to escape, making possession of an intact tail an honest signal of ability to escape without resorting to autotomy. Such a signal might affect mate choice.

Large, dominant males of the nonterritorial lacertid, *Psammodromus algirus*, decrease their activity and suffer contraction of home range size following autotomy during the breeding season due to abandonment of areas having less cover (Table 4-7) (Salvador, Martín, and López 1995). After autotomy, the smaller home ranges of males overlap with home ranges of fewer females, implying reduced opportunities for mating (Fig. 4-2) (Salvador, Martín, and López 1995; Salvador et al. 1996). As in *L. monticola*, tailless males of *P. algirus* stayed closer to refuges

Table 4-6   Mean numbers of courtships and copulations by tailed and tailless male and female *Lacerta
monticola.* Based on data in fig. 1 of Martín and
Salvador (1993a).

	Tailed	Tailless
Male		
Courtships	8.5	2.2
Copulations	3.3	0.7
Female		
Courtships	7.3	3.3
Copulations	3.5	0.4

Table 4-7   Movement and home range sizes of large adult male
Psammodromus algirus. Data taken from tables 2 and 3 of Salvador et
al. (1995) and presented as mean $\pm$ SE.

	Tails Intact	Tails Autotomized
Distance moved (m/min)		
Preautotomy period	0.33 ± 0.05	0.33 ± 0.04
Postautotomy period	0.38 ± 0.05	0.23 ± 0.05
Home range size (m²)		
Preautotomy period	99.2 ± 24.4	102.4 ± 13.1
Postautotomy period	104.2 ± 23.9	33.7 ± 5.5

(Martín and Salvador 1993b). The reduction in home range size and
tendency to remain near refuges again suggest that changes in spatial
occupation compensatory to increased predation risk entail a cost of
lower reproductive success.

Tail loss has potentially important consequences for socially mediated aspects of reproductive success, but its effects have been studied
only in the two lacertid species discussed. Research on lizards in a range
of families is needed to discover the generality and importance of reduced mating success after tail loss. It may be speculated that lizards in
taxa that move slowly, are cryptically colored, and move infrequently
may suffer less severe costs of autotomy associated with reduction in
courtship opportunities. However, preference for mates with intact tails
may be affected by other factors. The possibility that both sexes might
prefer mates having intact tails to tailless mates should be tested. Males
having intact tails might be preferred as mates because the tail is an honest signal of heritable escape ability. If the tail contributes to a female's

overall impression of male size, females in species that prefer large males (Cooper and Vitt 1993) might prefer males with intact tails over males of the same snout-vent length. For females, stored energy lost with tails can cause large reductions in clutch size and energetic content per egg (e.g., Dial and Fitzpatrick 1981; Taylor 1984; Wilson and Booth 1998). In species so affected, females with intact tails might be preferred over tailless ones as mates by males.

TAIL LOSS CAUSED BY INTRASPECIFIC FIGHTING

Tail loss during intraspecific fighting may have both social and antipredatory consequences. It presumably results in the same sorts of costs to dominance and ability to obtain mates associated with autotomy. Such costs would likely be most severe in species that use the tail as a weapon in intraspecific contests and in those slowed most by tail loss (e.g., Ballinger, Nietfeldt, and Krupa 1979; Punzo 1982; Martín and Avery 1998). Although such effects undoubtedly occur, the frequency and distribution among lizard taxa of tail loss resulting from intraspecific fighting is unknown. Although I do not intend to review the evidence, lizards in several families bite tails during agonistic bouts: Phrynosomatidae (Parker and Pianka 1973; Vitt et al. 1974), Gekkonidae (Brillet 1986), Eublepharidae (Greenberg 1943), Scincidae (Perrill 1973; Fitch 1954), Teiidae (Carpenter 1960), and Anguidae (Formanowicz, Brodie, and Campbell 1990). Because such biting sometimes causes tail loss, biting the tail might be a strategy for reducing the status of a competitor.

Discussion and Future Directions

Social and antipredatory behavior have long been prominent among the research topics of behaviorists, but classical ethologists rarely considered connections between them. Only when behavioral ecologists began to employ the cost-benefit analyses of optimality approaches did the likelihood of trade-offs between the reproductive consequences of social behavior and the survival benefits of antipredatory adaptations become clear. The few studies of relationships between social and escape behavior in lizards indicate that optimal escape theory provides a powerful means of predicting escape behavior in social situations. Its predictions were largely confirmed for studies of courtship, mate guarding, and agonistic behavior. Despite the great ecological differences between the nonterritorial, active foraging E. laticeps and the territorial, ambush foraging T. hispidus, considerations regarding costs and benefits permitted optimal escape theory to make similar predictions regarding the effects of intraspecific aggressive behavior on escape decisions. Their confirmation contributes to the impression that optimal escape theory has great potential for unraveling the complex interrelationships between social behavior and escape.

This review also makes it obvious that we remain more ignorant than knowledgeable, even about the topics treated here—the effects of courtship, mate guarding, and fighting on escape decisions, and relationships between tail loss and social behavior. The nexus between social behavior and antipredatory adaptations is broader than I have indicated. Pursuit-deterrent and social displays may be identical or very similar in some species (Leal and Rodríguez-Robles 1997), suggesting the possibilities of mutual evolutionary constraint and compromise. Social behavior may affect detectability to predators and increased detectability may, in turn, have consequences for selection of sites and times for social activities. Social behavior might also influence which of the available sets of escape tactics may be used. These examples suffice to indicate that there are many relationships between social behavior and defense that have been neglected and constitute prime topics for fruitful future research.

For relationships between social behavior and optimal escape behavior, I have noted some areas in need of research throughout the text. Here, I note further that new experiments designed to eliminate or control the effects of dilution of risk and reduced ability to detect approaching predators as alternate explanations for optimal escape behavior are highly desirable. Normally wary lizards of several species sometimes may be approached closely during courtship and copulation, while fighting or while guarding mates. Species normally difficult to approach may sometimes be collected easily by hand during this reproductive behavior (Cooper, unpublished observations). Although the animals appear alert and certain to have detected the potential predator, they do not attempt to escape or do so only at very close range. This suggests that the findings for *E. laticeps, P. algirus,* and *T. hispidus* may apply to a broad diversity of lizards.

Future studies should assess the effects of courtship on escape behavior in additional taxa that rely on escape to refuges and in slower species that rely more heavily on crypsis and immobility to avoid detection than on rapid escape. One obvious avenue that should be explored is the effect of courtship on flight initiation distance and reemergence times when escape occurs during an ongoing courtship or after copulation has occurred. Optimal escape theory predicts that flight initiation distance and time to reemergence should be shorter when escape interrupts courtship than after copulation in species that do not guard mates. It makes the same predictions in relation to that time in males that have not recently engaged in social behavior. The effect of egg guarding on escape decisions has not been studied formally.

I have already noted several topics regarding links between tail loss and social behavior that are in need of investigation. However, there is one other particularly important gap in our knowledge. Dominance is important to the ability of juveniles to compete for occupancy of the best habitats (Fox, Rose, and Myers 1981), which may affect their survival and growth, with ramifications for later reproductive success. If tail loss affects dominance by adults, it could strongly affect ability to hold a territory and to obtain matings. Examination of these relationships in adult territorial lizards is an important goal for future research.

Even in the restricted areas of relationships between social behavior, on the one hand, and escape behavior and autotomy, on the other, there are many interesting aspects remaining to be explored. The wider field of relationships between social behavior and antipredatory defense can provide more than enough material for a lifetime of research. It is my hope that this chapter will stimulate much-needed research on this fascinating topic.

Variation among Populations

Introduction

Gordon H. Orians

As demonstrated in part 1, experimental and observational studies of single populations in a particular environment can reveal much about the action of natural and sexual selection in animals. Nonetheless, those studies can demonstrate only the results of evolutionary agents at that particular site, at that particular time, and under those particular conditions. To determine the generality of those results and how they change with environmental conditions, comparative studies in different environments are needed. The three chapters in this section, which compare populations of a single species in different environments, focus on the role of sexual selection in producing sexual dimorphism in size, shape, color, and behavior. As first suggested by Darwin, sexual selection has two components: competition among males for access to females, and female choice of mating partners. Different outcomes are expected as a result of changes in the intensity of selection via these two components, as well as the nature and strength of environmental constraints that counteract the force of sexual selection. The two types of environmental constraints explored in the following chapters are predation and food supply. Because lizards provide no parental care once their eggs are laid, mate choice by female lizards may be simpler than choices in species such as birds in which substantial biparental care is believed to be the ancestral condition (Irwin 1994).

To interpret the results of their studies, the authors had to make several assumptions. The first is that the patterns represent real adaptations; that is, they are not by-products of other processes. For example, variations in food supply could influence growth rates, and growth rate variations would, in turn, affect the size distributions of individuals in the field. Similarly, differences in the thermal environments in which eggs develop or differences in egg sizes could influence size at hatching; subsequent growth rates of juveniles (Sinervo et al. 1992); and, as demonstrated for *Eumeces okadae* on the Izu Islands (chapter 6), age and size at which sexual maturity is reached.

The second assumption is that the observed phenotypic differences are at least in part heritable. Given that almost nothing is known about lizard genetics, the authors have employed the phenotypic gambit (Maynard Smith 1978; Grafen 1984). That is, they have assumed, as have many other investigators (e.g., Beletsky and Orians 1996), that there is a sufficient genetic basis for the morphological and behavioral traits they examined to provide a substrate on which natural selection could act.

Finally, the interpretations assume that enough time has elapsed since the different populations became isolated from one another for selection to have produced relatively stable new phenotypes. The isolation of the small islands from the adjacent larger islands in the Galápagos Archipelago is known to have occurred about 17,000 yr ago (chapter 7). The larger islands of the archipelago probably have always been separated from one another. The volcanic Izu Islands also have presumably been separated from one another throughout their existence (chapter 6). By contrast, the age of isolation of *Crotaphytus* populations in Oklahoma is unknown; some genetic exchange, but evidently not enough to prevent local adaptation, is still occurring today (chapter 5).

These assumptions cannot be established with certainty for any of the study populations. Nevertheless, some violation of them may not be critical. For example, even if sexual selection were still continuing to increase the degree of sexual dimorphism in some of the populations, the nature of the existing differences help us identify the trajectory being favored by selection. Similarly, the phenotypic gambit is helpful as long as the traits are at least partly heritable, which is usually the case.

Given these caveats, what have these intraspecific comparative studies of lizards demonstrated? The three studies suggest that differences in the potential for male-male competition and for female choice of mating partner do influence the evolution of sexual dimorphism. In chapter 5, Kelly McCoy, Troy Baird, and Stanley Fox demonstrate with laboratory studies that female collared lizards exhibit preferences among males. However, field studies show that males are not the most brightly colored where population structure should promote access by females to multiple males, perhaps because predation pressure counterselects against bright coloration in this population. The observed pattern emphasizes that how much choice among potential mating partners female lizards actually can and do exercise is uncertain, and that selection on traits that females may prefer may be constrained by other selection pressures. In chapter 6, Masami Hasegawa shows that home ranges of *E. okadae* females overlap those of many (40-50) males, indicating that the potential for mate choice is high on the Izu Islands. Opportunities to choose among potential mates appear to be much fewer for female *Crotaphytus collaris* in Oklahoma and for *Microlophus* lizards in the Galápagos Islands, the latter shown by Paul Stone, Howard Snell, and Heidi Snell in chapter 7. Rejection of courting males, which is observed by female *E. okadae*, may indicate choice, but it might also indicate that the female is at a stage in her reproductive cycle when she will not mate with any lizard.

Constraints on the action of sexual selection have been measured only indirectly in these studies. The presence of an observer virtually guarantees that predation will not be observed. Therefore, the authors have had to measure predation pressure by tallying the number of species of predators present in the environment, roughly estimating their abundances, and documenting evidence that survival rates are inversely correlated with richness of predator species. Although more precise measures would be desirable, these rough estimates probably serve the general purpose.

However, it is not a simple matter to predict the influence of predators on the evolution of the traits of their prey because lizard predators differ in the nature of the selective pressures they generate. For example, whereas avian predators are visual hunters, snakes and mustelids probably locate and pursue prey using primarily olfactory cues. Snakes, unlike the other predators, may be able to pursue lizards into their retreats. In addition, the structure of the environment strongly influences the visibility of the lizards, their escape routes, and their safety within their retreats. On the Izu Islands, predation apparently influences activity levels, which, in turn, influence the intensity of male-male competition for females (chapter 6). Complex outcomes may result from these differences. Not surprisingly, the authors' interpretations of the influence of predation pressures on the evolution of sexual dimorphism are tentative, and the conclusions differ among the studies.

The effects of food supply are also difficult to determine. Available energy may influence growth rates, time and size of sexual maturity, egg size, number of clutches produced per year, and time available for combat and mate choice. In general, we may expect less sexual dimorphism where food supply is poor than where food is abundant, which appears to be the case in the lava lizard of the Galápagos Islands (chapter 7).

These chapters illustrate what field investigators understand all too well—that untangling the influences of various evolutionary agents on patterns of phenotypic evolution is extremely difficult. The Galápagos Islands provide the most striking contrasts in food supplies, vegetation structure, and predation pressures (chapter 7). In addition, all the satellite islands have been isolated from the adjacent larger islands for approximately the same amount of time, which has been long enough for substantial phenotypic evolution to have occurred. Both the predictions about the influences of predation, escape, food availability, and sexual selection on the lava lizards and the patterns of variation are clearer than in the other two studies. The conclusion that predators select for small size on large islands, that food shortages select for small size on barren islands, and that sexual selection favors large size on diverse islands appears to account for the patterns observed among lizards on different types of islands.

Field investigations of the action of natural selection are difficult to conduct, and the results are generally difficult to interpret. Many of the key variables are difficult to measure, and all sites differ in multiple ways. Nevertheless, the ultimate goal of evolutionary studies is to understand what happens in the complex, messy world of nature. Laboratory studies can show us whether or not the study organisms possess, say, certain behavior patterns, but they cannot tell us how often they are used and how important they are in the lives of the organisms. The evolutionary outcomes in which we are really interested are the product of the total ecological situation in which the organisms live and the interactions among all the factors that influence their survival and reproductive success. The chapters in this section illustrate both the difficulties of conducting field studies and the important insights that can be derived from investigations that compare patterns and processes in environments in which evolutionary agents combine to yield variable results.

CHAPTER 5

Sexual Selection, Social Behavior, and the Environmental Potential for Polygyny

J. Kelly McCoy, Troy A. Baird, and Stanley F. Fox

Since Darwin (1871) first introduced the concept of sexual selection, the various mechanisms that result in selection for characters that convey an advantage in mating have been some of the most widely studied topics in evolutionary biology. It should be no surprise that this fascination with sexual selection has extended to the study of social behavior because for most vertebrates the vast majority of social behavior is involved, at least indirectly and usually directly, with sexual reproduction (Andersson 1994; Arnold and Duvall 1994). Most interactions with other individuals center around locating, identifying, courting, defending, and competing for mates. One might argue that interactions among individuals also include competition for resources, but even this is difficult to separate from reproductive behavior. Certainly mates often may be considered a resource for which individuals (usually males) compete (Bateman 1948), and even the defense of resources such as food, refugia, or other specific microhabitats may be mainly a mechanism for attracting potential mates (Halliday 1983; Krebs and Davies 1993; Baird and Liley 1989; Wade 1995).

Because sexual selection generally results in the evolution of sexual dimorphism, the study of social behavior is inextricably linked with the study of mechanisms that produce morphological differences between the sexes. Various models of selection that result in sexual dimorphism also result in specific mating systems that, in turn, produce certain patterns of social behavior (Krebs and Davies 1993; Andersson 1994). For example, strong intrasexual selection results in elaboration of characters that enhance male competitive ability and highly polygynous systems in which males compete to achieve control of multiple mates (Clutton-Brock et al. 1979; Cox 1981; Halliday 1983). Where intrasexual selection is strong, males attempt to maximize their fitness through competition with other males for access to females. In such a system, one would expect to find males displaying high rates of aggressive behavior toward other males and attempting to control either areas that contain many females or groups of females themselves (Clutton-Brock et al. 1979; Shine 1979; Stamps 1983b; Hews 1990). On the other hand, strong intersexual selection acts to enhance male characters that are attractive to females and may result in a mating system that involves much less competition among males. Rather than high levels of aggressive behavior, males gain an advantage through display behavior or expression of characteristics that females find attractive (Sullivan 1983; Borgia 1985; Baird, Fox, and McCoy 1997; Ryan 1997). Although this type of mating system may also involve strong territorial interactions (Arak 1983a; Stamps 1994), it generally results in high rates of display and courtship behavior aimed at attracting mates (Halliday 1983).

The link between sexual selection and social behavior is an important part of behavioral ecology. Sexual selection promotes certain types of behavior, but social behavior, like any character, is the result of all selective pressures acting together on that character. Social behavior is shaped not only by the requirements of mating, but by other selective forces such as predation pressure, the need to locate appropriate microhabitats, and foraging requirements. These selective forces may not be congruent and, indeed, may act in opposite directions (Arnold 1983). Numerous studies have demonstrated that the need to minimize risk of predation may act as a counterselective mechanism on social signals otherwise advanced by sexual selection (Semler 1971; Endler 1980, 1983; Ryan 1985). Any attempt to explain the evolution, or adaptive significance, of social behavior must therefore include an examination of all selective mechanisms that impinge on that behavior.

Many studies of sexual selection have been hampered by a relatively narrow approach. Despite (or perhaps as a result of) the recognition of multiple selective mechanisms within the general category of sexual selection, many studies have set out to test only a single, specific selective mechanism. The scope of such studies is limited because they are unable to examine the relative impact of both inter-and intrasexual selection on social behavior patterns. Overcoming this limitation is critically important because both inter-and intrasexual selection may act together to shape evolution of a species (Hews 1990; Moore 1990), and multiple selection models may be operating concurrently (McCoy 1995).

Although interspecific comparisons, especially within a phylogenetic framework, have provided enormous insight into broad patterns of the adaptive significance of social behavior (see chapters 9–11), such studies may not allow detailed examination of the effects of small differences in natural selection pressures or the relative effects of various types of sexual selection. The elucidation of such fine selective effects requires the study of populations with the same basic environmental requirements, similar phylogenetic histories, and comparable ecologies, yet subjected to differences in selection regime. In short, such studies will require comparisons among populations of the same species existing in habitats sufficiently different to provide differences in the selection pressures shaping social behavior. Chapters 3, 4, 6, 7, and 9–11 address this same issue.

Such a situation exists among populations of the collared lizard (*Crotaphytus collaris collaris*) in Oklahoma. *Crotaphytus collaris* is a large, diurnal lizard that is highly saxicolous (Fitch 1956); populations are typically restricted to outcroppings of exposed rock or boulders. Throughout much of its wide geographic range, *C. collaris* exists as small populations in pockets of suitable habitat, widely separated by large areas of unsuitable habitat. This presents a situation quite similar to that of island habitats where populations may experience differences in selection pressures (mediated by differences in habitat, habitat arrangement, food availability, or predation regime) and thus display different evolutionary trajectories (Brown and Lomolino 1998; Grant 1998; see also chapters 6, 7, and 11). One important difference between isolated populations of collared lizards and insular populations of other species is a small, but constant, degree of gene flow among collared lizard populations. This gene flow is sufficient to minimize the effects of genetic drift within populations but is unlikely to be sufficient to swamp the effects of local differences in selection pressure (Hranitz and Baird 2000).

Crotaphytus collaris is noted for its strong sexual dimorphism and its polygynous mating system (Fitch 1956; Bontrager 1980; Baird, Acree, and Sloan 1996). Males, which are much larger and more brightly colored than females, typically defend territories that may encompass the home ranges of several females (Fitch 1956; Bontrager 1980). Females are sometimes aggressive toward consexuals (Yedlin and Ferguson 1973; Bontrager 1980) and, under certain ecological conditions, may defend core areas within their home ranges (see chapter 1).

The observation of strong sexual dimorphism in body and head size in a territorial iguanian lizard might lead to the assumption that intrasexual selection has been the major force in shaping the evolution of social behavior in this species. Indeed, there are ample reports in the literature of aggressive male behavior associated with a polygynous mating system (e.g., Clutton-Brock et al. 1979; Cox 1981). Yet, the relationship between sexual dimorphism and behavior is particularly interesting in collared lizards because the degree of dimorphism differed significantly among lizards from three Oklahoma populations. Specifically, these populations differed in the degree of sexual dimorphism in body size, body shape, and coloration (McCoy, Fox, and Baird 1994; McCoy et al. 1997). However, the variation among populations did not show a simple pattern in which all characters were equally more or less dimorphic among the three populations. Rather, different populations showed the greatest or least extreme dimorphism in different characters (McCoy, Fox, and Baird 1994; McCoy et al. 1997). This lack of congruence in the degree of sexual dimorphism in various characters across populations of this species suggests the action of different selective mechanisms and that local differences in the environment may be producing differences in the mechanisms promoting sexual dimorphism in various characters.

The Study System

Behavioral field studies conducted from 1991 to 2000 on one of our study populations (Arcadia Lake [AL]) revealed the following information on the reproductive life history of *C. collaris;* the patterns are similar in our other two Oklahoma populations (McCoy and Fox, unpublished data) and to those reported for other areas (e.g., Ballinger and Hipp 1985). Adult *C. collaris* in Oklahoma are active from late March or early April through July or August. Females typically produce one or two clutches of eggs per season (Baird, Sloan, and Timanus 2001), although three clutches may be produced during good years (Baird, unpublished data). Hatchling collared lizards emerge from late July through September, during which time the activity by adults becomes increasingly diminished (Baird, Sloan, and Timanus 2001). Hatchlings remain active until decreasing temperatures drive them into hibernation, usually sometime in October.

Juvenile collared lizards usually attain sexual maturity shortly after emergence in the summer following their hatching. Although the smallest females sometimes do not reproduce during their first (yearling) year, most produce at least one clutch of eggs, albeit later than the older females (Baird, Sloan, and Timanus 2001). Males, however, show a different reproductive ontogeny. Yearling males, despite being sexually mature, have not yet developed the degree of sexual dimorphism displayed by older males (McCoy, Fox, and Baird 1994; McCoy et al. 1997), and yearling males usually exhibit markedly lower frequencies of the behavior patterns (e.g., advertisement displays and patrol) that older males use to control territories. Rather than defending territories and controlling access to females, yearling males use subordinate social tactics to avoid aggression from older territorial males (Baird, Acree, and Sloan 1996; Baird and Timanus 1998; see also chapter 1). Thus, the population of sexually mature collared lizards comprises three distinct classes of individuals: females, territorial adult males (2 yr old and older), and nonterritorial yearling males.

The populations studied were located at AL (Oklahoma County) in central Oklahoma, in the Glass Mountains (GM) (Major County) in northwestern Oklahoma, and at the Wichita Mountains National Wildlife Refuge (WM) (Comanche County) in southwestern Oklahoma. These populations are the same ones for which differences in sexual dimorphism were reported by McCoy and colleagues (1994, 1997). Although all populations display sexual dimorphism in color, body size, head length and width, limb length, and tail length, the interpopulation differences in the degree of sexual dimorphism are not congruent across these characters (Table 5-1). The GM population is most dichromatic, although only slightly more so than the WM population, whereas both are significantly more dichromatic than the AL population. The AL and WM populations are similar in the degree of sexual dimorphism in body size, but both are significantly more dimorphic than the GM population. Most of the other morphological characters follow the same pattern as body size, although the GM population is most dimorphic in head length, and the AL population shows little dimorphism in tail length.

We used the observed interpopulation differences in sexual dimorphism in our three populations to develop predictions regarding the relative strength of inter- and intrasexual selection in each population. From these expected differences in the strength of sexual selection, we predicted patterns of social behavior in each population, and differences among the populations. We tested our expected patterns of social behavior by conducting field observations on individuals of both sexes in all three populations, and by conducting controlled laboratory trials of intrasexual contests between males and female preferences of males.

Table 5-1 Summary of interpopulation differences in degree of
sexual dimorphism in various characters among three populations
of collared lizards from Oklahoma: Arcadia Lake (AL), Glass
Mountains (GM), and Wichita Mountains (WM).

Character	Differences among Populations
Body size	WM ≈ AL ≫ GM
Color	GM ≈ WM≫ AL
Head length	GM > WM ≈ AL
Head width, limb length	WM ≈ AL ≫ GM
Tail length	WM ≈ GM > AL

Methods

Field Studies

During 1991–93, we studied the behavioral ecology of collared lizards
in three populations in Oklahoma. We employed similar methods at all
three locations to collect data regarding space use and social behavior.
We established semipermanent study sites at all three locations (the AL
population is still under study), surveyed the study sites, and placed
permanent markers approximately every 10 m across each site. We
mapped these markers onto scale maps of the study sites. We captured
all lizards at each site and permanently marked them with unique toe
clips and temporarily with unique acrylic paint color codes. We regularly censused all study sites during the activity period for adult collared
lizards in Oklahoma (April–August), and we recorded point locations
of known individuals on the scale maps. We made focal observations
(20 min each) on known individuals to record rates of various types of
behavior. Additional details on methods may be found in McCoy (1995)
and Baird, Acree, and Sloan (1996)

   We used the point locations of known individuals from census data
to construct minimum convex polygon home ranges (Rose 1982) for
each lizard. From these home ranges, we calculated the percentage of
each individual's home range that was overlapped by different classes
of individuals. We also calculated for each individual the number of
overlaps with individuals of various classes. For the AL population, the
minimum number of points used to calculate home ranges for males
and females were 20 and 30, respectively (Baird, Acree, and Sloan 1996).
At WM and GM, we calculated home ranges for individuals for which
we had a minimum of five point locations. This number of points is
likely not sufficient to determine the full extent of home range area

(Rose 1982). Therefore, at WM and GM our minimum convex polygon technique likely underestimated the total area of home ranges. However, within each population we had similar numbers of sightings for the three classes of lizards studied. Thus, comparisons of space use among classes of lizards within populations are reasonable, even though the number of points per lizard was small in two populations. Comparisons of space use among populations, however, must be considered carefully in light of the underestimation of home range size at GM and WM.

During focal observations, we recorded all social and display behavior. For analysis, behavior was grouped into several categories and expressed as the rate at which a behavior was performed (behavior per hour). The categories of behavior that were compared among classes of individuals and across populations were as follows: push-ups (stereotypical push-ups as displayed by many iguanian lizards), displays (all other display behavior including lateral displays and dewlap extensions), aggressive behavior (clearly directed at another individual of the same sex and including approaches, displays, and physical confrontation and not including advertisement displays), and courtship encounters with members of the opposite sex. The push-ups and displays analyzed in this study are both forms of advertisement display. We included behavior in these categories only when it was clearly not directed at another individual. Behavior recorded as aggression was exhibited while within 1 m of another individual of the same sex and was clearly directed at that individual. Courtship encounters involved physical contact between a male and a female. Although many types of behavior were exhibited during these encounters, what we analyzed was the rate of encounters between males and females. Thus, each time a male approached and made contact with a female it was recorded as a courtship encounter. Although a pair might be in physical contact off and on over a fairly long period of time, only one courtship encounter between a male and a female was recorded unless the pair separated and clearly engaged in other types of behavior and then returned to contact each other. We analyzed behavior recorded during focal observations only for individuals with at least three focal observations on separate days.

Laboratory Trials

To test the possible roles of male coloration and size in female preferences for males and the outcomes of agonistic interactions between males, we separately conducted inter- and intrasexual trials in laboratory arenas. Detailed methods for these trials and our experimental treatments are described elsewhere (Baird, Fox, and McCoy 1997), and only the general methods are given here. Arenas each consisted of two chambers that were equal in size and positioned side by side but separated from one another by a movable, opaque partition. A chamber for a female spanned the front of each male's chamber, separated by a pane of glass such that the female could view each male equally. The female's chamber was fitted with a window covered by a blind with a small viewing port so that both the female and the males could be observed without disturbance.

We noosed males and females from each population, transported them to the laboratory, and measured their size (snout-vent length) and coloration using Munsell color charts. For each population, we then conducted three sets of trials using lizards from the same population (within-population design). In two separate heats, we first matched males for similar size but also maximal color difference, and then we matched males for similarity in color but also maximal difference in size. We randomly placed males paired for each of these treatments in the two side chambers of each arena. We then placed a female from the same population as the males, on the midline of the front chamber. On the following day, we recorded the position of the females at 10-min intervals for 3 h. We ran two heats of different females for each population ($n = 22$ females for GM and WM, $n = 21$ females for AL color-different trials, $n = 20$ for AL size-different trials). We calculated the percentage of observations that females positioned themselves on one side of their chamber or the other as our estimate of preference for brighter or larger males (Baird, Fox, and McCoy 1997). Following intersexual trials, we removed females from arenas.

On the day following intersexual trials, we staged intrasexual contests between males using mostly the same males in our pairings as for female preference trials. We ran a total of 32 replicates for size-matched, color-different trials and a total of 32 replicates for color-matched, size-different trials (for each treatment, $n = 11$ trials for both GM and WM, and $n = 10$ trials for AL). Male-male contests began when we removed the partitions and allowed the two males to interact directly. For 30 min we tallied behavior patterns of each male using a graded scoring system ranging in intensity from highly aggressive behavior (e.g., attacks, bites) to submissive behavior (e.g., fleeing). We then calculated for each male a graded agonistic score and compared the scores of brighter versus duller sized-matched pairs and larger versus smaller color-matched pairs (Baird, Fox, and McCoy 1997).

Predictions and Tests
Effects of Intrasexual Selection

Crotaphytus collaris is characteristically described as displaying a territorial, polygynous mating system. This mating system—along with the strong sexual dimorphism in body size and head size—suggests strong intrasexual selection acting on male traits. In this type of system, males maximize their fitness by controlling large areas (territories) that encompass the home range(s) of one or more females. The females in this system may also be territorial, may display dominance hierarchies, or may show little intrasexual agonism. Because female collared lizards display marked site fidelity and little tendency to wander over large areas, it is not possible to determine whether the polygyny is driven strictly by defense of resources or defense of females. Females are found in areas where resources are available and males are controlling those areas. A mating system consisting of territorial defense (by males) and strong polygyny makes several predictions regarding the social behavior both in terms of the use of space and in the type and rates of behavior displayed.

Because intrasexual selection in this species is largely driven by male-male competition for territories, we would expect to find evidence of strong interference competition for superior space among adult males. This evidence would come in the form of low overlap in territories among adult males, and high numbers of overlaps among adult males and females (as males attempt to control territories encompassing the home ranges of multiple females). We also would expect to find that male territories are substantially larger than the home ranges of females.

Strong intrasexual selection also implies certain aspects of social behavior. In this type of system, males would be expected to show high rates of aggressive behavior to other males. Because maintaining a territory (that includes the home ranges of several females) provides a large fitness benefit, males would also be expected to invest substantial energy in broadcast displays advertising possession of that territory.

The observation of a strongly polygynous system also allows us to make predictions regarding differences among males within a population. In a polygynous system, a small minority of the adult males will accomplish most of the reproduction, and many males will not be successful at mating (Wade 1979; Howard 1983; Clutton-Brock, Albon, and Guiness 1988; Le Boeuf and Reiter 1988; McVey 1988). This creates two distinct classes of males within a population. The successful (polygynous) males would be expected to show higher rates of agonistic behavior, more displays, and stronger territorial behavior than the unsuccessful males. There should be little overlap among the territories of the successful males, whereas the unsuccessful males may have much more of their home ranges overlapped by other individuals.

Because strong intrasexual selection indicates that males will maximize their fitness by successfully competing with other males, this selection should produce exaggeration of characters that confer an advantage in that competition (Howard 1979; Arnold 1983). Staged interactions among males (Baird, Fox, and McCoy 1997) demonstrated a significant advantage to larger males in all three populations, but no advantage of brighter color in two of the populations. Thus, we expect intrasexual selection to be strongest in those populations showing the greatest sexual dimorphism in body size (WM and AL).

Adult male collared lizards in all three populations displayed a pattern of space use supporting the hypothesis of strong intrasexual selection. At all three sites, the home ranges of adult males were substantially larger than those of adult females (Fig. 5-1). This difference was most marked at WM, where adult males used home ranges an average of five times as large as the home ranges of adult females. These data clearly fit the prediction for male superterritories expected in highly polygynous lizards (Stamps 1983b). There was also strong evidence of territorial behavior at all three populations. The home ranges of adult male lizards overlapped very little (Table 5-2), suggesting that these areas were actively defended and constituted territories. This contention was further supported by the much larger degree of overlap in the home ranges of yearling males, and the overlap of yearlings with adults (Table 5-2), indicating that yearlings were not actively defending their home ranges.

Although the analysis of space use shows no clear distinctions among the three populations, the rates of various types of social behavior provide greater discrimination. At all three populations, adult males performed displays and push-ups at fairly high rates, supporting the hypothesis of territorial defense (Fig. 5-2). Aggression directed at other adult males was likewise high at all three populations. However, adult males from the AL population showed a markedly higher rate of performing push-ups than males from the other populations, and adult males from both AL and WM displayed more frequently and exhibited higher rates of aggression than adult males at GM. There was also a distinct difference between classes of males at AL and WM that did not

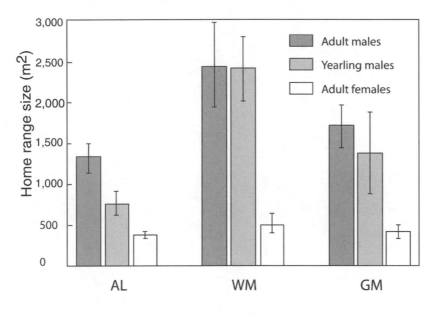

Figure 5-1 Home range sizes (mean ± 1 SE) for adult male, yearling male, and adult female collared lizards from three Oklahoma populations.

exist at GM. For populations at AL and WM, the rates of push-ups, displays, and aggressive behavior were significantly different between adult and yearling males (McCoy 1995; Baird, Acree, and Sloan 1996).

The pattern of differences among populations in the social behavior and space use of male lizards fits with the predictions based on the pattern of differences in sexual dimorphism in body size. Male lizards from all populations demonstrated territorial behavior and aggression toward other males, but the degree of territoriality was greater in those populations that showed the greatest sexual dimorphism in body size (AL and WM). Furthermore, in the populations in which intrasexual selection and sexual dimorphism in body size were greatest, there were two distinct classes of males based on their behavior. At GM, where sexual size dimorphism and intrasexual selection was the weakest, there was no significant difference between adult and yearling males in several categories of behavior.

Effects of Intersexual Selection

Although Stamps (1983b) has suggested that little opportunity for intersexual selection through female choice of mates exists in iguanian lizards, various aspects of the biology of collared lizards suggest that

Table 5-2 Home range overlaps for adult and yearling male collared lizards from three Oklahoma populations. Overlaps are expressed as the percentage of an individual's home range that is overlapped by other individuals (mean ± 1 SE).

	Adult Males			Yearling Males		
	AL (n = 15)	WM (n = 16)	GM (n = 22)	AL (n = 13)	WM (n = 18)	GM (n = 7)
Overlapped by adult male(s)	15.9 ± 4.3	8.4 ± 3.9	14.5 ± 4.8	84.3 ± 4.8	18.2 ± 4.7	37.7 ± 11.1
Overlapped by yearling male(s)	48.7 ± 4.1	33.3 ± 7.5	38.5 ± 15.9	35.6 ± 5.0	19.8 ± 6.5	17.1 ± 15.9

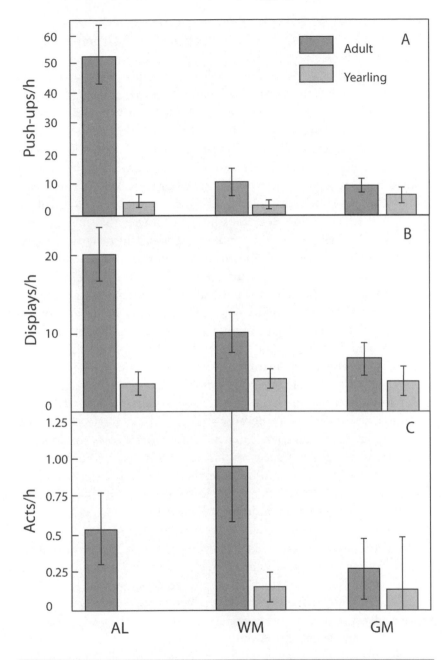

Figure 5-2 Rates of push-ups *(A)*, displays *(B)*, and aggressive acts *(C)* (mean ± 1 SE) for adult and yearling male collared lizards from three Oklahoma populations.

mate choice may indeed play an important role in sexual selection in this species. Unlike many other lizard species, the habitat and ecology of collared lizards frequently offer ample opportunity for females to see and assess the quality of multiple males. Collared lizards generally occupy large expanses of exposed rocks and boulders, and males display from prominent outcroppings. During these displays, males can frequently be seen from great distances. Additionally, although this species displays marked sexual dimorphism in color, staged agonistic encounters between size-matched males showed no advantage to more brightly colored males in two of these populations (GM and WM) (Baird, Fox, and McCoy 1997). Because color does not seem to play a major role in intrasexual selection, but does constitute a character that may be used to assess male quality visually, it is not unlikely that the evolution of sexual dichromatism may be promoted through intersexual selection.

If male coloration is a character that is selected through female mate choice, we would expect that the populations displaying the greatest degree of sexual dimorphism in color (GM and WM) would also be experiencing the greatest degree of intersexual selection. In these populations, we would expect to find high rates of advertisement displays (as in a system in which there is strong intrasexual selection), and we would also expect to find high rates of courtship behavior (Halliday 1983). This pattern of behavior would provide the greatest opportunity for females to assess male quality. Because strong intersexual selection would also act to increase the variance in male mating success, we would again expect to find large differences among successful males (those that are often chosen by females) and unsuccessful males (those that are rarely chosen) (Arnold 1983; Halliday 1983). If there are differences among populations in the strength of intersexual selection, we would again expect that pattern to be mirrored in the social behavior.

To distinguish behavior that might be promoted by intrasexual selection from behavior promoted through intersexual selection, we can compare only the rate at which males engage in behavior that includes contact with a female. This behavior is certainly not aggressive, does not involve advertisement to other males, and involves the cooperation of the female (female choice). Although advertisement displays by males may also play an important role in female choice of mates (see chapter 1), it is not possible to separate advertisement displays directed at females from advertisements directed at other males. Courtship encounters, however, are clearly important in intersexual selection. When we compared the rate of intersexual contact by males across the three populations, we found that adult males at AL and WM had far more frequent contact with females than males from GM (Fig. 5-3). Additionally, at AL and WM, there were significant differences between adult and yearling males, but no such differences at GM (McCoy 1995; Baird, Acree, and Sloan 1996).

In laboratory trials of female preference, females explored their chambers such that they detected both males, and females spent a larger percentage of their time positioned at the glass partition such that they could see the males. Females did not display a significant preference for larger over smaller color-matched males in any of the three populations (Table 5-3). By contrast, in size-matched color-different trials, AL females associated significantly more frequently with brighter males, whereas neither GM nor WM females associated preferentially with brighter or duller males (Table 5-3).

The results of staged male-male interactions further supported our predictions regarding the role of color in intrasexual selection. Males interacted agonistically in 94% of the 32 trials run with each of our two treatments. In color-matched size-different male contests, larger males had significantly higher agonistic scores in all populations (Table 5-4). By contrast, in the size-matched color-different treatment, only at AL did brighter males have significantly higher agonistic scores (Table 5-4) than males with duller coloration (Baird, Fox, and McCoy 1997).

Neither field observation on the differences in the rate of contact with females among male collared lizards from these three populations nor the outcome of laboratory trials on female mate choice coincide with our expectations based on the degree of sexual dichromatism. The GM population was the most dichromatic, yet the behavior of male lizards in the field and females in the laboratory seemed to indicate that intersexual selection in this population was the weakest. Conversely, both field observations and trials in the laboratory on the outcome of mate choice suggest that intersexual selection should be strongest at AL. Nonetheless, the AL population is the least dichromatic. Clearly, some additional explanation must be sought for the observed patterns.

Environmental Potential for Sexual Selection

Our data on the behavior and morphology of collared lizards from these three populations strongly suggest that the strength of sexual selection varies among populations. This suggests that some aspect (or aspects) of the environment affects sexual selection in these populations, or changes the environmental potential for polygyny (Emlen and Oring

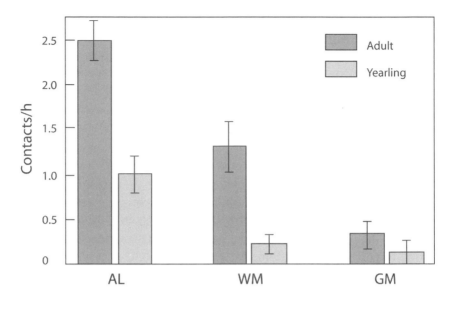

Figure 5-3 Rate of contact with females (mean ± 1 SE) for adult and yearling male collared lizards from three Oklahoma populations.

1977). It has been demonstrated for other species that the spatial distribution and arrangement of females affect the ability of males to control
access to females and thus affect the degree of polygyny (Vehrencamp
and Bradbury 1984). The greater the potential for polygyny, the greater
the strength of sexual selection. There is ample reason to suspect strong
differences in the environmental potential for polygyny and sexual selection among these populations. The habitats are markedly different,
and the localities differ particularly in the spatial arrangement of suitable microhabitat for collared lizards.

The WM location contains outcroppings of granite boulders widely
distributed across large areas. This area contains ample habitat for collared lizards, but the distribution of habitat is quite patchy. Areas of exposed boulders are often separated by small areas of sparse grass. Larger
areas that include exposed rock and sparse grass are widely separated by
dense grasslands and forested areas. At this site, females maintain relatively small home ranges on patches of exposed rock. The home ranges
of females are largely exclusive of other females' home ranges (Table
5-5) and may be defended to some degree. In this population, the home
range of each adult female overlaps an average of less than one other
adult female's home range and more than 81% of the area of the home

Table 5-3   Percentage of observations (mean ± 1 SE) that female collared lizards from three populations associated with males from their same population that were matched for size but differed in coloration, and males that were matched for coloration but differed in size. Data from Baird, Fox, and McCoy (1997).

Population	Size Matched		Color Matched	
	Duller	Brighter	Smaller	Larger
GM	59.4 ± 7.7	40.6 ± 7.7	54.0 ± 8.0	46.0 ± 8.0
AL	33.3 ± 6.7	66.7 ± 6.7	45.2 ± 5.6	54.8 ± 5.6
WM	44.2 ± 9.4	55.8 ± 9.4	41.9 ± 8.4	58.1 ± 8.4

Table 5-4   Graded agonistic scores (mean ± 1 SE) during dyadic contests between collared lizard males from each population matched for size but different in coloration, and matched for color but different in size. Data from Baird, Fox, and McCoy (1997).

Population	Size Matched		Color Matched	
	Duller	Brighter	Smaller	Larger
GM	41.7 ± 15.8	31.0 ± 13.4	13.5 ± 4.3	80.2 ± 22.0
AL	26.0 ± 7.2	67.2 ± 20.7	11.3 ± 3.9	69.3 ± 21.3
WM	29.4 ± 17.1	47.6 ± 11.5	5.4 ± 2.0	27.0 ± 8.8

range is not overlapped by other females. Adult females at WM also display a high rate of push-ups and advertisement displays (Fig. 5-4), indicating that females defend at least part of their home range, as has been documented for other populations (Yedlin and Ferguson 1973; Bontrager 1980). This spacing pattern of females on small patches of good habitat narrowly separated by less desirable habitat presents a situation in which adult males that are capable of defending a large territory may control access to several females at the same time and thus increase their fitness.

The habitat at GM is markedly different from that found at WM. At this site, collared lizards occupy narrow outcrops of gypsum at the top edges of steeply eroded buttes. Neither the flat tops of the buttes nor the steep clay banks below the gypsum offer suitable habitat for the collared lizards. Thus, the lizards in this population are confined to narrow (<20 m) bands of suitable habitat arranged in a nearly linear fashion. Females occupy home ranges nearly as large as those of females at WM (Fig. 5-1) and also display territorial behavior (Table 5-5, Fig. 5-4). However,

Table 5-5 Home range overlaps among adult female collared lizards in three Oklahoma populations (mean ± 1 SE).

	AL ($n = 23$)	WM ($n = 26$)	GM ($n - 19$)
Number of overlapping females	7.8 ± 0.5	0.9 ± 0.2	0.3 ± 0.1
Percentage of home range exclusive of overlaps	6.5 ± 3.5	81.5 ± 4.9	97.9 ± 1.1

this does not allow much opportunity for males to monopolize multiple females. A home range that completely includes the home range of several females would have a maximum linear dimension too large for a single male to patrol effectively. The spatial arrangement of habitat at the GM site results in a markedly reduced environmental potential for polygyny relative to WM.

Although the potential for polygyny is reduced at the GM site, the spatial arrangement of habitat does not preclude the action of intersexual selection through female choice. Although the home ranges of females are arranged almost linearly along the tops of the buttes, males display from prominent points and are visible from long distances, especially against the white background of the gypsum rock. Each female still has the opportunity to assess visually the quality of males. Females may also be assessing male quality and choosing mates using some threshold decision rule (Wittenberger 1983). Such mate choice may be maintaining the high degree of dichromatism found in this population. Although males from GM do not display the behavior that we would have expected for a population experiencing strong intersexual selection, estimates of male fitness (based on overlaps with females) do show a significant correlation with male color (McCoy 1995). The relatively low rate of courtship behavior by adult males may simply be an artifact of the spatial arrangement of the females and not an indication of the strength of selection. Males may be unable to reach enough different females to show a high rate of courtship.

The greatest environmental potential for polygyny is found at AL. This population exists on boulders imported to stabilize the spillway at the AL dam. The arrangement of these highly uniform boulders has created an environment that consists of small patches of highly homogeneous but excellent collared lizard habitat. Under these circumstances, females reach much higher densities than at the other two sites. There is extensive overlap among the home ranges of females with very little of each individual's home range exclusive to itself, and each female over-

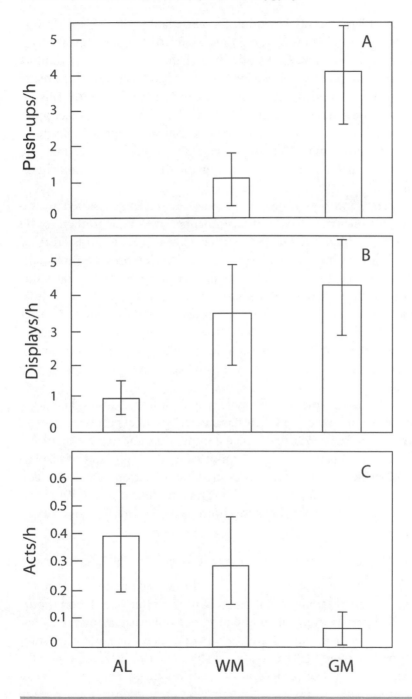

Figure 5-4 Rates of push-ups (*A*), displays (*B*), and aggressive acts (*C*) (mean ± 1 SE) for adult female collared lizards from three Oklahoma populations.

laps with a large number of other females (Table 5-5). This pattern of space use—combined with the observation that females engage in aggressive behavior more frequently than at the other sites but produce fewer advertisement displays (Fig. 5-4)—suggests that in this area of greater population density, females are not territorial. Clearly, this extremely dense spatial arrangement of females allows enormous opportunity for a male to control a territory that encompasses the home ranges of many females. The potential for polygyny, as well as the potential for sexual selection, is much higher in this population than in the other two.

The environmental potential for polygyny at AL has resulted in a situation in which there is enormous potential for a male to increase his fitness through competition with other males. Likewise, this creates a situation in which there may be enormous selection pressure on females to choose males that are successful in this competition. Females that choose poorly will produce males that inherit poor genes from their father and have very little chance of mating on reaching sexual maturity. On the other hand, females that choose a male possessing (genetic) characters that enhance his competitive ability may enjoy greatly enhanced fitness because their sons will also likely have high fitness (Fisher 1930).

Based on the potential for polygyny as inferred from the spatial arrangement of lizards and their habitat, we would expect strong sexual selection, through both male-male competition and female choice of mates, at AL. This prediction is supported by the high rates of displays and courtship among adult males in this population (Fig. 5-2). This prediction is also supported by the results of staged interactions that demonstrated strong female mate choice at AL (Baird, Fox, and McCoy 1997).

Natural Selection, Counterselection, and Other Possible Mechanisms

Given the apparent high strength of both intra- and intersexual selection at AL, the lack of strong sexual dichromatism in this population is puzzling. The lizards from this population not only display markedly less dichromatism than lizards from the other populations but are generally less brightly colored (Plates 6 and 7) (McCoy et al. 1997). Despite the strong sexual selection, lack of dichromatism and general lack of bright coloration may be a result of counterselection pressures. High rates of lizard injuries and sightings of potential predators strongly suggest that predation pressure at this site is much higher than at the other two sites

(Baird, Fox, and McCoy 1997) and may constitute a selective pressure that reduces the advantage of bright coloration. Brightly colored males may suffer higher mortality due to predation (especially on the uniformly gray boulders at this site) and thus experience a disadvantage through natural selection. The color dimorphism observed in this population is probably a balance between these two opposing selection pressures.

The degree of sexual dimorphism in head length in the GM population also does not fit with other observations. Although this is a character that might be important in male-male combat, there is no other evidence that such combat is a strong selective mechanism in this population. Head length, however, might also be related to the maximum size of prey that can be eaten. It is possible that sexual dimorphism in this character is promoted through the competition avoidance hypothesis (Selander 1966; Slatkin 1984). If greater sexual dimorphism in head length allows males and females to eat substantially different sizes of prey, the amount of competition for food between the sexes might be reduced. This could result in greater fitness for males and females sharing the same territory, and thus the same food resources. Comparisons of available prey between WM and GM showed lower prey density and greater diversity in size at GM (McCoy 1995). This suggests that the potential for selection through competition avoidance may be higher at GM. At least one study of collared lizard diet has indicated sexual differences (Best and Pfaffenberger 1987), although another failed to find such differences (Husak and McCoy 2000).

Conclusion

The pattern of sexual dimorphism, sexual dichromatism, and social behavior displayed by Oklahoma populations of collared lizards unveils a surprisingly complex evolutionary tapestry woven of the effects of various selection pressures. Both the morphology and the behavior of these lizards reveal the effects of intersexual, intrasexual, and natural selection. The current differences among these populations are likely the product of local differences in selection pressures overlain on the evolutionary history of the species.

Given the general observation that collared lizards are strongly sexually dimorphic and territorial, it is not unreasonable to hypothesize that strong intrasexual selection through resource defense polygyny has been the major driving force in the evolution of this species. The population at WM likely represents the selection regime typical for collared lizards through most of their evolutionary history. In this population and others similar to it (Sloan 1997; Baird and Sloan, unpublished data), females maintain small territories exclusive of one another. Males attempt to defend much larger territories that completely include the territories of several females. The fitness advantage gained by males that can control access to multiple mates through territorial defense has led to strong selection for characters that promote success in male-male agonistic encounters, or that enhance the ability of males to hold large territories. Thus, males at WM are much larger in body and relative head size than females and show a high level of advertisement displays. This social system also produces strong differences in behavior between males that hold territories (usually older males) and males that do not hold territories (usually yearlings).

At the other two locations, local differences in the spatial arrangement of suitable habitat have produced differences in selection that have promoted differences in sexual dimorphism. At GM, because lizards are distributed in long narrow bands, the ability of males to monopolize access to multiple females is strongly curtailed. This has produced a system in which polygyny and the strength of intrasexual selection are markedly reduced. Sexual dimorphism in body size is weaker in this population, and males show a reduced rate of territorial advertisement displays. The potential for color as a signal of male quality is still present, however, and this population displays strong sexual dichromatism. This population is also the most dimorphic in head length, a character that may enhance the ability of males and females to consume different sizes of prey. Low prey density and high diversity in prey size may be acting to increase the selection for sexual differences in diet and thus differences in characters that affect the size of prey consumed.

The population of collared lizards at AL also clearly shows the effect of local selection pressures. On this highly homogeneous and highly suitable habitat, collared lizards reach the highest density seen in any of these populations. Because critical resources are not limited at AL, females are not highly aggressive and do not defend territories (see chapter 1). The extremely high spatial overlap among females creates a situation in which males can monopolize access to a relatively large number of females. The strong fitness advantage to successful males in this population produces strong intra-and intersexual selection. Within this population, we find strong sexual dimorphism in body size, high rates of territorial and agonistic behavior among males, and evidence of female mate choice based on color (Baird, Fox, and McCoy 1997). However, this population also clearly displays the effects of counterselection. High predation rates in this population likely increase the cost of bright coloration. This results in a population that displays relatively little sexual dichromatism, despite female mate choice for the brightest colored males.

The study of sexual dimorphism, sexual dichromatism, and social behavior in populations of collared lizards not only reveals the interplay among various selection mechanisms, but supports the statement that every character is the evolutionary result of the totality of selection that impinges on that character. No behavioral or morphological trait is the product of a single selection mechanism. Thus, any study of the evolution of social behavior must consider all the possible mechanisms of selection that could modify the behavior of the organism being studied. Measuring the strength of a single selection mechanism will be insufficient to reveal the complex evolutionary mosaic that has produced the observed behavior.

Acknowledgments

We wish to thank J. Gosney and R. Parker for access to study sites in the Glass Mountains, the Army Corps of Engineers for access to the Arcadia Lake site, and the Fish and Wildlife Service for access to the Wichita Mountains National Wildlife Refuge. We also wish to thank the many students, graduate and undergraduate, who participated in the collection of data in the field. Funding for this research was provided in part by the Theodore Roosevelt Memorial Fund of the American Museum of Natural History, by the Office of Research and Grants at the University of Central Oklahoma, by the Zoology Department at Oklahoma State University, and by the Oklahoma Cooperative Fish and Wildlife Research Unit (Oklahoma State University, Oklahoma Department of Wildlife Conservation, U.S. National Biological Service and Wildlife Management Institute, Cooperating).

CHAPTER 6

Intraspecific Variation in Sexual Dimorphism and Mating System in Relation to Interisland Differences in Predation Pressure

Masami Hasegawa

Sexual selection, or selection for any character that confers an advantage in mating, often results in sexual dimorphism (Wade 1987; Hedrick and Temeles 1989; Andersson 1994). Dimorphism evolves by sexual selection when asymmetry in the development of one or more phenotypic characters confers an advantage in intrasexual competition for mates, or when there is nonrandom mate choice based on asymmetry in one or more phenotypic traits. Trivers (1972) pointed out that it is the difference of parental investment between the sexes that results in the different strategies that are usually observed in males and females for attaining reproductive success. Because male parental investment per gamete is usually low, males generally maximize reproductive success by mating with as many females as possible. The reproductive success of a given male is often highly dependent on his ability to outcompete other males for access to mates, and exaggerated male characters may be selected because they enhance male competitive ability. As a result, variance in male mating success is usually greater than that of females (Bateman 1948).

There is strong evidence in a variety of vertebrates that selection for intrasexual competitive ability sometimes results in the development of exaggerated male traits. In particular, increased body size and/or the size of particular structures are often selected in males because these promote the ability to prevail in intrasexual contests (e.g., Le Boeuf 1974; Trivers 1976; Clutton-Brock and Harvey 1977; Howard 1978a; Borgia 1979; Warner and Hoffman 1980; Packer 1983; Stamps 1983b; Arak 1988; Jarman 1988). Among reptiles, sexual dimorphism in male snakes and turtles is most strongly developed in species that exhibit male combat (Shine 1978; Berry and Shine 1980). Similarly, dimorphism in head and body size is more pronounced in lizard species in which males compete for mates directly, or for territories that overlap female home ranges, than in species in which male-male aggression is weak (Stamps 1983b; Carothers 1984; Anderson and Vitt 1990). Moreover, intraspecific studies have revealed that male reproductive success increases with male body size or the size of armaments used during intrasexual agonistic contests (Vitt and Cooper 1985a; Cooper and Vitt 1993; Hews 1990).

By contrast, female parental investment per gamete is generally higher and the reproductive success of individual females is usually not limited by their ability to compete intrasexually for access to males (Bateman 1948; Williams 1966; Trivers 1972; Alexander and Borgia 1979; Hoffman 1983). Rather than engaging in intrasexual contests, females generally maximize reproductive success in two other ways. They may maximize time and energy investments in the production of eggs and offspring (e.g., Hoffman 1983; Baird and Liley 1989). Maximizing parental investment might also involve choosing a mate that controls high-quality resources important for the production and survival of offspring. Females also may increase their reproductive success by mating preferentially with males that possess superior genetic qualities that are passed on to the offspring produced by those females (Fisher 1930; Trivers 1972). Although empirical study of female mate choice has lagged behind that of male competition, there is abundant evidence in all vertebrate classes to support the prediction that females may show preferences for mates on the basis of exaggerated characters (e.g., Sullivan 1983; Borgia 1985; Andersson 1986; Wells and Taigen 1986; Møller 1988b; Hoglund 1989; Houde and Endler 1990). Among lizards, females of some species associate preferentially with males that display bright, conspicuous coloration (Sigmund 1983; Baird, Fox, and McCoy 1997), whereas females of other species prefer males with enlarged heads (Vitt and Cooper 1985a; Cooper and Vitt 1993).

Whether selective advantages favoring the development of dimorphic male characters involve intrasexual competition, female mate choice, or both, the natural selection costs of such traits may have an important influence on the degree to which they develop. One potential cost involves the energetic investment required to develop morphological traits or to perform sustained sexual displays involving those traits (Garson and Hunter 1979; Wells and Taigen 1986; Klump and Gerhardt

1987; Reid 1987; Hoglund 1989; Vehrencamp, Bradbury, and Gibson 1989). Development of conspicuous characters may also increase predation risk. There is evidence from several studies that predation risk for males is increased by bright coloration (Endler 1983; Stoner and Breden 1988), conspicuous display patterns (Bradbury et al. 1989), and advertisement vocalizations (Ryan, Tuttle, and Rand 1982).

Different populations of the same species may experience marked variation in ecological factors that influence the strength of natural selection, such as predation pressure, and those that influence the potential for intra- and intersexual selection, such as the distribution and abundance of potential mates and same-sex competitors (Ryan, Tuttle, and Rand 1982; McCoy, Fox, and Baird 1994; Baird, Fox, and McCoy 1997; see chapters 3 and 5). Operational sex ratio (OSR), in particular, may influence the intensity of sexual selection (Emlen and Oring 1977). A heavily skewed OSR should promote intense sexual selection and result in higher variance in reproductive success among members of the more abundant sex (Wade and Arnold 1980). Especially in species with low dispersal ability, local differences in natural and sexual selection pressures may produce intraspecific differences in both the degree of sexual dimorphism (McCoy, Fox, and Baird 1994; McCoy et al. 1997) and the social behavior in local environments (see chapters 1, 3, 5, and 7). Although geographic variation in social behavior and sexual dimorphism is often considered adaptive, observed variation does not necessarily reflect adaptation to the conditions in a particular population. Gene flow among populations where conditions vary may prevent local populations from attaining adaptive maxima. However, if gene flow among populations is low, local variation in the environmental factors that influence natural and sexual selection may generate a variety of evolutionary outcomes (Thompson 1999). Because islands are isolated land masses where populations of the same species are sometimes subjected to markedly different ecological factors that can influence both natural and sexual selection (Foster and Endler 1999), insular populations offer a rich opportunity for the study of geographic variation in the evolution of social behavior and sexual dimorphism (see chapters 7 and 11).

Lizards are mostly insectivorous, relatively small vertebrates that are preyed on by a number of larger vertebrates, and lizard abundance on islands can be influenced markedly by predators. Lizards are especially abundant on remote oceanic islands or anciently isolated continents where carnivorous mammals are rare. Even though lizard populations on oceanic islands are often naturally dense, they are highly vulnerable to decimation by introduced mamalian predators (Case and Bolger 1991b). Not only because of introduced predators, but also because of faunal differences in the native predators among the islands of the same archipelago, lizards are expected to experience different predation regimes in different island populations. Lizards in populations suffering intense predation likely live at low densities with characteristically infrequent social interactions, whereas those in populations subjected to lax predation pressure are expected to live at higher densities with higher frequencies of social interactions. Insular lizard populations with different densities, therefore, are among the best model systems in which to examine the influence of variable predation regimes on the evolution of social behavior and sexual dimorphism.

The objective of this chapter is to review studies focusing on intraspecific variation in life history, social organization, and sexual dimorphism in insular populations of the lizard genus *Eumeces* under different predation regimes with the intent of unraveling the operation of sexual selection and its relationship to predation pressure and OSR. Specifically, I examine the role of sexual dimorphism in relative head size, male-male competition for females, mate guarding by males, and multiple copulation by females, as well as the influence of variation in these factors on the mating system under different predation regimes. Although there are a few studies with similar focus (e.g., Snell et al. 1988), the work that I review mostly focuses on the lizard genus *Eumeces*, specifically *E. okadae*, not only because there are many empirical studies of sexual dimorphism in north American *Eumeces* species (e.g., Vitt and Cooper 1985a), but also because I have collected a large demographic and life history data set on the insular populations of *E. okadae* (Hasegawa 1984, 1985, 1991, 1994a, b, 1997) with relevant studies to integrate ecological and behavioral interactions in this species with other animals within their food webs (Hasegawa 1999).

Sexual Dimorphism in *Eumeces* Lizards

Eumeces is the most primitive genus of the family Scincidae and is widely distributed in the northern hemisphere, from North Africa through West and East Asia to North America, even reaching the Bermuda Islands (Greer 1970). *Eumeces okadae* is one of the East Asian species belonging to the *fasciatus* species group. The lizards of this group are characterized by having a color pattern of dorsal stripes and a blue tail during the juvenile stage (Taylor 1935; Lieb 1985).

Life history, demography, behavior, and the social system have been studied for only a few species of *Eumeces* from North America and East Asia (Breckenridge 1943; Fitch 1954, 1956; Mount 1963; Hikida 1981; Guillette 1983; Hasegawa 1984, 1990a; Vitt and Cooper 1985a, b, 1986; Somma 1987; Ramirez-Bautista, Barba-Torres, and Vitt 1998). These studies indicate that *Eumeces* are relatively late-maturing, oviparous or viviparous, single-brooded, and small to medium lizards. One unique life history trait of *Eumeces* is that females brood their eggs (Fitch 1954; Hasegawa 1985; Vitt and Cooper 1989). Because a significant amount of energy and time are invested in egg brooding (Vitt and Cooper 1985b), females of some species are known to skip a year of reproduction (Fitch 1955; Hall 1971; Hasegawa 1984). Male lizards do not provide any paternal care. The significant investment by females in egg brooding may have a profound influence on the mating system of *Eumeces* lizards (Vitt and Cooper 1985a).

Many species of *Eumeces* show conspicuous sexual dimorphism in head size and coloration (Fitch 1954; Hikida 1978; Vitt and Cooper 1985a, 1986; Griffith 1991). In dimorphic species, adult males have a red or orange head coloration that is lacking in females (Cooper and Vitt 1988), and their heads are disproportionately larger (Vitt and Cooper 1985a). Male *Eumeces* aggressively fight with each other, mostly during the mating season (Fitch 1954, 1955; Mount 1963; Hikida 1978; Vitt and Cooper 1985a), and this intense male-male competition for receptive females is thought to be a major factor promoting sexual dimorphism in head size (Vitt and Cooper 1985a).

Interspecific comparisons of lizard social systems and sexual dimorphism have revealed a pattern of significant dimorphism of head and body size in species with strong male aggression, and little or no head size dimorphism in those species with weak male aggression (Stamps 1983b; Carothers 1984). Griffith (1991) investigated the evolution of sexual dimorphism within the North American members of the *fasciatus* group of *Eumeces* and found that the degree of sexual dimorphism differed among species. However, lack of detailed comparative data on the mating system of species within this group prevents statistical analysis of the correlation between the intensity of sexual selection and the degree of sexual dimorphism. Alternatively, within a species, instead of between species, it may be possible to detect a correlation between intraspecific variation in sexual dimorphism and the sociodemographic environment in order to evaluate how sexual or natural selection acts under particular environmental conditions. Along this line of thought,

I have studied the life history and mating system of *E. okadae* on the Izu Islands, Japan (Hasegawa 1994a, b).

Insular Environment and Predation Regime

The Izu Islands consist exclusively of volcanic islands and are situated linearly from north (Oh-shima: 32° 29′ N) to south (Aoga-shima: 34° 43′ N) over a distance of 230 km. Areas of inhabited major islands range from about 3 to 90 km². A uniformly mild climate (average air temperature = 16.2–17.9°C), under the influence of the warm temperate water of the Kuroshio Current, and abundant annual precipitation (about 3,000 mm/yr) have made these islands well vegetated with broad-leafed evergreen forests.

Populations of *E. okadae* in the Izu Islands were classified into three categories based on faunal differences of potential native predators: groups of populations on islands with just avian predators; those with avian and ophidian predators; and those with avian, ophidian, and mammalian predators (Table 6-1).

The first group consisted of populations on two islands (Miyake-jima and Aoga-shima) that were originally free of snake and weasel predators. The absence of these predators apparently allowed a high abundance of *E. okadae* and various insectivorous bird species (Hasegawa 1994b), with some of the birds (e.g., *Turdus celaenops* and *Butastur indicus*) acting as the only, albeit somewhat insignificant, lizard predators (Hasegawa 1990a, b). In time-constrained censuses conducted during the mating season in the late 1970s and early 1980s, 100 or more male lizards were seen per hour (Hasegawa 1994b).

The second group consisted of populations on To-shima, Mikura-jima, Kozu-shima, Nii-ima, Shikine-jima, and Hachijo-jima, which lacked only mammalian predators. These islands supported moderately high densities of the lizard. The numbers of male lizards seen in censuses were 20–60/h (Hasegawa 1994b). The snake *Elaphe quadrivirgata* was moderately abundant: sighting frequencies ranged from 0.5/h on Shikine-jima to 2.8/h on Kozu-shima during spring censuses (Hasegawa and Moriguchi 1989). Densities of insectivorous birds were low on these islands, probably as a result of the high densities of snakes, which also preyed on birds (Hasegawa and Moriguchi 1989; Hasegawa, unpublished data).

The third group consisted solely of the population on Oh-shima. This island accommodated the highest diversity of predators and the lowest density of *E. okadae* among the Izu Islands (less than six adult

Table 6-1 Correlates of life history traits and predation regime of three representative island populations of *Eumeces okadae* (after Hasegawa [1994a]).

Variable	Suites of Life History Traits		
	Early Maturity, Annual Reproduction, Many Small Eggs	Intermediate	Late Maturity, Biennial Reproduction, Few Large Eggs
Islands	Oh-Shima	Kozu-Shima	Miyake-jima
Major predator	*Mustela itatsi*	*Elaphe quadrivirgata*	*Turdus celaenops*
Vulnerable life history stages	All (intense)	All (mild)	Hatchling (intense)
Population density	Low	Intermediate	High
Major prey in summer—adults	Insect larvae	Annelids	Various
Major prey in summer—hatchlings	No data	Isopoda and Arachnida	Amphipoda
Prey abundance	Low	Intermediate	High
Feeding success	High	Intermediate	Low
Intraspecific competition	Mild	Mild	Severe

males seen per hour in the spring). Potential native predators included the weasel *Mustela itatsi* and five species of snakes.

Demography and Life History of *E. okadae*

The demographic characteristics of *E. okadae* relate strongly to local predation regime. Survivorship schedules differed greatly among the three groups of the Izu Islands. On Miyake-jima, where a long-term demographic study was conducted from 1978 to 1984 (Hasegawa 1990a, 1997), estimated annual survival rates were 36% for hatchlings, 56% for yearlings, 80% for 2-yr-olds, 63% for adult males, and 76% for adult females. Poor survival early in life with increasing survival with age was due mainly to size-specific predation by the endemic thrush *T. celaenops* on juveniles with a snout-vent length (SVL) <50 mm (Hasegawa 1990b). The cohort generation time was extremely long (7 yr). Thus, *E. okadae* on Miyake-jima was characterized by a combination of low natality and slow turnover (Hasegawa 1990a).

Consideration of predator fauna and density as well as the age composition of *E. okadae* on Kozu-shima indicates that, unlike the Miyake-jima population, all life history stages were vulnerable to predation by the snake *E. quadrivirgata,* and survival rates of juveniles and adults on this island were similar. On Oh-shima, where the weasel is native, more intense predation seemed to be acting on both juvenile and adult *E. okadae.*

Sexual Dimorphism in *E. okadae*

Mean SVL of adult male *E. okadae* ranged from 70 mm on Oh-shima to 86 mm on Aoga-shima, and mean SVL of adult females ranged from 78 mm on Oh-shima to 84 mm on Aoga-shima. In low-density populations, such as that on Oh-shima, males matured earlier (2 yr of age) and at a much smaller SVL (59 mm) than females (64 mm), and the mean SVL of adult males (70 mm) was smaller than that of females (78 mm). Conversely, in the high-density populations such as that on Aoga-shima, males and females matured at similar sizes (75 mm) and at later ages (3 to 4 yr of age), but maximum body size of males tended to be larger than that of females (Table 6-2).

Comparison of head width of mature males among different populations of *E. okadae* by analysis of covariance (with SVL as the covariate) indicated that head width adjusted by SVL varied significantly among the populations (Hasegawa 1994a). Although larger males within a population have larger relative heads than smaller males, the males in early

Table 6-2 Geographic variation in sexual dimorphism of SVL and head width in mature *Eumeces okadae* (after Hasegawa [1994b]).

Population	Male SVL (mm)			Female SVL (mm)			Head Width/SVL (males)		
	Minimum	Maximum	Mean	Minimum	Maximum	Mean	Mean	SD	n
Oh-shima	59	80	70	64	90	78	0.161	0.010	23
Kozu-shima	63	92	80	72	89	82	0.159	0.006	12
Mikura-jima	68	90	81	70	87	80	0.154	0.007	28
Shikine-jima	69	94	82	75	88	81	0.155	0.008	20
Miyake-jima	70	93	82	72	94	82	0.154	0.008	51
Aoga-shima	75	96	86	75	92	84	0.154	0.010	31

maturing populations at smaller body sizes (Oh-shima) had relatively larger heads than did the males on other islands (Table 6-2).

The pattern of variation in sexual dimorphism among *E. okadae* populations differs from that among the species within the *fasciatus* group. Although the largest species of this group (*E. laticeps*) is most dimorphic in relative head size (Griffith 1991), the population of largest individuals of *E. okadae* is least dimorphic (Hasegawa 1994a). Griffith (1991) postulated that accelerated growth up to the same age at maturity allowed greater dimorphism in body and head size (by rate hypermorphosis) within the *fasciatus* clade. On the contrary, variation in body size among different populations of *E. okadae* was mainly the result of delayed sexual maturity (Hasegawa 1994b).

Male-Male Aggression

The observed geographic variation in male head width may be a result of differences in the intensity of sexual selection through male-male aggression. A large and robust skull would be effective both for protection against biting by rival males and for delivery of a powerful bite by having massive jaw muscles (Hikida 1978; Carothers 1984; Vitt and Cooper 1985a). Focal observations of male-male aggression were conducted on Miyake-jima and Kozu-shima (Hasegawa [1994a] and unpublished data). On Miyake-jima, I spent a total of 781 min in observations during the mating season of 1983 and recorded 89 male-male interactions. Although interactions occurred on the average every 8.8 min, only 2 of 89 interactions escalated to head bites. The remaining interactions involved only threat postures, chasing and displacement from basking spots by dominant males, and flight by subordinate ones (Hasegawa 1994a). On Kozu-shima, I spent a total of 2,654 min in focal observations during 1988-94 but recorded only 13 interactions. Despite this low frequency of interactions (1 every 204 min), 3 of 13 interactions escalated to head bites. The proportion of male-male interactions escalating to head bites was nearly 10 times higher on Kozu-shima than Miyake-jima.

The proportion of males with bite wounds was compared for mature males captured at the peak of the mating season (late April–early May) between the Miyake-jima and Kozu-shima populations. Males in the Kozu-shima population showed a higher frequency of bite wounds on their heads (78.7%, $n = 108$) than those in the Miyake-jima population (54.5%, $n = 202$). Intensity of male-male aggression, therefore, could be indexed by the proportion of males with fresh wounds on their heads, because the more aggressive the males, the more wounds they had (Mount 1963; Cooper and Vitt 1987). In the case of *E. okadae* on the Izu Islands, both the proportion of males with fresh wounds on their heads and the relative head size of males varied significantly among the populations. As expected, the proportion of males with fresh wounds was positively correlated with relative head width, implying that more aggressive males had more robust heads (Fig. 6-1).

OSR in Relation to Predation Risk

Male *E. okadae* search for females active on the ground, and copulation takes place on the relatively exposed ground surface (Hasegawa 1994a). The sex ratio of mature lizards active on the ground surface is an important factor influencing frequency and success of copulation. The OSR (Emlen and Oring 1977) of the surface-active mature lizards was compared among island populations of *E. okadae* (Fig. 6-2). During the mating season, male *Eumeces* lizards are generally more surface active than females, and the OSR is usually male biased, whereas in the summer, when mating and egg brooding are completed, females are slightly more active (Fitch 1954, 1955; Fitch and von Achen 1977). The sexual difference in the pattern of seasonal surface activity suggests that males and females adjusted their activities in relation to costs and benefits. Limited surface activity of females suggests that they were time minimizers (at least during the mating season), so as to reduce predation risk associated with being gravid (Cooper et al. 1990). During the mating season, female *E. okadae* have well-developed yolked follicles about to ovulate (Hasegawa 1984), and their body mass is as heavy as those with oviductal eggs (Hasegawa 1997). Cooper et al. (1990) showed that reduced locomotory speed associated with being gravid was compensated by a behavioral shift from active escape by running to crypsis or secretive behavior in *E. laticeps*. Among-island variation in female activity in *E. okadae* during the mating season could reflect a difference in the risk of predation, and gravid females might be expected to be more wary in predator-rich environments than those in which predation pressure is relaxed.

Pattern of space use by both sexes of *E. okadae* was studied on Miyake-jima during the mating season of 1983 (Hasegawa 1994a). Eighty-five mature males and 68 mature females were registered within a small study site (0.35 ha), and home ranges were determined for 51 males and 12 females (6 gravid and 6 nonreproductive). Because of high density (Hasegawa 1990a) and lack of territoriality (Hasegawa 1994a), home ranges of both sexes overlapped extensively. Each of six gravid females

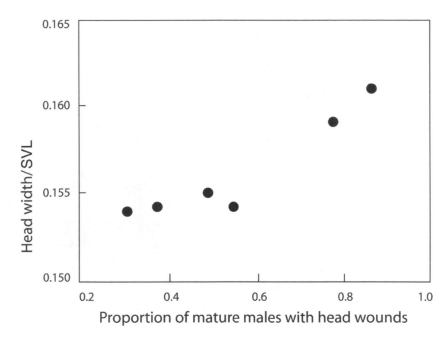

Figure 6-1 Geographic variation of sexual dimorphism in relative head size in relation to intensity of male-male aggression indexed by proportion of males with fresh head wounds.

had their home ranges overlapped by 40-50 males. The mean home range size of gravid females (88.1 m²) was slightly smaller than that of mature males (116.5 m²), but the difference was not statistically significant.

An attempt to study male-female spatial relationships was unsuccessful on the other islands where female *E. okadae* were relatively wary during the mating season. Instead, I relied on a fairly large number of predation records of *E. okadae* by the snake *E. quadrivirgata* on Kozushima to gain germane data indirectly. During the mating seasons of 1982–94, 82 adult *E. okadae* (23 gravid females and 59 males) were force regurgitated from the stomachs of *E. quadrivirgata*. Because the proportion of adult *E. okadae* active on the ground surface was strongly male biased during the corresponding period (92.9% males and 7.1% gravid females, *n* = 85), I suggest that gravid females were more susceptible to snake predation. During the summer months, males and females were equally surface active, but females were again taken more frequently by snakes—9 of 13 victims were females. These data suggest a higher vulnerability of female *E. okadae* to snake predation.

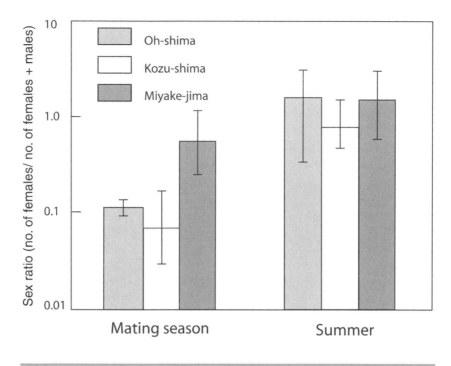

Figure 6-2 Geographic variation of apparent sex ratio of *Eumeces okadae* during the mating season in relation to predation regime.

Relaxed predation pressure on mature lizards should allow greater female activity in the Miyake-jima population. Apparent sex ratios (female/male + female) of surface-active *E. okadae* were distinctly higher on the islands free from both mammalian and snake predators (0.527 on Miyake-jima and 0.699 on Aoga-shima) than on islands with either snake (0.076 on Kozu-shima) or mammalian carnivores (0.083 on Oh-shima). On the latter islands, female *E. okadae* were highly secretive and rarely seen above ground. As a result, OSR differed greatly among the islands, although biennial reproduction of females on the high-density islands (Hasegawa 1984, 1994b) reduced the yearly availability of receptive females (Table 6-3).

Female Multiple Copulation and Male Mate Guarding

Mate guarding is a male behavior to defend females temporarily immediately before or after copulation. Guarding of females by males during the mating season has been reported for *E. laticeps* by Vitt and Cooper (1985a) and Cooper and Vitt (1987) and was suggested for *E. obsoletus*

Table 6-3   Geographic variation in OSR of *Eumeces okadae* during mating season (after Hasegawa [1994b]).

Population	Female/ (male + female)	Reproductive females (%)	Estimated OSR
Oh-shima	0.114	96.4	0.110
Kozu-shima	0.074	93.8	0.069
Mikura-jima	0.286	86.7	0.248
Shikine-jima	0.139	77.8	0.108
Miyake-jima	0.551	51.8	0.285
Aoga-shima	0.669	44.2	0.309

by Fitch (1955). Goin (1957) observed that a male *E. laticeps* moved along and followed a female before and after copulation. Another lizard species reported to exhibit mate guarding is *Lacerta agilis* (Olsson 1993a). After copulation, the male stays temporarily with the mated female and vigorously attacks approaching males; thereafter, he resumes mate searching (Olsson 1993a). Studies of *L. agilis* suggest that mate guarding is a male tactic to ensure reproductive success under the situation that females can mate multiple times with one or several males (Olsson, Gullberg, and Tegelstrom 1994).

Female *E. okadae* also mate multiply in a given mating season (Hasegawa 1994a). During copulation, the male bites the female on the neck, leaving a small wound that can later be detected as a scar. The average number of copulation scars per female was 2.4 ($n = 46$) on Miyake-jima and 2.4 ($n = 18$) on Kozu-shima, indicating that females from both islands copulate more than once per season. Mean copulation durations were 224 s for Miyake-jima pairs ($n = 5$) and 116 s for Kozu-shima pairs ($n = 6$). Because *E. okadae* displays female multiple mating and because female multiple mating and male mate guarding are associated in *L. agilis* (Olsson 1993a; Olsson, Gullberg, and Tegelstrom 1994), I suggest that male *E. okadae* should exhibit mate guarding. Field studies for a decade, however, provide little evidence of mate guarding by male *E. okadae* on Miyake-jima. Intensive focal observations did not yield evidence of mate-guarding behavior as shown by *E. laticeps* (Goin 1957; Vitt and Cooper 1985a).

On the other hand, I observed several circumstantial incidents of mate guarding on Kozu-shima. Because females are highly secretive on this island, I sometimes tried to capture gravid females by excavating their burrow systems. In the mating season of 1991, I found four cases in which pairs of lizards were resting in the same burrows, and males were always in front of females. Those females had several copulation scars on their necks, suggesting that they had been mated previously. I also observed similar pairs in burrows for a population of *E. latiscutatus* on the Japanese main islands (Hasegawa, unpublished observation).

Until recently, whether female lizards copulate multiply with different males was considered little (Olsson and Madsen 1998). Once its occurrence was confirmed, its adaptive significance received considerable attention (Dugan and Wiewandt 1982; Hicks and Trivers 1983; Madsen et al. 1992; Olsson 1993a; Olsson, Gullberg, and Tegelstrom 1994). Although there are several presumed benefits to the female of accepting multiple copulations (Olsson and Madsen 1998), Olsson and Madsen (1995) conclusively presented evidence to show that female *L. agilis* obtain good genes for their young through multiple matings, thereby avoiding costs associated with mate choice on the basis of external traits of males.

Interpretation of Geographic Variation

Consideration of both OSR and the degree of male aggression sheds light on the adaptive significance of geographic variation in sexual dimorphism from the perspective of sexual selection. Vitt and Cooper (1985a) suggested that intense male-male competition for receptive females is a principal factor promoting sexual dimorphism in relative head size and body size of *E. laticeps*. In *E. okadae* on the Izu Islands, the lower OSR seen on certain islands is correlated with higher male aggression (Fig. 6-3) and greater sexual dimorphism in head size. However, on islands where predation is relaxed, OSR is relatively high, males are not so aggressive (Fig. 6-3), head size is relatively small, but mature males are larger than females. This pattern suggests that sexual dimorphism in head size and absolute body size are subject to different intensities of sexual and natural selection.

On Oh-shima and Kozu-shima, where predation pressure is high, direct male-male competition (intrasexual selection) predominates. On these islands, females are highly secretive during the mating season, leading to a low OSR, and males aggressively compete for these secretive and sedentary females. A similar situation was reported for males of *E. laticeps*, which roam over a large area during much of the mating season (Vitt and Cooper 1985a), whereas females are generally secretive, typically remaining in trees with appropriate nest sites. Like male *E. laticeps*, which temporarily guard receptive females, male *E. okadae* on the

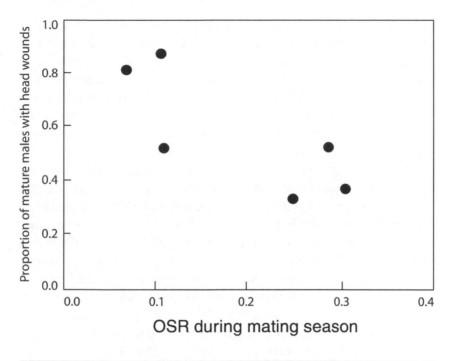

Figure 6-3 Inverse relationship between OSR and proportion of males with fresh head wounds among populations of *Eumeces okadae*.

islands with higher predation pressure appear to exhibit pre- and post-copulatory mate guarding.

Under conditions of high predation intensity, mere survival and aggressive superiority over rival males might justify a male's quality as a mating partner. Sedentary females would simply accept copulations by the winner of male-male aggressive interactions. Even if females wished to copulate multiply, sedentary females have few opportunities for overt choice due to the low density of males. On these islands, therefore, males can increase their reproductive success by physically outcompeting other males. In doing so, they develop conspicuous and robust heads, which help them win aggressive encounters, although their body sizes are small due to early maturation as an adaptation to higher predation pressure (Hasegawa 1994a, b).

By contrast, on islands where predation pressure is largely relaxed, females are not as secretive as on islands with higher predation pressure. On Miyake-jima, for example, during the mating season reproductive females move through as large an area as do mature males, and their home ranges overlap with those of as many as 40 males. These differences in female behavioral ecology make the OSR on Miyake-jima three to four times greater than on Oh-shima and Kozu-shima. However, females do not simply passively accept male courtship. Only 2 of 13 courtships resulted in successful copulations (Hasegawa 1994a). This suggests that female choice may be in operation, although it is uncertain on what basis females choose mates. Under conditions of female choice, intense male-male aggression alone may no longer be advantageous for increasing the reproductive success of male lizards.

Despite the fact that males do not develop robust heads, maximum or asymptotic body size of males on these high-density islands is larger than females. Limited resources in the crowded population might explain smaller body size in females. On Miyake-jima, almost every mature female reproduced biennially (Hasegawa 1984), and females ceased growth during reproduction. Although growth of females resumed slightly between reproductive bouts, males were always larger than females of the same age (Hasegawa 1990a). Small differences in mean adult SVL between the sexes might be attributable to a lower annual survival rate of adult males compared with females (Hasegawa 1990a). The lower survival rate of mature males may partly be due to a heavy parasitic load during the mating season by the female tick *Ixodes asanumai* (Hayashi and Hasegawa 1984a, b).

The principal message garnered from my studies of *E. okadae* is that environmental factors influencing intensity of sexual selection vary greatly among these Japanese islands. Of the putatively important conditions promoting sexual selection, variation in the OSR among populations with different risks of predation seems most influential. Quite simply, risk of predation determines the level of female activity, which, in turn, influences the intensity of male-male competition for females.

Interestingly, the effect of predation on the strength of sexual selection and the evolution of sexual dimorphism in *E. okadae* is opposite that seen in other species. In other species, increased predation risk has been associated with reduced sexual selection and selection against exaggerated male characters or displays (Endler 1980, 1983; Baird, Fox, and McCoy 1997; see also chapters 5 and 7). In these species, predation acts to increase the cost of showy male characters and displays, thus acting counter to sexual selection. Among populations of *E. okadae*, however, increased predation pressure is associated with increased sexual selection and sexual dimorphism. In this system, predation acts indirectly on male behavior and morphology through its action on female behavior. Because female *E. okadae* are more susceptible to predation, females in populations with greater predation pressure are far more secretive than are females in other populations. This highly secretive behavior results in greatly altered OSRs and greater competition among males for the few females that are active at any one time. The intense competition among males produces a great advantage for increased head size. Thus, in *E. okadae,* predation acts to promote the evolution of sexual dimorphism.

Of course, future research on the phylogeny of *Eumeces* (Murphy, Cooper, and Richardson 1983; Lieb 1985; Hikida 1993; Kato, Ota, and Hikida 1994) can offer an opportunity to evaluate the evolutionary history of sexual dimorphism among the species of the genus (e.g., Griffith 1991). However, comprehensive field and laboratory studies of life history, mating system, and sexual dimorphism of *Eumeces* lizards are still in their infancy except for a few well-studied species (Vitt and Cooper 1985a, b; Cooper and Vitt 1993; Hasegawa 1994a).

CHAPTER 7

Island Biogeography of Morphology and Social Behavior in the Lava Lizards of the Galápagos Islands

Paul A. Stone, Howard L. Snell, and Heidi M. Snell

Species assemblages on islands and mainlands are very different. Compared to the nearest mainland, islands are characterized by low species diversity, particularly of species with low dispersal abilities and/or specialized niches, and high population densities of species with good dispersal abilities (MacArthur and Wilson 1967; Brown and Lomolino 1998; Grant 1998). MacArthur and Wilson (1967) considered colonization and extinction rates as the core of their theory of island biogeography. They argued that colonization rates are influenced primarily by the degree of isolation of islands, with distant islands receiving colonists at low rates (MacArthur and Wilson 1967). Among colonists, extinction rates are influenced by island size, with small islands experiencing extinctions at high rates (MacArthur and Wilson 1967). These patterns can be illustrated using examples from the Galápagos Islands, the study area in this chapter. Lava lizards (*Microlophus* spp.), the study animal herein, are found on most islands in the Galápagos; the most notable exceptions are Genovesa, Darwin, and Wolf, three of the most isolated islands in the Galápagos (Van Denburgh and Slevin 1913; Carpenter 1966; Grant and Grant 1989; Snell, Stone, and Snell 1996). It is likely that lava lizards would thrive on these islands but have never successfully colonized. Galápagos tortoises (*Geochelone elephantopus*) historically were found on 11 islands (de Vries 1984), the smallest of which is the fourteenth largest island in the archipelago (Snell, Stone, and Snell 1996). It is likely that smaller islands lack the resources to support tortoises.

Because assemblages on islands constitute only a subset of the source (mainland) assemblages, successful colonists may experience ecological release from predation, parasitism, and interspecific competition once they are established on an island (Carlquist 1974; Brown and Lomolino 1998). Ecological release may be the ultimate factor explaining the dramatic adaptive radiations of *Drosophilia* in Hawaii, *Anolis* lizards in the West Indies, and Darwin's finches in the Galápagos (Throckmorton 1975; Williams 1983; Grant 1986). Ecological release also may have contributed to predictable morphological and behavioral differences between island and mainland populations, including insular giantism (Carlquist 1974; Case 1982), flightlessness, and tameness (Carlquist 1974). A third result of ecological release on islands is the high population densities, relative to mainlands, of successful colonists (Nilsson 1977; Schoener and Schoener 1980; Stamps and Buechner 1985). Ecological release appears to be strongest on the smaller islands within an archipelago, where population densities (Nilsson 1977; Wright 1981; Wright, Kimsey, and Campbell 1984) and survivorship (Schoener and Schoener 1978, 1982a; McLaughlin and Roughgarden 1989) are often highest.

The successful colonization of an island by a species may depend on the colonization and extinction rates of its competitors, predators, parasites, and prey. Numerous studies have shown that ecologically similar species occur together on islands less often than expected by chance, suggesting that interspecific competition has a central role in the composition of island assemblages (Diamond 1975; Brown and Lomolino 1998). In the Galápagos, the distribution of red-footed boobies (*Sula sula*) may be proximately constrained by the distribution of its main predator, the Galápagos hawk (*Buteo galapagoensis*), and ultimately constrained by the distribution of lava lizards, the main prey of the hawk (Anderson 1991). Variation in the intensity of ecological conditions may also lead to the evolution of interisland differences in morphology and social behavior. In multispecies assemblages on islands, character displacement resulting from interspecific competition may influence the expression of morphological traits such as beak size (Grant and Grant 1989) and body size (Schoener 1969b, 1970; Williams 1969). Similarly, the high population densities of insular populations, coupled with the inability to disperse, can lead to intense intraspecific competition, which can affect the social system of insular populations. Taxa that are territorial on mainlands may form dominance hierarchies or have undefended home ranges in insular populations (Stamps and Buechner 1985; Lott 1991). In addition, interpopulational variation in predation pressure can be correlated with interpopulational variation in a variety of morphological and behavioral traits (Moodie 1972; Bantock and Bayley 1973; Endler 1980; Reznick and Endler 1982; Steadman 1986; Snell et al. 1988; Stone, Snell, and Snell 1994; Wellborn 1995).

In this chapter, we attempt to take an assemblage-level process (island biogeography) and link it to the evolution of interisland differences in morphology and social behavior in Galápagos lava lizards (*Microlophus* spp.). The literature is rife with hypotheses about potential effects of ecological conditions on insular populations. We first summarize these hypotheses, then develop predictions that are tested using the distributions of lava lizard predators, parasites, prey, and conspecifics. We test these predictions using two different data sets. First, we compare body size, limb length, and chin-patch length among males from 37 populations of four species of lava lizards and ask if variation in lava lizard morphology is explained by the distribution and abundance of lava lizard predators, parasites, and prey. Second, we describe the social systems of four populations of lava lizards that occur on small islands at high population densities and compare them to published descriptions of lava lizard social systems from large islands at lower population densities, in an attempt to describe the effect of population density on the social system and the intensity of sexual selection.

Study Animal and Study Area

Lava lizards are sexually dimorphic in pattern (Van Denburgh and Slevin 1913; Werner 1978) (Plates 8 and 9) and size, with males always larger than females (Carpenter 1966). Male lava lizards have black chin patches that can cover most of the anterior ventral surface of the animal (Plates 9 and 10). The chin patch may function as an ornament in lava lizards (Werner 1978) and closely related lizards (Cooper and Greenberg 1992). Lava lizards, like most iguanian lizards, are saxicolous, sit-and-wait predators, and adults of both sexes are thought to be territorial (Carpenter 1966; Stebbins, Lowenstein, and Cohen 1967; Werner 1978). Although the systematics of lava lizards needs revision (Wright 1983; Frost 1992), there are seven recognized species, all of which are allopatric and confined to the Galápagos. Six of the seven species are restricted to one large island and associated islets (Carpenter 1966). The seventh species, *M. albemarlensis,* is found in the center of the archipelago on five large islands and associated satellites (Carpenter 1966). Lava lizards are the only member of the family Tropiduridae present in the Galápagos; thus, interspecific competition among lava lizards and other lizards is probably weak.

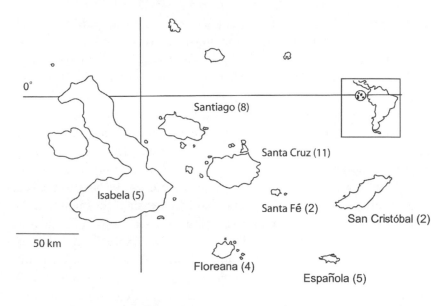

Figure 7-1 Map of Galápagos Islands. Numbers in parentheses refer to the number of islands sampled in a particular island group, with the large island in the group to the left of the parentheses.

The Galápagos Islands straddle the equator about 800 km west of Ecuador (Fig. 7-1). The islands are volcanic in origin and have never been connected to the mainland. The islands that are currently emerged are relatively young (3–5 million years old) (Simkin 1984). There are 123 islands in the archipelago, and data are available on the area and isolation of each (Snell, Stone, and Snell 1996). The flora of the Galápagos is primarily of South American origin (Porter 1976). The native vertebrates are typical of oceanic archipelagos, consisting mostly of birds and reptiles, with few mammals (three families, all with specialized dispersal ability) and no freshwater fish or amphibians (Bowman 1966; Perry 1984).

The 37 islands on which lava lizards were sampled can be classified as one of three types: islands large enough and high enough to create their own weather; small, diverse islets with woody vegetation that are usually near a large island but always share a similar flora with the coastal regions of the nearest large island; and small, barren islets that lack woody vegetation and are always a considerable distance from the nearest large island (Figs. 7-2 and 7-3). There are seven large islands, the smallest of which is 24.1 km² (Figs. 7-1 and 7-2). These islands are the wettest of the three types of islands (Grant and Boag 1980). Our study sites on all of the large islands were coastal and within the arid zone (Grant 1986). Each large island is associated with at least one satellite islet (Fig. 7-1). The 30 satellite islets are small, ranging in size from 0.00235 to 1.24 km² (Fig. 7-2). There are 9 barren islets and 21 diverse islets (Fig. 7-2).

During the most recent ice age (about 17,000 yr ago) sea levels were approximately 120 m lower than they are today (Fairbanks 1989). Geist (1996) constructed a tentative map of the "Glacial Galápagos" using depth contours from oceanographic maps. From this map, it can be inferred that none of the seven large islands in our sample was connected to each other during the ice age, and that most or all of the satellite islets were connected to the nearest large island. As sea levels subsequently rose, these seven "super" islands must have fragmented, separating 7 large populations of lava lizards into 37 smaller populations. At this point, it is likely that the newly separated populations began to differentiate from each other due to both stochastic and deterministic factors. We can look at differences among populations within island groups and see morphological and behavioral changes that have occurred since these islands were separated from each other. If the differences we see are consistently correlated with ecological conditions on the different types of islands within an island group, this would be evidence that local adaptation or phenotypic plasticity was responsible for the differences. On the other hand, if the differences are not consistently correlated with local ecological conditions, we could conclude that stochastic factors, such as genetic drift or founder effects, were responsible.

Hypotheses and Predictions
Body Size

Among iguanian lizards, body size in insular populations is usually larger than in mainland relatives (Case 1978). Several hypotheses have been developed and tested in attempts to explain this phenomenon (Schoener 1969a; Case 1978, 1982; Dunham, Tinkle, and Gibbons 1978). Much attention has been devoted to hypotheses involving interspecific competition (Schoener 1969b, 1970; Williams 1969; Losos 1990c; Case and Bolger 1991a; Miles and Dunham 1996). However, interspecific competition should be relatively unimportant in the depauperate lizard fauna of the Galápagos. We adapt the remaining hypotheses to interpopulational comparisons of body size among lava lizards.

Large body size in insular lizards may evolve in response to reduced predation on islands (Case 1982; but see chapter 6). Several mechanisms,

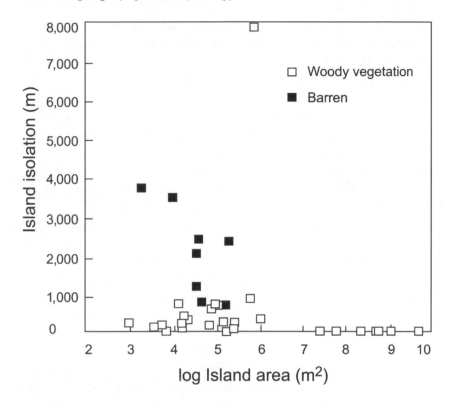

Figure 7-2 Plot of island isolation versus log island area for barren islets, diverse islets, and large islands (all seven of which had woody vegetation). Data on island isolation and area are taken from Snell, Stone, and Snell 1996. Data on the presence or absence of woody vegetation are from direct observations.

both selective and plastic, could account for this, including delayed sexual maturity (Reznick and Bryga 1987; Gibbons and Lovich 1990), increased life span, niche expansion (Case 1982), and size-selective foraging (Bantock and Bayley 1973; McLaughlin and Roughgarden 1989; Külling and Milinski 1992; Wellborn 1995). Within an archipelago, each mechanism would be predicted to result in large body size on islands with relatively low predation pressure. In the Galápagos, Schluter (1984) discounted the role of predation in the evolution of differences in body size among lava lizards on large islands. However, on two large islands with introduced cats and rats, lava lizards are smaller compared with those on satellite islets that are free of exotic predators (Grant 1975; Steadman 1986), and on one island, lava lizards are smaller than lava lizard fossils that predate the exotic predators (Steadman 1986). This supports the *predation hypothesis,* and we ask, Does this pattern exist within a larger sample of islands, and can it be extended to include native predators such as hawks and snakes (Table 7-1)?

Case (1982) also suggested size-limited predation as a possible mechanism of the predation hypothesis. However, when adapted as a prediction of variation in body size within an archipelago, this mechanism, identified here as the *escape hypothesis,* predicts a different pattern from the predation hypothesis. If individuals can escape predation by attaining large body size because the predator is gape limited, for example, we would expect strong natural selection for large body size and rapid growth rates on islands where gape-limited predators were present (Table 7-1) (Case 1978, 1982). As a result, we would expect larger lava lizards on islands where predation pressure was highest (Table 7-1). Such a pattern has been noted in threespine stickleback (*Gasterosteus aculeatus*) from lakes that vary in the presence of potentially size-limited predators (Moodie 1972). Among predators of lava lizards, snakes appear to be the only one that is potentially gape limited.

Large body size in insular lizards may be the result of increased mean food availability or increased variation in food availability on islands (Case 1978, 1982). Within an archipelago, this hypothesis is actually two hypotheses that predict different patterns. First, reduced predation and reduced interspecific competition on islands, coupled with a territorial spacing system, may create conditions of superabundant resources for insular populations (Licht 1974; Case 1982; Stamps and Buechner 1985). Under such conditions, large body size should evolve. The *food availability hypothesis* predicts that body size will be directly correlated with mean food availability (Table 7-1). Supporting this hypothesis are studies that have shown increased growth rates in populations with natural or supplemental increases in food availability (Licht 1974; Dunham 1978; Stamps and Tanaka 1981a; Guyer 1988a, b). However, this hypothesis is complicated by data that show mean food availability is often lower, per unit area and per individual, on islands compared to the nearest mainland, and that food is limiting on islands relative to mainlands (Licht 1974; Andrews 1979; McLaughlin and Roughgarden 1989).

Second, Case (1978, 1982) argued that variation in food availability should be higher on islands than on mainlands, and that during drastic food shortages on islands, larger individuals should survive better than smaller individuals. This hypothesis is complicated by data showing that large individuals starve sooner than small individuals during episodes of starvation (Laurie and Brown 1990; Wikelski 1994). However, one

Table 7-1 Predictions about the relationship of traits to ecological conditions. For simplicity, predictions are presented as the trait expression expected when ecological conditions are harsh. Hypotheses and predictions are described more fully in the text.

Hypothesis	Prediction
Predation	High predation/small body size
Escape	High predation/large body size
Food availability	Low food/small body size
Herbivory	Low food/large body size
Sexual selection	High parasites or high density/large body size
Speed	High predation/long limbs
Crypticity	High predation/small ornaments
Hamilton-Zuk	High parasitism/large ornaments

version of this hypothesis, the *herbivory hypothesis* (Schluter 1984), has some empirical support from our study system. Comparison among large islands shows that lava lizards are largest on islands where arthropod abundance is lowest (Schluter 1984). To explain this pattern, Schluter (1984) proposed that lava lizards on islands with low food availability had shifted to a more herbivorous diet, and that selection for large body size had occurred in order to maximize digestive efficiency (Pough 1973; Iverson 1982). Within an archipelago, this hypothesis predicts that body size will be largest on islands where chronic arthropod shortages occur (Table 7-1).

In contrast to the multiple predictions that stem from hypotheses about predation and food availability, the *sexual selection hypothesis* yields a single prediction: large body size should evolve as a direct result of sexual selection. High population densities on islands should lead to increased intensity of both intersexual and intrasexual selection (Andrews 1979; Case 1982; Stamps and Buechner 1985; Baird, Fox, and McCoy 1997). Under such conditions, there is a clear fitness advantage for larger individuals (Table 7-1) (Wilbur, Rubenstein, and Fairchild 1978; Thornhill and Alcock 1983; Trivers 1985; Cooper and Vitt 1993; Wikelski 1994; Lewis, Tirado, and Sepulveda 2000). Because we lack basic information on mating success from most of the 37 islands in the study, we cannot directly test this prediction with our data. However, both population density and rates of parasitism are expected to be directly correlated with the intensity of sexual selection (discussed later). Therefore, a positive correlation between body size and either lava lizard density or parasite prevalence would be indirect support of the prediction (Table 7-1). In addition, this hypothesis will be revisited when we analyze the behavioral data later in the chapter.

Limb Length

Avoiding predation is a complex process that may involve a variety of traits, including wariness, sprint speed, and degree of ornamentation (Endler 1978; Greene 1988; see also chapter 4). Several studies have found a positive correlation between limb dimensions and sprint speed (Garland 1984; Snell et al. 1988; Losos 1990b; Sinervo and Losos 1991; Miles 1994). On Plaza Sur in the Galápagos, female lava lizards had longer approach distances, shorter hindlimbs, and slower sprint speeds than males (Snell et al. 1988). The differences in approach distances were attributed to sexual selection in males against premature territory abandonment for flight (for similar patterns see chapter 4). However, because males allowed closer approach than females, there was an increased need for speed in males once they did flee; this was achieved via natural selection for increased limb length (Snell et al. 1988). Wariness in lava lizards is positively correlated with predation pressure, at both the intrapopulational (Snell et al. 1988) and interpopulational levels (Stone, Snell, and Snell 1994). If long limbs help lizards evade predators, we would expect variation in limb length to parallel variation in predation pressure (Snell et al. 1988). Here, the *speed hypothesis* is tested by asking, Does limb length positively correlate with predation pressure at the interpopulational level (Table 7-1)?

Ornamentation

In environments where rates of predation are high, natural selection for crypsis may be stronger than sexual selection for ornamentation. In such environments, conspicuous mating behavior and time-consuming mate choice decisions may be avoided (Magnhagen 1991; Berglund 1993; Hedrick and Dill 1993), and the expression of sexually selected ornaments may be reduced (Moodie 1972; Endler 1980; Reznick and Bryga 1987; Wellborn 1995). Because the chin patch and the displays in which it is featured may attract the attention of predators, the *crypticity hypothesis* predicts an inverse relationship between predation pressure and chin-patch length (Table 7-1).

Parasites differ from predators in the type of selection pressure they exert on their hosts. Because parasites do not normally kill their hosts, mortality selection exerted by parasites is typically weaker than that exerted by predators (Begon, Harper, and Townsend 1990). However, sexual selection is expected to be strong in environments where rates of parasitism are high, because parasite-resistant individuals defeat diseased individuals in intrasexual competition (Schall and Dearing 1987) and are favored as mates via intersexual mate choice (Hamilton and Zuk 1982). Mating success is often correlated with expression of ornaments (Andersson 1982; Borgia 1985; Kodric-Brown 1989; Zuk et al. 1990). Hamilton and Zuk (1982) argued that ornaments revealed information about the genetic resistance to parasites of prospective mates. The *Hamilton-Zuk hypothesis* presented an interspecific prediction that ornament expression should be positively correlated with historical parasite burdens (Hamilton and Zuk 1982). The logic of this prediction is that in species with historically high rates of parasitism, there has been great evolutionary incentive for parasite-resistant individuals to advertise their resistance and for prospective mates to heed these advertisements (Hamilton and Zuk 1982). Analyses of this prediction have been confounded by phylogenetic effects and by disagreements over how to rank the degree of ornamentation of different species (Hamilton and Zuk 1982; Read and Harvey 1989; Read and Weary 1990). We believe that our data provide an opportunity to extend the interspecific prediction to the interpopulational level, and that the measurement of variation in ornaments should be less confusing at this level. Thus, we predict that ornamentation across populations of lava lizards should be positively correlated with parasite prevalence (Table 7-1).

Social Behavior

Iguanian lizards on mainlands almost always have territorial social systems characterized by male territoriality (Stamps 1983a), resource defense polygyny (Emlen and Oring 1977), sexual size dimorphism favoring males, and characteristic male display behavior (Stamps 1983a). Territoriality is characterized by tenacious and exclusive (or nearly so) defense of a well-defined space (Noble 1939). On islands, at high population densities, territorial systems often change as the cost of territorial defense becomes prohibitively expensive (Stamps and Buechner 1985; Trivers 1985). Observed changes include smaller home ranges, more home range overlap, reduced rates of aggressive behavior, and shifting patterns of space use (Stamps and Buechner 1985). Despite numerous examples of such changes in insular populations of lizards, birds, and rodents (Gliwicz 1980; Stamps and Buechner 1985; Lott 1991), data from Galápagos lava lizards suggest a typical mainland-like social system. Werner (1978) studied the mating system of *M. delano-*

nis on Isla Española and described territorial displays, low home range overlap, and distinct classes of territorial and nonterritorial males. Similarly, Stebbins, Lowenstein, and Cohen (1967) inferred low home range overlap and other characteristics of territoriality from a limited number of observations on *M. albemarlensis* at the Charles Darwin Research Station on Isla Santa Cruz. Carpenter (1966) collected behavioral and morphological data from 12 populations of lava lizards, all on large islands, and wrote that "extensive observation on wild populations of *Tropidurus* (*Microlophus*) on many islands indicates that territoriality is well developed in both males and females." However, all these studies were conducted on large islands in the Galápagos, where population densities are lower than on many of the small islets we studied (discussed later). Social systems of lava lizards on small islets may be dramatically affected by high population density. Indeed, we predict that these islets will have less stable territorial systems and more intense sexual selection than the larger islands studied by previous researchers (Carpenter 1966; Stebbins, Lowenstein, and Cohen 1967; Werner 1978).

Ecological Conditions
Predators

Lava lizards are exposed to a variety of potential predators, including the Galápagos hawk (*B. galapagoensis*), Galápagos snake (*Alsophis* spp.), herons and egrets (as many as seven species), introduced cats (*Felis cattus*), introduced black rats (*Rattus rattus*), and native rice rats (*Oryzomys bauri*). Of these predators, we considered only hawks, rats, cats, and snakes, because they are thought to be important predators on lava lizards (Kruuk 1979; Kramer 1984; Konecny 1987; Stone, Snell, and Snell 1994; Stone and Snell, unpublished data), and because they have variable distributions among the study islands (Eckhardt 1972; de Vries and Black 1983; Hoeck 1984; Kramer 1984; Stone and Snell, unpublished data). We used the number of predator species present on a given island as an index of predation pressure on that island. The presence or absence of each predator on each island was determined from the literature (Eckhardt 1972; de Vries and Black 1983; Hoeck 1984; Kramer 1984) and our field notes. We realize that this index fails to account for variation in the density of predators on islands, and that such variation could create important differences in predation pressure. However, information on predator densities on these 37 islands is unavailable and unlikely to become available. Using the presence/absence of predator

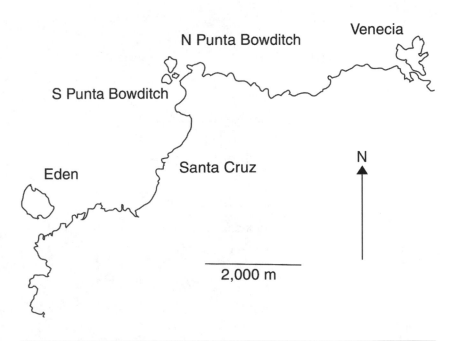

Figure 7-3   Map of northwestern Santa Cruz, showing four barren islets (Guy Fawkes) and four diverse islets.

species precludes the use of herons and egrets in the index, despite data that suggest herons and egrets are important predators on lava lizards (Snell et al. 1988). There is a heron/egret component to the fauna of all our study islands, and, therefore, the group adds no discriminatory power to the index.

There is considerable variation in the diversity of predators on our 37 study islands, and this variation is consistent with the predictions of island biogeography (MacArthur and Wilson 1967). Predator diversity was positively correlated with island area and negatively correlated with

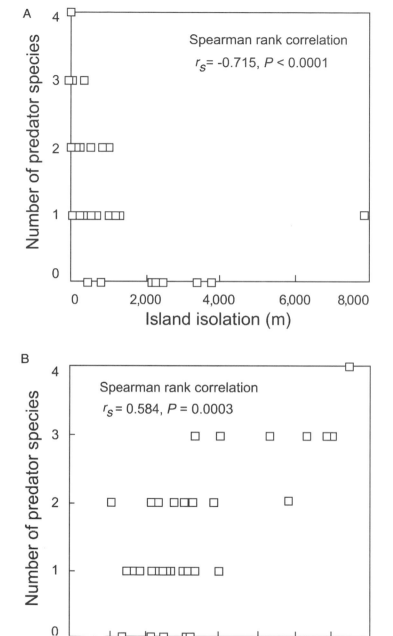

Figure 7-4   Plot of predator diversity versus island isolation (*A*) and log island area (*B*).

Table 7-2 Comparison of predator diversity by island type. Mean ranks are from Kruskal-Wallis nonparametric analysis of variance (ANOVA) $H_{corr} = 20.8$, $P < 0.001$.

Island Type	Mean Rank	Modal Number of Predator Species
Large	32.4	3
Diverse	19.0	1
Barren	8.4	0

island isolation (Fig. 7-4). Predators were most diverse on large islands, and least diverse on barren islets (Table 7-2). The number of predator species on an island ranged from zero to four (Table 7-3). The distribution of cats is especially constrained by human distribution; the four islands where cats occur are the four islands in the sample that are inhabited by humans (Table 7-3). Introduced rats appear to be the least constrained by biogeographic phenomena (after being transported to multiple islands by humans); black rats occur on 20 islands, including four of the barren islets (Table 7-3). Native rice rats, by contrast, are found on only one island in our sample: the large island of Santa Fé (Table 7-3). Snakes are found on 15 islands, including the 12 largest islands in the sample, and are absent from barren islets. Hawks have the most idiosyncratic distribution, being restricted to four island groups, but ubiquitous within the large islands and diverse islets of these groups (Table 7-3). Hawks have been extirpated (or nearly so) on three islands, all inhabited by humans, cats, and rats (Table 7-3) (Anderson 1991; Snell, unpublished data).

Parasites

Several different parasites are known to be associated with lava lizards (Ayala and Hutchings 1974; Aquino-Shuster, Duszynski, and Snell 1990; Couch et al. 1996). We sampled four types of parasites, documenting both presence/absence and prevalence on each island. In 1987 and 1990–92, fecal samples were collected opportunistically, and the presence or absence of intestinal coccidia was recorded as outlined in Aquino-Shuster, Duszynski, and Snell (1990). A total of 593 samples were processed from 17 populations. This is the same data set published by Couch et al. (1996), except that 1993 samples from Plaza Sur were not included because only that island was sampled that year. During 1990–92 and 1994, ectoparasite data were collected from all 37 popula-

Table 7-3 Distribution of predators on study islands.

Island Group (*Microlophus* sp.)	Island	Island Type	Snakes	Rats	Hawks	Cats	Total
Isabela (*albemarlensis*)	Isabela	Large	1	1	1	1	4
	Cráter Beagle 1	Diverse	0	1	1	0	2
	Cráter Beagle 2	Diverse	0	1	1	0	2
	Cráter Beagle 5	Diverse	0	1	1	0	2
	Marielas Sur	Diverse	0	1	1	0	2
Santa Cruz (*albemarlensis*)	Santa Cruz	Large	1	1	[a]	1	3
	Eden	Diverse	1	1	0	0	2
	Plaza Norte	Diverse	0	0	0	0	0
	Plaza Sur	Diverse	0	0	0	0	0
	N Punta Bowditch	Diverse	0	1	0	0	1
	S Punta Bowditch	Diverse	0	1	0	0	1
	Venecia	Diverse	1	1	0	0	2
	N Guy Fawkes	Barren	0	0	0	0	0
	S Guy Fawkes	Barren	0	0	0	0	0
	E Guy Fawkes	Barren	0	0	0	0	0
	W Guy Fawkes	Barren	0	0	0	0	0
Santa Fé (*albemarlensis*)	Santa Fé	Large	1	1[b]	1	0	3
	Santa Fé Islote	Diverse	0	0	1	0	1
Santiago (*albemarlensis*)	Santiago	Large	1	1	1	0	3
	Albany	Diverse	0	0	1	0	1

Table 7-3 *(Continued)*

Island Group (*Microlophus* sp.)	Island	Island Type	Snakes	Rats	Hawks	Cats	Total
	Bartolomé	Diverse	1	1	1	0	3
	Sombrero Chino	Diverse	1	1	1	0	3
	Roca Bainbridge 3	Barren	0	1	0	0	1
	Roca Bainbridge 4	Barren	0	1	0	0	1
	Roca Bainbridge 5	Barren	0	1	0	0	1
	Roca Bainbridge 6	Barren	0	1	0	0	1
Española (*delanonis*)	Española	Large	1	0	1	0	2
	Gardner (Española)	Diverse	1	0	1	0	2
	Islote Oeste	Diverse	0	0	1	0	1
	Osborn	Diverse	0	0	1	0	1
	Xarifa	Diverse	0	0	1	0	1
San Cristóbal (*bivittatus*)	San Cristóbal	Large	1	1	[a]	1	3
	Isla Lobos	Diverse	1	1			2
Floreana (*grayi*)	Floreana	Large	1	1	[a]	1	3
	Champion	Diverse	1	0	0	0	1
	Gardner (Floreana)	Diverse	1	0	0	0	1
	Enderby	Barren	0	0	0	0	0
Total			15	21	16	4	

[a]Hawks extirpated (Floreana, San Cristóbal) or nearly so (Santa Cruz).
[b]Native rice rats.

tions (Plate 8). Lizards were carefully inspected, and the number of ticks were counted ($n = 7,741$ lizards) and the number of mites ($n = 5,336$ lizards) were assigned to one of six categories (0, 1–10, 11–30, 31–70, 71–150, >150). During 1992 and 1994, blood smears were made from 418 lizards in 34 populations. These smears were stained with Wright-Giemsa, and blood parasites, probably hemogregarines (Ayala and Hutchings 1974), were counted during 10-min scans under oil emersion at 1,000–1,200×.

The prevalence (percentage of infected individuals in a population) of each parasite in each population was calculated using data from all sex and age classes, and with individual lizards represented more than once in the data set. We feel this best represents the prevalence of parasites in the population, despite potential biases. As an index of the intensity of parasitism in each population, we used the average prevalence for all types of parasites sampled in each population. The 17 populations for which fecal samples were collected included all but three of the populations in which male morphology was sampled in 1992. The average parasite prevalence in these 17 populations was calculated using all four types of parasites. Average parasite prevalence for the other three populations sampled in 1992, and for 14 of the 17 populations sampled in 1994, was calculated using the ectoparasite and blood parasite data. In the remaining three populations, average prevalence was calculated using data from only ectoparasites.

Parasite diversity was negatively correlated with island isolation, but not with island area (Fig. 7-5). Parasites were more likely to be absent from samples of barren islets than from the other two types of islands (2 × 3 χ^2 contingency table, $\chi^2 = 18.13$, $P < 0.001$) (Fig. 7-6). On two particularly well-sampled barren islets (N Guy Fawkes and E Guy Fawkes), we have never encountered a parasite of any of the four types we sampled (total samples = 646). This pattern suggests that island isolation may have prevented parasites from establishing on distant islets. Ticks were the most likely parasite to be recorded as absent from an island (Fig. 7-6), even though they were the most intensively sampled parasite.

Parasite prevalence was also a function of island type. Prevalences were highest on diverse islets in all island groups except the Santiago group, where lizards on the large island (Santiago) had the highest parasite prevalences (Tables 7-4 and 7-5). Parasite prevalence was lowest on barren islets (Tables 7-4 and 7-5).

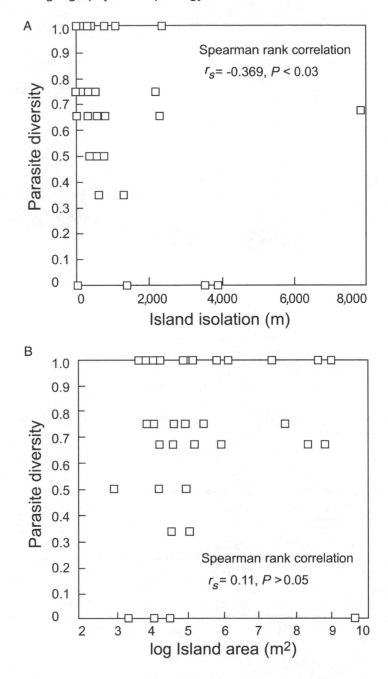

Figure 7-5 Plot of parasite diversity versus island isolation (*A*) and log island area (*B*). Because all types of parasites were not sampled on all islands, *y*-axis values are proportions of parasites sampled that are present on an island.

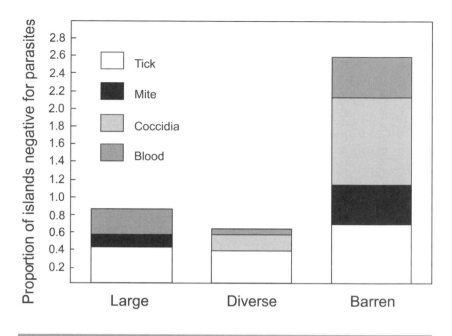

Figure 7-6 Plot of proportion of population samples that were negative for a given parasite. The stacked bars show the cumulative effect of each parasite on overall parasite diversity. For example, a value of one on the y-axis (large islands $\simeq$ 1) indicates that, on average, islands of a given type are missing one of the four types of parasites. A value of three (barren islets $\simeq$ 3) indicates that islands of a given type are missing three of the four types of parasites.

Table 7-4   Comparison of prevalence of parasites by island type. Mean ranks are from Kruskal-Wallis nonparametric ANOVA (H_{corr} = 13.4, $P < 0.005$).

Island Type	Mean Rank	Mean Summed Parasite Prevalence
Large	18.2	0.2972
Diverse	23.9	0.3889
Barren	8.2	0.0963

Table 7-5 Prevalence of parasites in 37 populations of lava lizards.

Island Group (*Microlophus* sp.)	Island	Island Type	Mite Prevalence (n)	Tick Prevalence (n)	Coccidia Prevalence (n)	Blood Prevalence (n)	Mean Prevalence
Isabela	Isabela	Large	0 (26)	0 (26)	—	0 (9)	0
(*albemarlensis*)	Cráter Beagle 1	Diverse	0.7500 (24)	0 (24)	—	0.1818 (11)	0.3106
	Cráter Beagle 2	Diverse	0.5455 (11)	0 (11)	—	—	0.2728
	Cráter Beagle 5	Diverse	0.7778 (9)	0 (9)	—	—	0.3889
	Marielas Sur	Diverse	0.8654 (52)	0.0385 (52)	—	0.1250 (16)	0.3430
Santa Cruz	Santa Cruz	Large	1.0000 (58)	0.0247 (162)	0.3077 (13)	0.4444 (18)	0.4442
(*albemarlensis*)	Eden	Diverse	0.1154 (104)	0.2697 (152)	0 (5)	0.4706 (17)	0.2139
	Plaza Norte	Diverse	1.0000 (93)	0.0691 (217)	0.3846 (13)	0.5625 (16)	0.5041
	Plaza Sur	Diverse	0.9828 (720)	0.8498 (1171)	0.6000 (20)	0.6316 (19)	0.7660
	N Punta Bowditch	Diverse	0.7530 (327)	0.1103 (408)	0.2000 (5)	0.8387 (31)	0.4755
	S Punta Bowditch	Diverse	0.8910 (155)	0.2251 (191)	—	0.4444 (18)	0.5202
	Venecia	Diverse	0.9301 (471)	0.0632 (570)	0.6667 (6)	1 (25)	0.6650
	N Guy Fawkes	Barren	0 (124)	0 (124)	0 (11)	0 (8)	0
	S Guy Fawkes	Barren	0.0192 (52)	0.3590 (78)	0 (2)	0.8000 (5)	0.2946
	E Guy Fawkes	Barren	0 (152)	0 (211)	0 (5)	0 (11)	0
	W Guy Fawkes	Barren	0.3119 (109)	0.1410 (156)	—	0.6000 (5)	0.3510
Santa Fe	Santa Fé	Large	0.4717 (53)	0.0738 (149)	0.2000 (15)	0.1667 (6)	0.2280
(*albemarlensis*)	Santa Fé Islote	Diverse	0.8389 (180)	0.5062 (320)	0 (11)	0.1667 (12)	0.3780
Santiago	Santiago	Large	0.8235 (51)	0.2674 (86)	—	0.4615 (13)	0.5175
(*albemarlensis*)	Albany	Diverse	0.0909 (33)	0 (33)	—	0 (11)	0.0303

Table 7-5 (Continued)

Island Group (Microlophus sp.)	Island	Island Type	Mite Prevalence (n)	Tick Prevalence (n)	Coccidia Prevalence (n)	Blood Prevalence (n)	Mean Prevalence
	Bartolomé	Diverse	0.4681 (47)	0.2979 (47)	—	0.2000 (10)	0.3220
	Sombrero Chino	Diverse	0.9754 (121)	0.1965 (173)	—	0.3500 (20)	0.5073
	Roca Bainbridge 3	Barren	0.4630 (54)	0 (89)	—	0.3000 (10)	0.2543
	Roca Bainbridge 4	Barren	0 (20)	0 (20)	—	0 (5)	0
	Roca Bainbridge 5	Barren	0.0370 (27)	0 (27)	—	0 (9)	0.0124
	Roca Bainbridge 6	Barren	0 (45)	0.0889 (45)	—	0.2000 (10)	0.0963
Española	Española	Large	0.9563 (686)	0.0252 (753)	0.2013 (154)	0 (20)	0.2957
(delanonis)	Gardner (Española)	Diverse	0.9906 (426)	0.0018 (552)	0.1408 (142)	0.0526 (19)	0.2965
	Islote Oeste	Diverse	0.9786 (139)	0.0280 (143)	0.4000 (10)	0.1250 (8)	0.3829
	Osborn	Diverse	0.9116 (328)	0 (442)	0.7294 (85)	0.1111 (18)	0.4380
	Xarifa	Diverse	0.8920 (423)	0.1910 (555)	0.7375 (80)	0.2000 (10)	0.5051
San Cristóbal	San Cristóbal	Large	0.8710 (31)	0 (75)	—	0.4167 (12)	0.4290
(bivittatus)	Isla Lobos	Diverse	0.8269 (104)	0 (277)	0.2500 (16)	0.6875 (16)	0.4411
Floreana	Floreana	Large	0.4062 (32)	0 (32)	—	0.2500 (11)	0.2187
(grayi)	Champion	Diverse	0.8571 (21)	0 (21)	—	—	0.4286
	Gardner (Floreana)	Diverse	0.8333 (6)	0 (6)	—	0.2500 (4)	0.3611
	Enderby	Barren	0.4595 (37)	0 (37)	—	0.1579 (19)	0.2058
Overall mean (n)			0.5971 (5351)	0.1034 (7444)	0.2834 (593)	0.2998 (452)	

Plate 1 Male *Platysaurus broadleyi*. Males are polymorphic for front leg color, which may be yellow, orange, or an orange-yellow mixture. Photograph by M. J. Whiting.

Plate 2 Ventral view of a male *Platysaurus broadleyi* showing abdominal status signaling-badge (light color). Badges may be orange, yellow, or an orange-yellow mixture. Photograph by M. J. Whiting.

Plate 3 Male *Platysaurus broadleyi* flashing his badge at a rival (termed a ventral display). Males also simultaneously expand their throats. Photograph by M. J. Whiting.

Plate 4 Status signaling can break down during competition for high-quality resources and result in fighting, which is preceded by back arching and lunging. Photograph by M. J. Whiting.

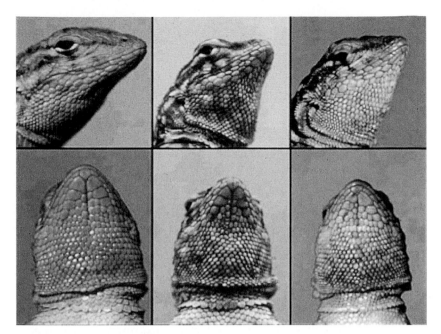

Plate 5   Male side-blotched lizards *(Uta stansburiana)* in a California population adopt alternative mating strategies that differ in behavior and morphology. These three male types include highly aggressive orange-throated males *(left)*, mate-guarding blue-throated males *(center)*, and yellow-throated "sneaker" males *(right)*. Photographs by B. Sinervo.

Plate 6   Adult male collared lizard *(Crotaphytus collaris)* from Wichita Mountains, Oklahoma, showing very bright coloration typical of this population of lizards. Photograph by J. K. McCoy.

Plate 7 Adult male collared lizard *(Crotaphytus collaris)* from Arcadia Lake, Oklahoma, showing subdued coloration typical of this population of lizards. The white and red dots are paint for identification of individuals. Photograph by T. A. Baird.

Plate 8 Female *Microlophus albemarlensis* from Isla Plaza Sur. Note the tick just anterior to the forelimb. Photograph by H. M. Snell.

Plate 9 Male *Microlophus albemarlensis* from Isla Plaza Sur, showing a side view of the black chin patch. Photograph by H. M. Snell.

Plate 10 Male *Microlophus delanonis* from Isla Española, showing a side view of the black chin patch. Photograph by H. M. Snell.

Plate 11 Abdominal coloration of male *(left)* and female *(right) Scelo-porus undulatus consobrinus.* This is an example of a typical, sexually dichromatic species and is one of our study species. Photograph by D. Hews.

Plate 12 *Liolaemus bellii,* a high-elevation (2,100–3,000 m), viviparous lizard of central Chile with negligible sexual dimorphism and reduced aggression that reaches very high densities. Photograph by J. Jiménez.

Plate 13 *Liolaemus leopardinus,* a high-elevation (2,100–3,000 m), large, viviparous lizard of central Chile with negligible sexual dimorphism and reduced aggression. Photograph by J. Jiménez.

Plate 14 *Liolaemus schroederi,* a moderately high-elevation (1,800–2,590 m), shy, viviparous lizard of central Chile that often climbs into bushes. Photograph by J. Jiménez.

Plate 15 *Liolaemus monticola,* a saxicolous, oviparous lizard of central Chile that inhabits low to moderate elevations (1,500–2,450 m) and shows obvious sexual dimorphism and elevated aggression. Photograph by S. Fox.

Plate 16 *Liolaemus nitidus,* a ground and rock-dwelling, large, oviparous lizard of central Chile that inhabits a wide range of elevations (0–3,000 m) and shows notable sexual dimorphism, with bigger, more colorful males. Photograph by S. Fox.

Food Availability

Previous studies have documented variation in arthropod abundance on large islands in the Galápagos and suggested that food availability is important in the evolution of body size in lava lizards (Schluter 1984). We provide a coarse-grained analysis of food availability in lava lizards using data from different microhabitats on Plaza Sur (see Stone [1995] for description of methods). These data show that food availability is strongly correlated with microhabitat, and that food is most abundant around a few species of woody vegetation (Table 7-6), and least abundant around *Sesuvium,* an abundant succulent that is the dominant vegetation on barren islets. In addition, food was abundant around dead marine animals (sea lions, fish, and sea birds). Living marine animals also contribute to food availability for lizards by providing organic matter (feces, vomit, tissue) that attracts potential prey and/or can be fed on directly by lizards (Hews 1993; Stone 1995). We catagorized the study areas of each island according to two criteria: the presence of woody vegetation and the presence of breeding colonies of sea lions (*Zalophus californianus*) or boobies (*Sula* spp.). We then assigned each of our study islands to one of three categories of food availability: low (islands without woody vegetation or breeding sea lions/boobies), medium (islands with woody vegetation or breeding colonies of sea lions/boobies), and high (islands with woody vegetation and breeding colonies of sea lions/boobies).

Of the 37 study islands, 7 had low food availability, 23 had medium food availability, and 7 had high food availability. The seven islands with low food availability were all barren islets (two barren islands harbored large populations of breeding sea lions and/or boobies), and all except one of the islands with high food availability were diverse islets, with the exception of the large island of Santa Fé.

Population Density

Previous studies have suggested that population densities of lava lizards on large islands with introduced predators are lower than on associated islets without introduced predators (Steadman 1986). We calculated population densities for six islets (five diverse, one barren) for which we had accurate maps and large samples of lava lizards (Table 7-7). Five of the six islets are associated with two large islands for which population densities of lava lizards have been reported in the literature (Stebbins, Lowenstein, and Cohen 1967; Werner 1978) (Table 7-7). We also included qualitative estimates of population density from the Floreana is-

Table 7-6 Food availability by microhabitat on Isla Plaza Sur. Each 4-m^2 plot was searched for 10 min and potential lizard prey were counted. Prey items were converted into ant equivalents based on prey size. Mean rank values are from Kruskal-Wallis nonparametric ANOVA, which revealed significant differences among microhabitats (H = 21.65, P < 0.01). See Stone (1995) for a more complete description.

Microhabitat	Number of Ant Equivalents (range)	Mean Rank	Plots Sampled
Dead *Opuntia*	328–1,908	42.5	8
Dead marine vertebrate	222–5,367	33.8	4
Maytenus	212–915	32.8	13
Parkinsonia	143–835	28.5	6
Intertidal	299–798	28.2	8
Opuntia	159–1077	21.6	8
Grabowskia	111–570	10.8	6
Sesuvium	39.5–116	2.0	2

Table 7-7 **Population densities of lava lizards.** Data are from this study, unless otherwise indicated.

Island Group	Island	Island Type	Population Density (males/ha)
Santa Cruz	Santa Cruz[a]	Large	31
Santa Cruz	Plaza Sur	Diverse	216
Santa Cruz	N Punta Bowditch	Diverse	266
Santa Cruz	N Guy Fawkes	Barren	106
Santa Fé	Santa Fé Islote	Diverse	211
Española	Española[b]	Large	42
Española	Xarifa	Diverse	160
Española	Oeste	Diverse	32
Floreana	Floreana[c]	Large	Extremely rare
Floreana	Champion[c]	Diverse	Common

[a]Data from Stebbins, Lowenstein, and Cohen (1967).
[b]Data from Werner (1978).
[c]Data from Steadman (1986).

land group (Steadman 1986). With one exception, population densities on diverse islets were much higher than on associated large islands (Table 7-7). The exception, Islote Oeste, is one of the smallest islands in the Galápagos, and the low density of lava lizards might be associated with an absence of suitable nesting habitat on the island (M. A. Jordan, personal communication). The one barren islet in this sample had a population density that was intermediate compared to diverse islets and large islands (Table 7-7).

Synopsis

Island size and island isolation appear to influence the population density of lava lizards and the distributions of their predators, parasites, and prey. On large islands, predation is probably more important than on small islets. Predation may explain the relatively low population densities of lava lizards on large islands. In turn, the low population densities of lava lizards on large islands may reduce rates of parasite transmission and lessen intraspecific competition for food and mates. Thus, large islands may experience selection regimes that are driven by predation pressure, with natural selection favoring morphology and behavior that facilitate avoidance of predators (Table 7-8). The ecological characteristics of small islets may depend on the degree of isolation of the islet. Islets that are distant from a large island tend to be barren of woody vegetation and have low diversity of predators and parasites, as well as low food availability. On such islets, food availability may be more important than on other islands, and natural selection may favor traits that facilitate resource acquisition and/or defense (Table 7-8). Small islets with woody vegetation appear to provide environments with reduced predation (compared to large islands) and increased food availability (compared to barren islets). This combination may explain the high population densities of lava lizards on diverse islets. The high densities of lava lizards in turn may increase rates of parasite transmission and the intensity of intraspecific competition. On such islets, sexual selection may favor traits that confer resistance to parasitism, attractiveness to potential mates, and prowess in intrasexual competition (Table 7-8).

Correlations between Ecological Conditions and Morphology

The 37 populations studied include four species, *M. albemarlensis, M. delanonis, M. grayii,* and *M. bivittatus* (Table 7-9, Fig. 7-1). In 1992, we sampled *M. delanonis, M. bivittatus,* and 13 populations of *M. albemar-*

Table 7-8   Predictions about relationship of traits to island type. Predictions are presented as the expected trait expression on island types where specific ecological conditions are harsh. Note that the sexual selection hypothesis predicts the same pattern as the predation hypothesis or the food availability hypothesis, and that the crypticity and Hamilton-Zuk hypothesis predict the same pattern. Hypothesis and predictions are described more fully in the text.

Hypothesis	Prediction
Predation	Large islands/small body size
Escape	Large islands/large body size
Food availability	Barren islets/small body size
Herbivory	Barren islets/large body size
Sexual selection	Diverse islets/large body size
Speed	Large islands/long limbs
Crypticity	Large islands/small ornaments
Hamilton-Zuk	Diverse islets/large ornaments

Table 7-9   Summary of morphology sample.

Species	Number of Islands Sampled	Males Sampled per Island (mean ± 1 SD [range])	Year Sampled
M. albemarlensis	26	24 ± 26.8 (2–140)	1992, 1994
M. delanonis	5	40 ± 29.4 (8–88)	1992
M. bivittatus	2	27 ± 9.9 (20–34)	1992
M. grayii	4	8 ± 5.6 (2–15)	1994
Totals	37	898	

lensis. The patterns this sample revealed were suggestive enough to convince us to add another species (*M. grayii*) and 13 more populations of *M. albemarlensis* to the sample in 1994. The two years appeared to have similar patterns of rainfall (normal wet years), and the sampling periods in the two years were congruent (May–July).

Using dial calipers, we measured snout-vent length (SVL), chin-patch length, and hindlimb length from a total of 898 adult male lizards. SVL was measured as the straight-line distance from the tip of a lizard's snout to the posterior edge of the anterior lip of its cloaca. Chin-patch length was measured as the greatest midline patch of uninterrupted black scales on the ventrum. Lizards in the San Cristóbal group (*M.*

bivittatus) lacked chin patches and were excluded from chin-patch analyses. Limb length was measured as the straight-line distance on a stretched lizard from the tip of the third (longest) toe to the ball of the acetabulum.

We evaluated the relationship between ecological conditions and morphology of male lava lizards using two types of analysis. First, we tested predictions involving biotic variables (parasites, predators, and prey) and morphology of lava lizards (Table 7-1), using Spearman rank correlation analyses or Kruskal-Wallis nonparametric ANOVAs. For each test, a morphological variable was the dependent variable, a biotic variable was the independent variable, and the island mean was the sampling unit. Second, we tested predictions about variation among island types (Table 7-8). We performed an ANOVA for each island group, with island as the categorical variable, morphological variables as the dependent variables, and individual lizards as sample units. We then looked for differences between island types using Fisher PLSD multiple comparisons. Although many of these analyses involve related tests, we decided not to adjust our *P* values for multiple comparisons (Rice 1990) because of the diffuse nature of the data set and the already high potential for making Type II errors in the interpretation of the analyses.

Body Size

Table 7-10 gives the results of multiple comparisons of islands within island groups. The predictions of body size (Table 7-8) agree with the observed pattern, with one major exception: the largest males in the Española island group were on the large island, Española (Tables 7-10 and 7-11). Otherwise, most (76 and 74% for columns 2 and 3, respectively, of Table 7-10) of the pairwise comparisons showed that diverse islets had significantly larger lizards than either large islands or barren islets (Tables 7-10 and 7-11). This pattern is consistent with the predation, food availability, and sexual selection hypotheses and would be expected if predators select for small size on large islands, chronic food shortages select for small size on barren islets, and sexual selection favors large size on diverse islets.

Correlations between predation pressure and body size are consistent with this interpretation. Body size was largest on islands with intermediate predator diversity, and smallest on islands with zero or four predators (Fig. 7-7). We grouped islands according to the presence or absence of individual predator species and performed *t*-tests to identify the predator with the largest influence on body size. Lizards were

Table 7-10 Morphological differences within island groups. A plus sign indicates significant differences in the direction of the prediction at the head of the column, a minus sign indicates significant differences in the opposite direction of the prediction at the head of the column, and a zero indicates no differences among island types. The number of symbols in each cell corresponds to the number of diverse islets in a particular island group. For each symbol, a diverse islet is compared to a large island (or a barren islet) using Fisher PLSD multiple comparisons. The commas in the third column separate barren islets. NA, not available.

Island Group	Small Lizards on Large Islands	Small Lizards on Barren Islets	Large Limbs on Large Islands	Small Chin Patches on Large Islands
Isabela	++++	NA	++00	+000
Santa Cruz	+++++0	0+++++, 0+++++, −+++++, −+++++	+++++−	+++000
Santiago	+00	0++, 0++, 0++, 00+	0−−	+00
Santa Fé	+	NA	0	0
Española	−−−−	NA	++++	+00−
San Cristóbal	+	NA	0	NA
Floreana	+0	+0	00	00

smaller on islands with rats or cats (Table 7-12), which provides general confirmation of isolated reports from specific islands suggesting that introduced predators were associated with reduced body size in lava lizards (Grant 1975; Steadman 1986). This pattern cannot be extended to include predation by native snakes and hawks. There was very little difference in mean body size on islands with or without snakes, and lizards were significantly larger on islands with hawks (Table 7-12). It is unlikely that lava lizards can escape predation from hawks by attaining large body size. Hawks routinely take prey that are larger than lava lizards (Laurie and Brown 1990; Anderson 1991), and hawks may even selectively prey on large lava lizards (Werner 1978). Instead, the large size of lava lizards on islands with hawks may be a spurious correlation caused by the correlation of both lizard size and hawk presence with the presence of cats. The escape hypothesis, though generally unsupported by our data, is consistent with the pattern of variation in body size ob-

Table 7-11 Variation in morphology in 37 populations of lava lizards. Values are mean ± 1 SD (*n*). Values for hindlimb length and chin-patch length are residuals from regressions with SVL as the independent variable.

Island Group (*Microlophus* sp.)	Island	Island Type	SVL (cm)	Residual Hindlimb Length (cm)	Residual Chin-Patch Length (cm)
Isabela (*albemarlensis*)	Isabela	Large	8.05 ± 0.424 (13)	0.14 ± 0.155 (13)	−0.80 ± 0.443 (13)
	Cráter Beagle 1	Diverse	10.07 ± 0.382 (11)	−0.10 ± 0.256 (10)	−0.77 ± 0.791 (11)
	Cráter Beagle 2	Diverse	9.74 ± 0.233 (2)	−0.31 ± 0.163 (2)	−0.24 ± 0.249 (2)
	Cráter Beagle 5	Diverse	9.62 ± 0.612 (5)	−0.10 ± 0.240 (4)	0.88 ± 0.920 (5)
	Marielas Sur	Diverse	10.14 ± 0.478 (27)	0.094 ± 0.230 (27)	−0.49 ± 0.799 (27)
Santa Cruz (*albemarlensis*)	Santa Cruz	Large	8.27 ± 0.417 (24)	0.058 ± 0.199 (23)	−0.15 ± 0.532 (24)
	Eden	Diverse	8.21 ± 0.516 (19)	−0.16 ± 0.144 (19)	−0.32 ± 0.507 (19)
	Plaza Norte	Diverse	9.08 ± 0.309 (40)	−0.26 ± 0.106 (36)	0.33 ± 0.486 (40)
	Plaza Sur	Diverse	9.18 ± 0.366 (140)	−0.029 ± 0.195 (131)	0.50 ± 0.432 (138)
	N Punta Bowditch	Diverse	9.52 ± 0.329 (43)	−0.26 ± 0.188 (40)	0.10 ± 0.732 (43)
	S Punta Bowditch	Diverse	9.26 ± 0.311 (20)	−0.084 ± 0.204 (18)	0.25 ± 0.402 (20)
	Venecia	Diverse	9.12 ± 0.412 (43)	0.17 ± 0.221 (40)	0.026 ± 0.681 (42)
	N Guy Fawkes	Barren	8.69 ± 0.337 (19)	−0.31 ± 0.230 (17)	0.11 ± 0.745 (19)
	S Guy Fawkes	Barren	8.37 ± 0.308 (10)	0.058 ± 0.166 (10)	−0.44 ± 0.423 (10)
	E Guy Fawkes	Barren	8.49 ± 0.318 (47)	−0.38 ± 0.217 (46)	0.30 ± 0.448 (47)
	W Guy Fawkes	Barren	8.18 ± 0.230 (18)	−0.11 ± 0.129 (17)	−1.12 ± 0.268 (18)
Santa Fé (*albemarlensis*)	Santa Fé	Large	10.19 ± 0.344 (13)	0.33 ± 0.152 (12)	−1.40 ± 0.411 (13)
	Santa Fé Islote	Diverse	11.53 ± 0.553 (26)	0.34 ± 0.2 (26)	−1.49 ± 0.515 (19)

Table 7-11 (Continued)

Island Group (*Microlophus* sp.)	Island	Island Type	SVL (cm)	Residual Hindlimb Length (cm)	Residual Chin-Patch Length (cm)
Santiago (*albemarlensis*)	Santiago	Large	9.69 ± 0.480 (16)	0.049 ± 0.216 (16)	−0.76 ± 0.498 (16)
	Albany	Diverse	9.74 ± 0.239 (12)	0.08 ± 0.152 (12)	−0.76 ± 0.542 (12)
	Bartolomé	Diverse	9.29 ± 0.702 (4)	0.37 ± 0.059 (4)	−0.11 ± 0.615 (4)
	Sombrero Chino	Diverse	10.07 ± 0.543 (18)	0.18 ± 0.190 (18)	−0.69 ± 0.674 (17)
	Roca Bainbridge 3	Barren	8.94 ± 0.234 (9)	0.24 ± 0.102 (8)	0.97 ± 0.360 (9)
	Roca Bainbridge 4	Barren	9.57 ± 0.293 (4)	0.24 ± 0.216 (4)	0.32 ± 0.391 (4)
	Roca Bainbridge 5	Barren	9.09 ± 0.153 (11)	0.30 ± 0.226 (10)	0.40 ± 0.253 (11)
	Roca Bainbridge 6	Barren	9.31 ± 0.389 (18)	0.32 ± 0.136 (16)	0.41 ± 0.550 (18)
Española (*delanonis*)	Española	Large	13.55 ± 0.595 (88)	0.12 ± 0.287 (86)	−0.073 ± 1.363 (88)
	Gardner (Española)	Diverse	12.95 ± 0.511 (40)	−0.18 ± 0.270 (37)	0.07 ± 0.902 (40)
	Islote Oeste	Diverse	12.85 ± 0.398 (8)	−0.25 ± 0.177 (7)	−1.317 ± 0.454 (8)
	Osborn	Diverse	12.88 ± 0.352 (35)	−0.16 ± 0.272 (27)	−0.45 ± 0.814 (35)
	Xarifa	Diverse	13.05 ± 0.513 (30)	−0.079 ± 0.260 (21)	1.30 ± 1.188 (30)
San Cristóbal (*bivittatus*)	San Cristóbal	Large	7.91 ± 0.55 (20)	0.20 ± 0.132 (20)	
	Isla Lobos	Diverse	9.99 ± 0.79 (34)	0.22 ± 0.216 (34)	

Table 7-11 (Continued)

Island Group (*Microlophus* sp.)	Island	Island Type	SVL (cm)	Residual Hindlimb Length (cm)	Residual Chin-Patch Length (cm)
Floreana (*grayi*)	Floreana	Large	8.79 ± 0.569 (5)	0.27 ± 0.209 (5)	−0.025 ± 0.494 (5)
	Champion	Diverse	10.68 ± 0.442 (8)	0.11 ± 0.247 (5)	−0.53 ± 0.567 (8)
	Gardner (Floreana)	Diverse	9.50 ± 0.219 (2)	0.32 ± 0.038 (2)	0.45 ± 0.249 (2)
	Enderby	Barren	9.88 ± 0.402 (15)	0.12 ± 0.203 (14)	0.38 ± 0.540 (15)

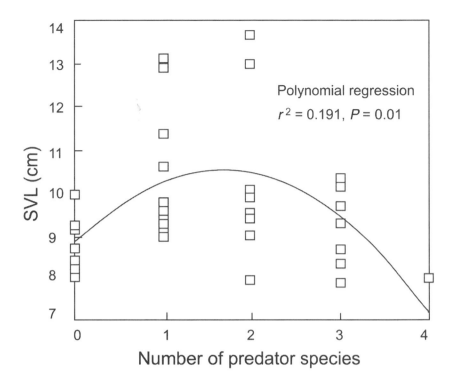

Figure 7-7   Plot of SVL versus number of predator species on an island.

served within the Española group, where the largest lizards occur on the large island, where snakes are present.

The food availability hypothesis is the most likely explanation for the small body size of lizards on barren islets. However, ANOVA revealed no significant relationship between food availability and SVL ($F_{2,36} = 1.34$, $P = 0.27$), but the trend was in the predicted direction. Similarly, indirect tests of the sexual selection hypothesis (Table 7-1) were not conclusive. Body size was not correlated with parasite prevalence across the entire sample of islands ($r_S = 0.204$, $P > 0.05$), and population density appeared to be unrelated to body size in three populations from the Española island group (Fig. 7-8). However, there was a remarkably strong positive correlation between population density and SVL in four populations from the Santa Cruz island group (Fig. 7-8). The herbivory hypothesis, advanced by Schluter (1984) to explain variation in body size between lava lizards on large islands, received no support in our analyses and apparently does not explain variation in body size within island groups or among island types.

Table 7-12   Relationships between distributions of specific predators and lava lizard morphology. Values are differences in means between islands with the predator and islands without the predator. Negative values indicate differences in means that are opposite of predicted differences. The predicted pattern is that lizards on islands with a particular predator will have small SVL, long limbs, and small chin patches, relative to islands without the predator. NS, not significant.

Predator	Mean Difference SVL (cm)	Mean Difference Residual Limb Length (cm)	Mean Difference Residual Chin-Patch Length (cm)
Cats	1.75*	0.143 NS	0.19 NS
Hawks	−1.79***	−0.016 NS	0.54*
Rats	1.26**	0.138*	−0.04 NS
Snakes	0.12 NS	0.179*	0.292 NS

*$P < 0.05$.

**$P < 0.01$.

***$P < 0.0001$.

Chin-Patch Length

There was a significant relationship between chin-patch length and SVL, but SVL accounted for only a small amount of the variation in chin-patch length ($df = 831$, $r^2 = 0.2$, $P = 0.0001$). After the effects of SVL were removed, residual chin-patch length was not correlated with parasite prevalence across the entire sample (Fig. 7-9). This lack of correlation is partly due to the unexpected cluster of most of the barren islets in the low parasitism/large chin-patch quadrant of the graph (Fig. 7-9). When the barren islets are removed from the analysis, the relationship between chin-patch length and parasite prevalence is positive and significant ($r_S = 0.409$, $P < 0.05$) (Fig. 7-9). There was also a significant relationship between food availability and residual chin-patch length ($F_{2,32} = 4.39$, $P = 0.02$); lizards on islands with low food availability had relatively large chin patches. The seven islands with low food availability were the same seven barren islets that appeared to confound the relationship between parasite prevalence and chin-patch length (Fig. 7-9). Thus, chin patches were large when either parasite prevalence was high or food availability was low. This provides support for the Hamilton-Zuk hypothesis but also suggests a more general role of ornaments as useful signals in harsh environments.

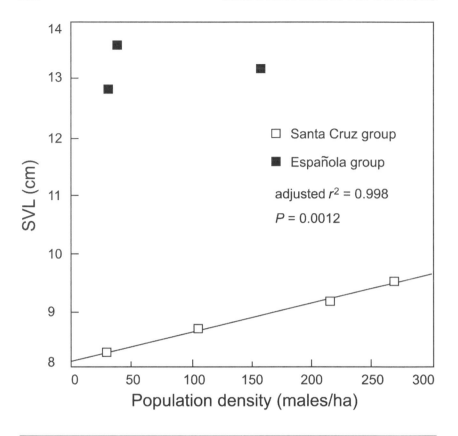

Figure 7-8   Plot of SVL versus population density on an island.

The crypticity hypothesis was supported by the negative correlation between residual chin-patch length and the number of predator species on an island (Fig. 7-10). Separate t-tests involving individual predator species revealed that chin patches were smaller on islands with hawks (Table 7-12).

Comparisons within island groups generally did not support the results just described. Residual chin-patch length was not significantly different in 67% of the pairwise comparisons, and in no island group was a majority of the comparisons significant in the predicted direction (Tables 7-10 and 7-11).

Limb Length

A hindlimb length versus SVL regression, for all males combined, was highly significant ($df = 836, r^2 = 0.947, P = 0.0001$). After removing the effects of SVL, residual hindlimb length and the number of predator

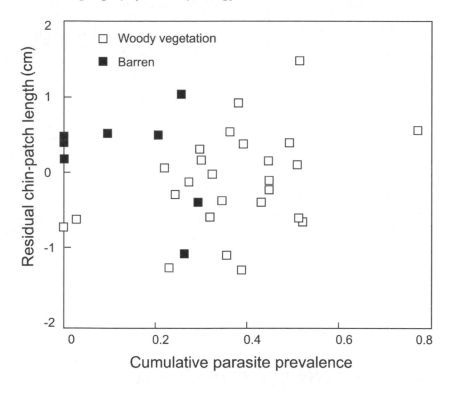

Figure 7-9 Plot of residual chin-patch length versus cumulative parasite prevalence on an island.

species on an island were positively correlated, providing support for the speed hypothesis (Fig. 7-11). Separate *t*-tests further revealed that lizards had longer limbs on islands with rats or snakes (Table 7-12). Interestingly, Stone, Snell, and Snell (1994) showed that the presence of cats, but not snakes, rats, or hawks, was positively correlated with wariness of lava lizards. This suggests that lava lizards may respond to different predators in different ways, increasing wariness in the presence of introduced cats, increasing limb length in the presence of native snakes, and decreasing chin-patch length in the presence of hawks.

Comparisons within island groups provided partial support for the speed hypothesis. Overall, only 50% of pairwise comparisons of residual limb length to island type were significant in the predicted direction (Tables 7-10 and 7-11). However, in the two island groups that were most intensively sampled (Española and Santa Cruz), lizards on large islands had significantly longer limbs than lizards on diverse islets in all but one comparison (Tables 7-10 and 7-11).

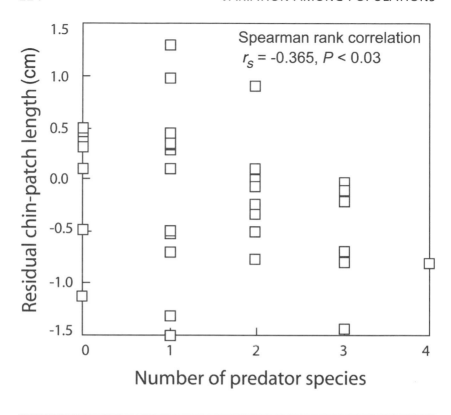

Figure 7-10   Plot of residual chin-patch length versus number of predator species on an island.

Correlations between Ecological Conditions and Social Behavior

In 1993, we spent 3 mo studying the mating system of *M. albemarlensis* on Isla Plaza Sur (Stone 1995). In 1994, using a similar protocol, we collected data over shorter time periods from three other populations of *M. albemarlensis* (two diverse islets, one barren islet). Our study areas covered relatively large areas on these small islets (Table 7-13), and we therefore monitored a relatively large percentage of the males in each population. On N Guy Fawkes, we censused the entire island and marked every male on the island except one, an elusive male that lived on the cliff. All four study areas were bordered by the intertidal zone and were typical of the available habitat on each islet. At the beginning of each study, we mapped each study area and captured every adult lizard within study area boundaries. We marked lizards permanently by toe

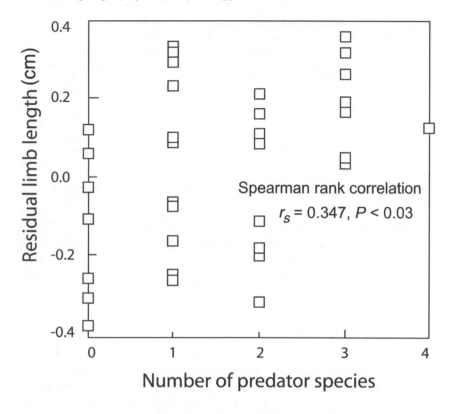

Figure 7-11 Plot of residual hindlimb length versus number of predator species on an island.

Table 7-13   Attributes of islets in behavior sample.

Island	Island Size (ha)	Island Isolation (m)	Size of Study Area (ha)
Plaza Sur	11.900	302	0.394
Santa Fé Islote	0.671	17	0.123
N Guy Fawkes	0.235	3,870	0.235
N Punta Bowditch	2.918	272	0.150

clipping, and semipermanently by painting a unique mark (numbers or letters) on the dorsum with yellow, oil-based paint. After processing, we released lizards at the site of capture. After a few days of intense capture efforts, we conducted censuses in the study areas. During censuses, we noted the location and behavior of marked lizards, and we captured any unmarked lizards that had eluded initial capture. Reference points were

Table 7-14   Attributes of behavior sample.

Island	Males Sampled	Sightings Recorded	Sightings per Male (range)	Duration of Sampling
Plaza Sur	79	6,403	1–256	19 Jan–24 Apr 1993
Santa Fé Islote	26	981	3–85	29 Mar–6 Apr 1994
N Guy Fawkes	25	327	1–34	14 Feb–16 Feb 1994
N Punta Bowditch	34	294	1–24	1 Feb–15 Feb 1994

concentrated enough so that most lizard sightings were within 5 m of a reference point.

Censusing effort varied from island to island (Table 7-14). On N Guy Fawkes and Santa Fé Islote, censuses were conducted around the clock during daylight hours over the course of a few days to 1 wk (Table 7-14). On N Punta Bowditch, censuses were spread out over a 2-wk period, with fewer censuses each day (Table 7-14). The sample on Plaza Sur was spread out over 3 mo and concentrated on morning (0700–1000 h) and afternoon (1400–1800 h) censuses (Table 7-14). Sample sizes were affected by censusing effort. On Plaza Sur, sample sizes were large and covered most of a breeding season. On Santa Fé Islote, sample sizes per male appear adequate, but these samples occurred over a short time period. On N Guy Fawkes and N Punta Bowditch, sample sizes per male and study duration were small. Our data on Plaza Sur, where sampling effort was spread out over time, suggest little variation in rates of behavior within the breeding season. This may mitigate the short duration of sampling on the other three islands. The low number of sightings per male on N Guy Fawkes and N Punta Bowditch probably increases the error associated with the data we present for these islands.

From the census data, we calculated the proportion of sightings that individuals performed four types of behavior: rejection, bobbing, courtship, and agonism. The rejection postures and signature bobs we observed were typical of iguanian lizards (Carpenter and Ferguson 1977; Werner 1978). We recognized three distinct types of courtship behavior: refuging, circling, and pairing. These are described elsewhere (Carpenter and Ferguson 1977; Stone 1995). We recorded four behavioral patterns associated with agonism: displacement, challenge, fight, and chase (Carpenter and Ferguson 1977).

In addition to calculating the rates of agonism, we also created a dominance index (DI). For each male, we calculated the DI by first scoring each male-male interaction as a win (displacing, chasing or winning a fight), loss (being displaced, chased, or losing a fight), or draw (challenge that does not escalate into a fight, fight with no clear outcome) for each interacting lizard. We arranged interacting pairs of males in dyads and recorded the outcome of agonistic interactions within all dyads. We accumulated the record of each member of a dyad using the following formula: (wins − losses)/(wins + losses + ties). This yields scores for each member of a dyad that are identical in magnitude but have opposite sign. Scores could range from −1 (all losses within a particular dyad) to 1 (all wins). We considered an individual as dominant within a dyad if its score was >0.33, as subordinate if its score was <−0.33, and as neutral if its score was between 0.33 and −0.33. The DI for each male was: dyads won − dyads lost.

We estimated home range area and home range overlap between males from the census data from three islands. We estimated home range area using the minimum convex polygon method (Rose 1982) and home range overlap as the sum of all individual overlaps with a given male, expressed as a percentage of home range size. These estimates are sensitive to sample size; both home range area and home range overlap tend to be underestimated when sample size is small (Rose 1982). For Santa Fé Islote and Plaza Sur, sample sizes were large (Table 7-14), and for most lizards we probably had enough sightings to map home ranges accurately. For these islands, we mapped the home ranges of all lizards with more than 10 sightings. For N Guy Fawkes, we mapped home ranges of all lizards with greater than six sightings but stipulate that home range area and overlap are probably underestimated because of small sample sizes (Table 7-14). For N Punta Bowditch, the number of sightings for each male (Table 7-14) was probably too low to map home ranges accurately.

Home Range Area

Home ranges on islets were smaller than reported home ranges on large islands (Table 7-15). However, differences in home range area were much less than differences in population density (Table 7-7). Home ranges reported in the literature are probably underestimated because of small samples (Stebbins et al. 1967) or nonstandard techniques (Werner 1978), which means that differences in home range area may be larger than reported here. Thus, differences in population density between small islets and large islands appear to have led to differences in home range area in lava lizards, and these differences are consistent with

Table 7-15   Summary of space use in males. Values are mean ± 1 SD.

Island	Home Ranges Plotted	Home Range Area (m²)	Home Range Overlap (%)	Number of Overlappers
Plaza Sur	67	175 ± 88.7	209 ± 106.7	8.7 ± 4.0
Santa Fé Islote	22	156 ± 136.4	243 ± 161.4	10 ± 4.8
N Guy Fawkes	23	68 ± 52.5	70 ± 56.4	4 ± 2.0
Española[a]	—	374	—	—
Santa Cruz[b]	—	252	—	—

[a]Data from Werner (1978).

[b]Data from Stebbins, Lowenstein, and Cohen (1967).

previous comparisons of island and mainland social systems (Stamps and Buechner 1985).

Home Range Overlap

Home range overlap was much higher on small islets than would be expected in a typical territorial system (Table 7-15). Home range overlap was >200% on the most heavily sampled islands, Plaza Sur and Santa Fé Islote (Table 7-15, Figs. 7-12 and 7-13), suggesting that male lizards on these islets had little if any space to occupy exclusive of other males. Moreover, on these two islets, males on average had home ranges that overlapped with 9 to 10 other males (Table 7-15, Figs. 7-12 and 7-13). With a larger sample on N Guy Fawkes, a similar pattern would not be surprising (Fig. 7-14). On large islands, home range overlaps are much lower (Stebbins, Lowenstein, and Cohen 1967; Werner 1978), as is typical in mainland populations (Stamps 1983a). Thus, it seems clear that high population densities on small islets have resulted in extensive overlap of home ranges, a pattern consistent with expectations (Stamps and Buechner 1985).

Rates of Behavior

Previous studies have not reported rates of specific behavior in lava lizards (Carpenter 1966; Stebbins, Lowenstein, and Cohen 1967; Werner 1978). As a result, we cannot directly compare the behavior of lava lizards on large and small islands. However, by comparing among the islets we sampled, we can identify behavior associated with the social systems on large islands that is performed at a low rate in specific islet populations, and this would support predictions about reduced aggression in high-density populations (Stamps and Buechner 1985). For ex-

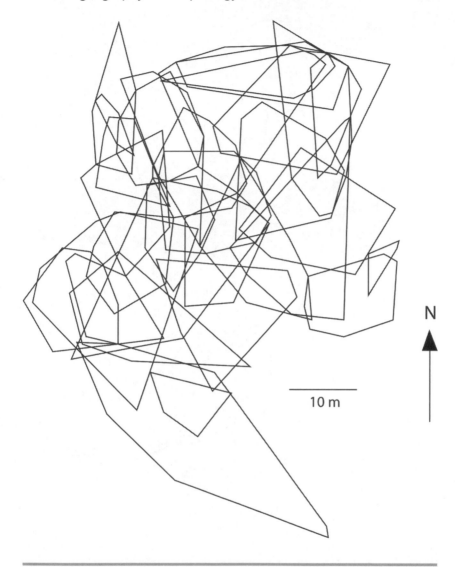

Figure 7-12 Home range map for male lava lizards on Plaza Sur showing all males from one part of study area that was sampled.

ample, rates of rejection were particularly low on N Punta Bowditch and Santa Fé Islote (Table 7-16). On Santa Fé Islote, rejection displays were not observed despite almost 1,000 sightings of males. On N Punta Bowditch, one female rejected a male one time (294 sightings of males). By contrast, females on Plaza Sur and N Guy Fawkes performed rejection displays regularly (Table 7-16), as do females on large islands

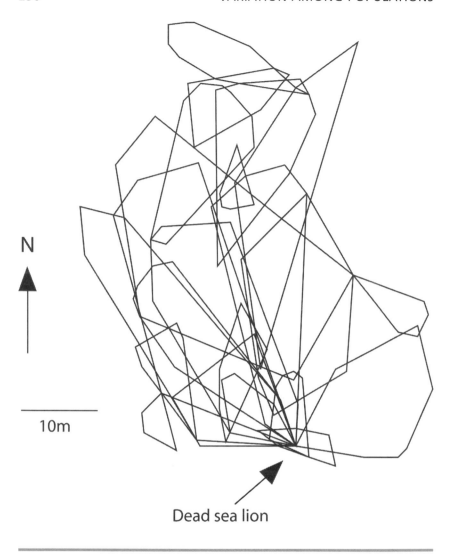

N

10m

Dead sea lion

Figure 7-13   Home range map for male lava lizards on Santa Fé Islote. Note the location of the dead sea lion and the convergence of home ranges in that area.

(Stebbins, Lowenstein, and Cohen 1967; Werner 1978). Similarly, rates of agonistic behavior were low on Plaza Sur and N Punta Bowditch, relative to the other two islets (Table 7-16). There was a similar pattern, not significant, for rates of bobbing (Table 7-16). The majority of male-male interactions had neutral outcomes, except on Santa Fé Islote, where male-male interactions with winners and losers slightly outnumbered those with neutral outcomes. Neutral intrasexual interactions

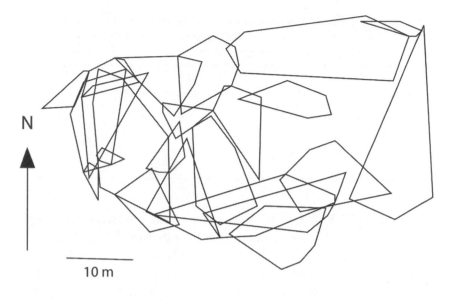

Figure 7-14   Home range map for male lava lizards on N Guy Fawkes.

were the rule on N Punta Bowditch, where only 2 of 34 marked males won fights or chases, and no marked male lost a fight or was chased (the two dominant males interacted with unmarked males). Courtship was common on each island, and there were no interisland differences in courtship frequency (Table 7-16). Thus, despite high lizard densities, small home ranges, and high home range overlap, in certain instances aggressive social behavior, rejection and agonism, appeared to occur at decreased frequencies, suggesting the possible erosion of the territorial system.

Shifts in Space Use

The high levels of agonism on Santa Fé Islote may be the result of the death of a large sea lion on the beach in the study area on the first day of censusing. As the sea lion decomposed over the next several days (the length of our censusing period), thousands of flies and maggots became available as prey. The dead sea lion was visited by almost half (12/26) of all marked males, and by more than half (16/30) of all marked females. Many of these lizards moved to the dead sea lion from home ranges that were centered on the opposite side of the study area (Fig. 7-13). In addition, several unmarked males and females and many juveniles were observed at the dead sea lion. Since

Table 7-16   Interpopulation comparison of social behavior. Values for each island are mean ranks from Kruskal-Wallis nonparametric ANOVA. *P* values are from Kruskal-Wallis nonparametric ANOVA.

Variable	Plaza Sur	Santa Fé Islote	N Guy Fawkes	N Punta Bowditch	P
Rejection frequency	93	64	91	67	<0.001
Bobbing frequency	76	91	101	77	<0.10
Agonism frequency	76	97	114	64	<0.001
Courtship frequency	89	72	68	85	<0.10

all the adults in the study area were marked, the unmarked adults observed at the dead sea lion must have migrated to the sea lion from outside the study area. Many of the lizards sighted around the sea lion were observed feeding on flies, maggots, and sea lion flesh. The dead sea lion was also the site of about 40% of all agonistic interactions between male lava lizards, and many of these interactions escalated into fights that had winners and losers. Thus, it seems clear that the death of this sea lion temporarily created a dramatic change in the pattern of space use and probably the levels of aggression in this population. Hews (1993) conducted food supplementation experiments involving an island iguanian, *Uta palmeri*, and found that females shifted their home ranges in response to the increased food resources, but that males did not. This suggests that the risk of losing a territory outweighs the benefits of leaving the territory to feed in males but not in females. On Santa Fé Islote, the movement of both males and females from their home ranges to the dead sea lion may indicate a weakened territorial system.

Social Classes

In contrast to these obvious differences between the social systems of lava lizards on large islands and small islets, there was one obvious similarity: two distinct classes of males were apparent on all four islets (Fig. 7-15). These classes were identified on the basis of the strong intercorrelations among dominance, courtship frequency, and bobbing frequency on all four islets (Table 7-17). The correlations were all positive, which means that dominant lizards frequently bobbed and courted females. Such a pattern has been noted in a variety of social systems (Temeles 1990; Baird, Acree, and Sloan 1996; Iguchi and Hino 1996), including the lava lizards on the large island of Española (Werner 1978).

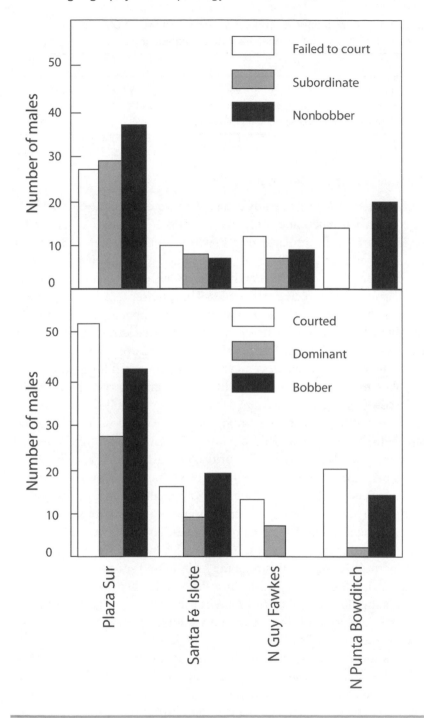

Figure 7-15   Histograms showing two distinct classes of males.

Table 7-17   Correlations among behavioral variables. Values are from Spearman rank correlation analyses with all males combined.

Spearman Rank Correlation	r_s	P
Dominance index versus courtship frequency	0.319	<0.001
Dominance index versus bobbing frequency	0.432	<0.001
Courtship frequency versus bobbing frequency	0.458	<0.001

Potential for Sexual Selection

In Werner's (1978) main study areas, there were 15 territorial males, 16 nonterritorial males, and 19 females. In 1994, on another part of Española, we found a similarly male-biased sex ratio (31 males:25 females, unpublished data). On Plaza Sur, we recorded 42 males that performed bobbing displays, 37 males that did not (roughly equivalent to Werner's two classes of males), and 94 females. Thus, there were similar proportions of dominant and subordinate males on the two islands, but the number of females per male was more than twice as high on Plaza Sur. Females on Española are territorial, with low territory overlap (Werner 1978); this disperses females and reduces the potential for males to monopolize multiple females. In addition, the larger territories of males often completely encompass female territories, probably limiting the potential for females to choose among prospective mates. Territory overlaps between females and dominant males on Española indicate that most individuals overlapped one or two potential mates, and that some males overlapped zero females. On the other hand, on Plaza Sur, among males that bobbed, the modal number of females overlapped was nine, with one male overlapping 21 female home ranges, and only two males overlapping less than 5 female home ranges. This suggests a greater opportunity for males to court multiple females, and for females to choose among multiple males. Although the male-biased sex ratio on Española suggests that females may be the limiting sex, the dispersed territories of females should greatly reduce the environmental potential for polygamy, as well as reduce both the mean and variance of male mating success. Sexual selection in lava lizards on Española should be relatively weak, and should operate primarily through intrasexual competition, with mate choice being relatively unimportant. On the other hand, the female-biased sex ratio and dense overlapping distributions of females on Plaza Sur create a high potential for polygamy and probably greatly increase both the mean and variance of male mating success. Under these conditions, sexual selection should be strong; male-male competition should not be oriented toward maintaining a large, exclusive territory; and the opportunity for female choice should be high.

Synopsis

We found evidence of the collapse of territoriality described by Stamps and Buechner (1985) for island populations of lizards, birds, and rodents. Specifically, we found reduced levels of aggression, shifts in home ranges in response to shifts in food availability, extensive home range overlap, and reduced home range area on some or all of the four islets. These data suggest that trends found in mainland-to-island comparisons of social behavior are valid within an archipelago, with large islands having mainland-like characteristics, when compared to small islets. The high population densities and observed changes in the social systems of lava lizards on small islets probably result in stronger inter- and intrasexual selection relative to large islands.

A Biogeographical Hypothesis for Evolution in Lava Lizards

Seventeen thousand years ago, as sea levels began to recede, 7 large populations of lava lizards were gradually fragmented into 37 populations. The initial fragmentation of each large population may have introduced interpopulational variation among the newly formed populations via founder effects. Subsequent genetic drift, particularly on the smaller islets, may have led to further differentiation of the new populations. In addition to these stochastic effects, the new populations probably differentiated in response to deterministic factors, such as natural selection, sexual selection, and phenotypic plasticity, that varied in intensity from island to island because of interisland variation in ecological conditions.

We developed and evaluated a series of predictions about the nature of these deterministic effects. The results suggest that the pattern of interpopulational variation in lava lizards is predictable from biogeographical data. In Fig. 7-16, we propose a biogeographical hypothesis for the evolution of interpopulational variation in lava lizards. Although our data support many elements of this hypothesis, it nevertheless remains speculative, and there are doubtless other evolutionary forces that affect variation in lava lizards (e.g., Jordan 1999).

Predation appears to have a central role in the evolution of lava lizards (Fig. 7-16). Body size, chin-patch length, and hindlimb length were all correlated with the diversity or distribution of lava lizard pred-

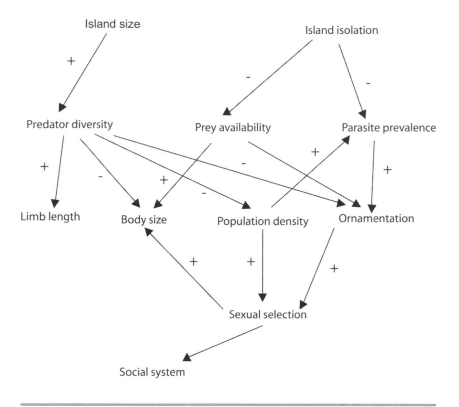

Figure 7-16 General hypothesis for evolutionary forces affecting lava lizard morphology and social behavior. Plus and minus signs indicate whether the effect is directly (+) or inversely (−) correlated with the trait. See text for discussion.

ators. Predation may also exert indirect effects on the behavior and morphology of lava lizards by influencing the density of these lizards and their parasites. Predation pressure was correlated with decreased population density, which may have led to changes in the social systems of lava lizards. Rates of parasitism were highest in dense populations, perhaps because of increased rates of parasite transmission. Parasite prevalence, in turn, was correlated with chin-patch length.

Ultimately, the distribution of predators in the Galápagos appears to be a function of island size, with small islets providing lava lizards with habitats that have reduced predator diversity. On the other hand, parasite diversity and food availability appeared to be a function of island isolation, with isolated islets having reduced parasite diversity and an absence of woody vegetation. These two biogeographical variables, island size and island isolation, are distributed in such a way in the Galá-pagos that we can recognize three distinctly different types of islands: large islands, diverse islets, and barren islets. The different types of islands present different types of ecological challenges to lava lizards. On large islands, high rates of predation, low population densities of lava lizards, intermediate rates of parasitism, and abundant food for lava lizards may create a selective environment in which natural selection for predator avoidance is stronger than sexual selection. As a result, lava lizards on large islands are relatively small and have long limbs and small chin patches. The social system on large islands appears similar to that of typical mainland iguanian lizards, with male territoriality and low, invariant mating success. On the other hand, on diverse islets, intermediate rates of predation, high population densities of lava lizards, high rates of parasitism, and high food availability may create a selective environment in which sexual selection is more important than natural selection. Lava lizards on diverse islets are larger and have short limbs, large chin patches, and a social system characterized by the breakdown of territoriality, with high, variable mating success. Finally, on barren islets, there is evidence that both natural and sexual selection may be important. Despite very low rates of predation and parasitism, population densities of lava lizards are intermediate, and body size is small, perhaps because of low food availability. However, chin patches on barren islets are larger than expected, and the social system appears to be similar to that on diverse islets, a degenerated form of the typical territorial system on large islands and mainlands.

Future Directions

The data presented herein took 4.5 people four seasons of full-time field work to collect. Despite these efforts, the dataset is meager with respect to the goals of the analyses. Indeed, every variable we measured could have been measured better, with larger sample sizes or more direct measurements. Furthermore, there are doubtless numerous other important variables that we failed to measure that have important roles in this system. However, we have described in broad strokes the pattern of morphology and behavior of lava lizards, as well as the ecological conditions experienced by lava lizards, across a large sample of islands. This pattern leads to numerous interesting questions, many of which can be addressed with discrete experiments on smaller subsets of islands. Here are some of these questions:

1. What factors contribute to male mating success on diverse islets? Our coarse description of space use and social behavior of lava lizards on diverse islets suggests that mating systems may be dramatically affected by high population density. Indeed, the mating system appears to be quite unlike the typical iguanian social system, and perhaps more like the scramble competition seen in explosively breeding anurans. Under these conditions, it is likely that sexual selection will be intense, and the phenotypic traits that contribute to male mating success may therefore be critical to male fitness.

2. Does the importance of intra- and intersexual selection change across the different island types? On large islands, there appears to be little opportunity for female choice, but male-male competition for territories containing females may be important. On islets, where females are overlapped by multiple males, female choice may be more important, whereas high population density may make the cost of territorial defense prohibitively expensive, thereby reducing the importance of male-male competition.

3. Can males on Española escape predation by snakes by attaining large body size? In six of seven island groups, lizards on large islands were smaller or not significantly different in body size compared with lizards on neighboring diverse islets. In the seventh group, Española, lizards on the large island were significantly larger than on all four satellite islets. This reversal suggests a different pattern of evolution on Española, and this difference merits further exploration, with the escape hypothesis as a possible starting point.

4. Are population density and body size of lava lizards on barren islets limited by food availability? Our data suggest that lizards on barren islets are smaller and less densely spaced compared with lizards on diverse islets. Many of the barren islets are tiny, and the potential for population-level manipulations of food availability is therefore high. Such manipulations could provide direct tests of the effects of food availability on growth and survival of lava lizards.

5. Are dead sea lions an important food resource for lava lizards? The dead sea lion on Santa Fé Islote appeared to cause a radical change in space use and behavior in all sex and age classes of lava lizards. How this may affect the survival, growth, and fecundity of lava lizards is unclear but merits further investigation.

Acknowledgments

We wish to thank Don Duszynski, Peter Grant, Diane Marshall, and Randy Thornhill for reviewing early drafts of the manuscript, and Mark Jordan and Don Miles for reviewing a later version. We also thank Brent Hinkle, Terry Moslander, and Robert Spencer for assisting with construction of home range maps. Numerous people contributed to data collection and camp living, including Sandra Andaluz, Denise Anderson, Lee Couch, Don Duszynski, Eduardo Espinoza, Lee Fitzgerald, Mark Jordan, Don Miles, Billy Schaedla, Don Sias, but especially Marco Altamirano and Mimi Wolok. Several people in Quito and the Galápagos deserve thanks for their logistical and moral support. In Quito, Alfredo Carrasco, Gonzalo Cerón, and the staff of the Charles Darwin Foundation made smooth what would otherwise have been a logistical nightmare. In the Galápagos, Arthur Lee, Linda Cayot, Pat Walsh, Fionnuala Walsh, Daniel Palacios, Gayle Davis-Merlen, Don Ramos, Chantal Blanton, the staff of the Charles Darwin Research Station, the staff of the Galápagos National Park Service, and the staff of the Rincón del Alma provided support. Especially important were Bernardo and Fermín Gutiérrez and the crew of the San Juan, a sight for sore eyes on many thirsty, lonely days. The Galápagos National Park Service provided permits for the field work portions of this research. This research was supported by grants from the University of New Mexico, the Latin American Institute, the Society for the Study of Amphibians and Reptiles, the Agency for International Development, the U.S. Man and the Biosphere Program, the National Science Foundation (IBN-9207895), FUNDACYT of Ecuador, and by TAME airlines and Prima.

Variation among Species

Introduction

George W. Barlow

Most people are convinced that science heightens our understanding of the world at an almost dizzying pace. You cannot argue with that conclusion, at least when the comparison is with prescientific progress. Think of the race to decipher the nature of the gene. A giant step forward. Later, the discovery of the polymerase chain reaction unleashed a massive effort to map the human genome and, more relevant to this book, provided a potent tool for creating molecular phylogenies. But between the first of these discoveries and the second, a substantial part of the twentieth century went by. Given that thousands of scientists were laboring in molecular biology during that time, that is a modest tempo.

Glacially slower, in comparison, have been fundamental advances in traditional evolutionary biology, in which the steps have been moderate for the most part, and for understandable reasons: the number of scientists conducting such research has always been many times smaller than that in biomedical research, and society has provided evolutionary biologists with far fewer resources. The subset of investigators analyzing the evolution of animal behavior is yet smaller—much smaller. Understandably, then, traditional evolutionary biology has not seemed to change much with time, at least at first glance.

Looking back, I see the field has nonetheless progressed, even when only through the gradual but steady accretion of findings, resolving now this explicit question and now that. As a result, issues that earlier lay in the realm of conjecture have acquired a solid empirical basis through systematic observations and well-planned experiments. At times, important ideas, large steps, have burst on the scene, and these have accelerated discoveries by providing new questions and with them new research paradigms. A quick review of some of these landmark ideas will help put into context the chapters in this section.

The modern study of animal behavior started with the writings of the incomparable Charles Darwin. Many of his insights in *The Expression of the Emotions in Man and the Animals* (Darwin 1872) became cornerstone concepts in the field of ethology, roughly 70 years later. Especially relevant to ethology were Darwin's interest in the form of behavior and its underlying motivation. This gave ethologists a springboard into the evolutionary origin of behavior, such as that used in signals.

The contributions of Konrad Lorenz, the father of ethology, arose from what was, for an evolutionary biologist, an unusual, though not unheard of, education. He got a medical degree in Vienna. But in those days a doctor's schooling included original research. For Lorenz, this meant delving into avian anatomy, how wings work and the likes, probably because he was already interested in birds and their evolution. His medical training biased him toward explanations of behavior that lay in neurobiology. A year abroad as a young student at Columbia University introduced him to the new field of genetics. He combined genetics with anatomy, physiology, and bird behavior and embarked on a comparative study of the spatiotemporal "morphology" of displays and their ornamentation, following the lead provided by C.O. Whitman (1919).

This branch of Lorenz's investigations produced a revised phylogeny of ducks based on their displays (Lorenz 1941) (I'll return later to discuss a problem inherent in his approach). Less important than his revised classification were the inferences he drew from the phyletic tree he constructed. He hypothesized evolutionary paths for the origins and modification of displays and their ornaments. The genesis of displays was explained in motivation models of conflicting behavior, the neurological bias, such as conflicting impulses in a goose both to attack and to flee from its opponent.

Others followed Lorenz's leadership, intensively studying certain taxa comparatively and reasoning how displays evolved (e.g., Kortlandt 1940; Tinbergen 1959; Nelson 1978). Regrettably, such comparative studies fell out of favor, though a few continued to be published (e.g., Zimmermann and Zimmermann 1990).

Those analyses all shared the comparative approach in the genuine meaning of the word: they concentrated on a group of related species, often a family, or as we see in some of the chapters to follow, a genus or a few kindred genera. That was in contradistinction to how comparative psychology was practiced during that period.

I remember well, in the 1950s or thereabouts, reading an article in which maze solving was compared in chickens, goldfish, and ants. The conclusion was that ants are the smartest of the three. Most biologists would not be surprised by that inference because they know that ants are adapted to solving a maze whereas the other two species are not. One wonders what was learned from that research. Certainly nothing about the evolution of behavior. Teasingly, Lorenz delighted in saying, "The *Journal of Comparative Psychology*, in which no comparative article has ever been published."

Strategically planned comparative studies illuminate the evolution of behavior. In the last decade, such studies have reemerged but now employ a more potent paradigm. Recall that Lorenz erected a phylogeny based on duck displays, then used that phylogeny to deduce evolutionary pathways. That tautological method is susceptible to pitfalls, in particular the difficulty of sorting out convergence and genetic relationship.

Far better is the current approach in which an independent phylogeny is first constructed, utilizing one of several molecular/statistical techniques (Stepien and Kocher 1997). Then the behavior is mapped onto that phylogeny. From this synthesis, one can infer the course of behavioral evolution, often revealing parallel and convergent evolution (Ryan 1994; Goodwin, Balshine-Earn, and Reynolds 1998).

The ethological approach to behaviorally constructed phylogenies differed in another way and one that appears to have been little noticed. In the early studies, the behavior used tended to be acts, often modal action patterns (Barlow 1977), such as the pioneering studies of Charles Carpenter on lizard displays (Carpenter 1978a, b). A few modern studies are still based on discrete behavioral traits, such as the structure of frog vocalizations (Ryan and Rand 1993).

Today the behavioral elements of choice are most often outcomes, also called consequences (Hinde 1970), of behavioral mechanisms. Some examples are dominance relationships, territoriality, home range, habitat occupied, spacing of retreats, and activity level. Outcomes are more remote from the genome than are the less variant stereotypical motor patterns (Dawkins 1982). Dominance relationships are a step further removed from their behavioral mechanisms because they result from the interactions of separate individuals. Therefore, talking about the heritability of behavior that exists only in the interaction of two self-contained individuals, who are separate genetic entities, becomes a controversial endeavor (Barrette 1987; Dewsbury 1990; Nol, Cheng, and Nichols 1996). Aggressive interactions provide illustrative and germane examples (Francis 1988).

Compared to consequences, therefore, behavioral mechanisms are more clearly developmental expressions of their genetic substrate and are characteristically conservative in an evolutionary sense (Barlow 1981; Kavanau 1990). Outcomes are typically more plastic, being continuously variable adjustments to changing environments while using the same behavioral tools, that is, patterned motor output. Consider territoriality, which may wax and wane with the economic defensibility of a resource (Emlen 1980; Carpenter 1987; Stamps and Krishnan 1998).

The two mechanisms producing territoriality are the tendency to remain and to fight (Tinbergen 1956). The decisions to stay on territory and the readiness to defend it fluctuate depending on a high-level assessment of the possible costs and benefits to the actor. The underlying behavioral mechanisms that carry out the decision, however, such as threat display, fight or flee, are invariant by comparison. Thus, when selecting elements of behavior for tracking their evolution in a phylogeny, behavioral acts should, in general, be a better choice than their consequences.

I oversimplify, of course. Small differences in behavioral mechanisms can change the likelihood that one or the other consequence will happen. A one-gene mutation can alter the form of a display just by making its expression more or less complete; the change is produced by an alteration of thresholds of the components of the act (Bastock 1956). The readiness to respond aggressively, a change in threshold, can vary greatly in range of expression from one individual to another within a species (Barlow, Rogers, and Fraley 1986), and that provides variation on which selection can act.

Across related species, the probability of territoriality might be higher in one species than in another as a consequence of small differences in the threshold of the readiness to respond aggressively. Thus, even though dominance per se is not heritable, the probability of an individual becoming dominant in a given environment can be acted on by natural selection (Dewsbury 1990). Social behavior can similarly be selected for indirectly. For instance, cooperative breeding in birds, which is the outcome of a suite of behavioral mechanisms, can characterize higher taxa (Edwards and Naeem 1993).

Evolutionary analysis is not restricted to comparative studies, of course. Adaptation is key to understanding the process of evolution (Williams 1966). In the early days of ethology, adaptiveness was assumed, and adaptive explanations tended to be circular, or just-so stories; in many instances they were covertly group selectionist (Huxley 1963; Lorenz 1966). Niko Tinbergen (1964) wrote that a particular behavior was done "for the benefit of the species." But that must have been penned in a moment of hurried thinking. During that time, he and his students were investigating gulls in the field, demonstrating the adaptiveness of individual differences in behavior (e.g., Tinbergen et al. 1962; Kruuk 1964; Patterson 1965).

This tack more clearly tied the evolution of behavior to its adaptiveness. The advance was augmented by David Lack's comprehensive research on the natural history of birds (Lack 1968). His student John Crook then extended the program to elaborate the leitmotif that bird social behavior and mating systems are shaped by their ecology (Crook 1964) (this is described in some detail in chapter 10). Crook's research is noteworthy for the way it melded the comparative and adaptive approaches.

The major selective forces acting on social behavior, as epitomized by mating systems, were identified as limiting resources and exposure to predation. When theorizing, attention tended to zero in on the distribution in space and time of resources such as food and places of refuge such as nest sites and escape refuges (Emlen and Oring 1977). Greatly simplified, males map onto the distribution of females. To a large extent, the mating system was regarded as a consequence of whether females were clumped or dispersed and to what degree.

As we will see in the chapters to follow, lizards offer what amounts to a constructive replication, hence a test, of these generalizations, which until recently derived primarily from studies of other vertebrates and insects. Refuges are a critical resource to several lizards, as chapter 9 demonstrates.

However, where lizards differ most conspicuously from other vertebrates resides in their need to tune their body temperature to some thermal optimum in the face of daily and seasonal swings in ambient temperature. This is especially the case for lizards dwelling in habitats with widely fluctuating and extreme temperatures such as characterize deserts of mid-latitudes, and elevated mountainous terrain.

Lizards attune their body temperature behaviorally, moving in and out of sun or shade, and positioning their bodies so as to either minimize or maximize the amount of radiant energy striking their bodies. They also adjust the extent of contact with the substrate to conduct heat to or from themselves. Their chosen body temperature is often near the upper range of heat that they can tolerate, even in extremely hot desert habitats (e.g., Heath 1965). Those lizards pursue life on the thermal edge, where a small increase in body temperature can quickly curb their activities, potentially permanently.

Lizards' social behavior, therefore, could be shaped by the way their thermal environment promotes or constrains their activities. When too hot, they are forced to forsake productive activities such as foraging, defending territory, or mating. When too cool, lizards become sluggish and vulnerable to predation (Smith 1997). Lizard behavior is hence squeezed by thermal boundaries that vary in space and time.

Even though so-called homeothermic animals are more thermally labile than is commonly appreciated, the extreme dependence of many lizards on thermal conditions sets them apart from other vertebrates, although some freshwater fishes exercise a degree of behavioral control over their body temperature (e.g., Barlow 1958). The thermal aspect of lizard behavior provides a certain glue when comparing species inhabiting different environments, and from different genetic lineages. The need to regulate body temperature is a recurrent theme in what follows, but now we need to examine another major force in the evolution of behavior in sexually reproducing animals.

The progression described so far has been much like that in more traditional areas of evolutionary biology. A new perspective, and a major step in the development of evolutionary theory, was the strict adherence to individual selection when explaining social behavior, as enunciated so clearly by G. C. Williams (1966).

Individual males and females have opposing reproductive interests (Williams 1975). Females bear comparatively few, large eggs, and the prudent female permits only the best male to fertilize them. Males have abundant sperm and seek to fertilize as many eggs as possible, so they need pay little regard to the quality of the mother. Sperm are readily available to females, so they need not compete aggressively with other females for sperm. Because they hold the trump card—the scarce eggs—females are the limiting sex.

Males have a different role in life. Because a single male can fertilize the eggs of an immense number of females, most males are superfluous. If an individual male is to reproduce, it must compete with other males for access to females. In that competition, some males fail to reproduce at all whereas others mate multiply. By contrast, all females have the capacity to reproduce. The result is sexual selection of males. This is expressed in two ways: a male displays his qualities to females to persuade them to choose him, and males aggressively exclude rival males, often in direct combat.

But, I oversimplify the large picture. When parental care is present, father and mother may come into conflict over how much care each will invest in the offspring, and that produces a variety of evolutionary scenarios (Maynard Smith 1977). Rarely, the male is even the limiting sex, and then sex roles are reversed (Vincent 1992; Swenson 1999). These complications do not arise in lizard social systems, which generally lack parental behavior, and that lessens the complexity of the analysis of the adaptiveness of their social systems.

One more complication needs mentioning, even though it is again a complication that probably does not arise among lizards. The emphasis on individual selection presented what at first seemed a paradox, the existence of cooperative behavior displayed at some cost to the actor. Here Hamilton developed the idea that individuals should confer benefits on other individuals in direct proportion to their degree of kinship (Hamilton 1964; see also Williams and Williams [1957]). This concept—inclusive fitness—is relevant only to animals that live in close contact with kin and nonkin and where they have the potential of conferring, or withholding, benefits from them.

This might seem to exclude lizards, but in one lizard the young can recognize the odor of their mother, and when given a choice, they select the burrow bearing the odor of kin (Lena and De Fraipont 1998). The young of other ectothermic vertebrates recognize kin and behave differently toward them (Waldman 1984; Sadler and Elgar 1994; Brown and Brown 1996; Pouyaud et al. 1999). Clustering of juvenile lizards is not typical, but when it occurs they might be expected to gather preferentially with kin.

The common lack of behavior obviously organized around kin helps further delimit the boundaries of lizard social behavior. As the authors remind us, one of the advantages of studying lizards is that some of the more complex aspects of social behavior do not arise to cloud the issues in lizards—only limited parental behavior and no known kin-related altruism, at least so far. The other themes mentioned here are strung through most of the four chapters to follow. Some authors emphasize this or that element, but the interlacing of themes is evident.

Chapter 8, by Diana Hews and Vanessa Quinn, manifests a research strategy somewhat different from the other chapters of part 3. In keeping with the other chapters, species of lizards are compared. Most are in the genus *Sceloporus,* the familiar fence lizards, but some are in the allied genus *Urosaurus*. The point of departure is the presence of color patches on the body (for now, head and throat coloration are ignored). The overarching goal is clearly integrative. The authors start with the presence or absence of sexual dichromatism, the patches of color, in different species.

Species differ in ways that suggest dissimilarity in masculinity and femininity. In some species, the male has a colorful patch, the female not, and the male is the more aggressive of the two. In another set of species, neither male nor female has the patch of color, and both sexes are relatively unaggressive. In the third situation, both sexes have colorful patches, and both are aggressive. Are color and aggression linked, perhaps two outcomes of, say, higher levels of testosterone? This looks like a wonderful natural experiment for investigating the relationship among sex hormones, dichromatism, and behavior.

The ultimate goal appears to be to connect differences in behavioral biology and dichromatism to differences in endocrinological substrates. Hormonal levels do not vary as much as one might have predicted, so differences in target-tissue receptivity become the focus. A step further along the path of integration, how does this relate to the organization of the central nervous system? This is an ambitious and synthetic program that raises many questions.

Chapter 9, by Paul Gier, is a close comparison of two species of iguanid lizards from separate genera. The plan followed was to select related species that occupy radically different environments. How do they differ in their use of the environment and how does that shape sexual dimorphism and their mating system? Sexual differences in the size and shape of head became the focus of the analysis of dimorphism, and I found myself thinking of antlers on deer as a parallel.

In this comparison, the key ecological resource for the Central American species appears to be refuges where females obtain some protection from predators. The refuges are clustered, enabling males to practice resource polygyny by defending territories that encompass the refuges of more than one female. Temperature is not a major constraint in this benign thermal climate, though basking sites play some role.

The other species occurs in the desert in our Southwest. The thermal regime there is harsh, reaching life-threatening high temperatures. At times, shade is crucial and the intense heat interrupts normal activity. The mating system is again polygyny, but in this species, rather than being territorial, males follow and defend females to secure matings.

More questions were raised than answered. The reality of cranial sexual dimorphism remains unresolved, as does its importance in dominance interactions. Fights in both species were rare but displays frequent. Might the displays be significant in female assessment of male quality?

In chapter 10, Stanley Fox and Paul Shipman report in detail their long-term study of seven species of *Liolaemus* living at different elevations in the Andes. The thermal regimes differ with elevation, as do many other aspects of the environments, such as exposure to predation and abundance of food. Some provocative patterns emerge, notably that the high-elevation species live in a mild thermal regime with abundant food and few predators. The low-elevation species occur in a harsher thermal environment with more predators and apparently less food. The species differ in aggressiveness. The "high" species are more abundant and less aggressive than the "low" ones. In addition, they are more interactive, suggesting a richer social system. The study is amply quantified and analyzed with regard to ecology and phylogeny, yet much remains to be done. The actual nature of the social systems is unknown, but the authors provide plausible hypotheses to test.

Chapter 11, by Jonathan Losos, Marguerite Butler, and Thomas Schoener, turns to one of nature's truly grand experiments—the repeated radiation of anoline lizards on Caribbean Islands, with different lineages converging on the same ecotypes. The chapter is limited to one aspect of sexual dimorphism: body size and shape. This is done in the context of adapting to different parts of the habitat, which is partitioned into vertical position in the forest and where on the tree, or in the grass nearby, a given species resides. Further, another potential cause of sexual dimorphism is introduced: males and females that might exploit different resources in the environment.

In a lovely demonstration of the role of the environment in shaping sexual dimorphism, these authors discovered that the type of habitat, not phyletic relationship, was the best predictor of size and shape in the two sexes. Teasing out the separate effects of overall size from relative size, to obtain a confident assessment of dimorphism per se, required extensive statistical analysis. Other studies on, for example, sexually dimorphic and dichromatic dewlaps are left for another day.

These chapters offer a rich pallette of research questions for future investigations. As John Maynard Smith (1989, p. 179) wrote, "More scientists are held up in their research by a shortage of ideas to test than by an inability to test the ideas they have got."

CHAPTER 8

Endocrinology of Species Differences in Sexually Dichromatic Signals

Using the Organization and Activation Model in a Phylogenetic Framework

Diana K. Hews and Vanessa S. Quinn

Many animals have conspicuous social signals. Often these signals are expressed in one sex and function in the context of mate choice, intrasexual competition, or both (Andersson 1994; Bradbury and Vehrencamp 1998). A more complete understanding of sex-specific signals will come from integrative studies within a phylogenetic context (Ryan, Autumn, and Wake 1998). Integrative studies document the action of natural and sexual selection on signalers and receivers; determine the mechanistic basis of signals, signal perception and processing; and use historical perspectives to ask how sensory systems evolve (Endler 1992; Ryan 1997). Although much is known about ecological and evolutionary aspects of sexually selected traits (Andersson 1994), considerably less is known about their developmental basis and how selection has acted on these developmental mechanisms.

This chapter focuses on our proximate physiological work. Specifically, we examine sex steroids and how they control both sexual signaling morphology and signaling behavior. We focus on endocrine mechanisms because sex steroid hormones play fundamental roles in the development and expression of sexual differences in vertebrates (Becker, Breedlove, and Crews 1992). Evolutionary biologists increasingly recognize the key role endocrine systems can play in the expression of correlated suites of life history traits (e.g., Moore 1991; Moore 1995; Mousseau and Fox 1998; Sinervo and Svensson 1998). In particular, the endocrine system can result in the coupling of display morphology and display behavior.

Currently, we are assessing the roles of sex steroid hormones, their receptors, and their cellular actions in mediating expression of a sexual color signal and of the behavior involved in signaling. Our long-term goal is to study key species pairs in the lizard genus *Sceloporus* that vary in the degree of sexual dichromatism in a signaling trait. The signaling trait we study is a pair of large, brilliant, blue patches of abdominal skin on an otherwise white background. The patches are exposed to conspecifics in stereotyped postural displays during aggression and courtship in *Sceloporus* and in some species of its sister genus, *Urosaurus* (tree lizards) (Fig. 8-1). Differences in expression of abdominal patches in *Sceloporus* appear to covary with differences in aggression (Vinegar 1975; Quinn and Hews 2000; Hews and Benard, in press). We have begun our endocrine work by focusing on two closely related species and a third more distantly related species of *Sceloporus*. When results are available for multiple species within a clade it will be possible to apply statistically rigorous phylogenetic analyses, such as independent contrasts (Martins and Hansen 1996). The goal will be to determine whether there are common endocrine mechanisms underlying independent evolutionary state transitions of sexual dichromatism.

There is great potential with *Sceloporus* lizards for using explicit comparative methods to examine evolutionary patterns in the mechanistic bases of sexually dimorphic signaling traits. The group has a well-corroborated phylogeny (Reeder and Wiens 1996; Wiens and Reeder 1997), with a number of independent losses of sexual dichromatism in abdominal patches, due to either loss of the patches in males or gains of them in females (Wiens 1999). Thus, our work highlights a poorly studied but potentially important aspect of sexual selection: the evolutionary *loss* of sexual signals. Phylogenetic studies suggest that such losses may be common in a number of vertebrate clades (Peterson 1996; Price and Birch 1996; Omland 1997; Burns 1998; Wiens 1999). Determining the evolutionary forces producing losses and the various physiological bases of such evolutionary change will increase our general understanding of the evolution of sexually dimorphic signals (Emerson 1994, 1996; Reynolds and Harvey 1994).

Sexual Dimorphism in Color Signals and Aggressive Behavior in *Sceloporus* Lizards

The blue abdominal patches are involved in sex recognition and in aggression (Cooper and Burns 1987; Quinn and Hews 2000), develop with sexual maturation, and show little seasonal variation in their expression. Phylogenetic analyses (Wiens 1999) strongly support the hypotheses that (1) sexual dichromatism in the abdominal patches is the ancestral state in this genus, and (2) that monochromatism represents evolutionary loss

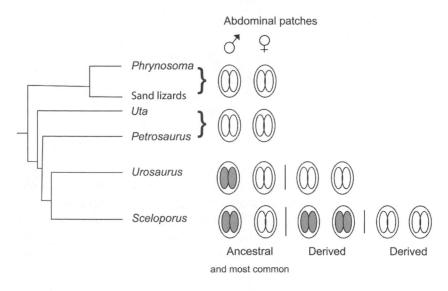

Abdominal patches

Figure 8-1 Simplified phylogeny of *Phrynosomatidae* (based on Reeder and Wiens [1996]). For each genus (or the clade of *Phrynosoma* and the sand lizards: *Uma, Callisaurus, Cophosaurus, Holbrookia*) the adjacent illustrations indicate character states for abdominal coloration of males and females (based on Wiens [1999]). Variation within a genus, such as in *Sceloporus*, is indicated by the presence of more than one set of male-female character states. Solid ovals = blue abdominal patches present; open ovals = blue abdominal patches absent.

of sexual dichromatism in the abdominal patches, either by trait gain in females or by trait loss in males. There are at least 10 independent evolutionary losses of abdominal patches in *Sceloporus* males, and 8 independent gains of them in females. *Sceloporus* males exhibit high levels of territorial aggression compared with females (reviews in Stamps [1977b, 1983b], Carpenter [1978a], and Martins [1994]). Thus, territorial aggression is typically sexually dimorphic in this genus. The behavior patterns involved in aggressive displays are well described for many *Sceloporus* species (e.g., Carpenter 1978a; Ruby 1978; Moore 1987). Growing evidence (discussed later) suggests that increased aggression and occurrence of the abdominal patches covary in this genus.

Our research focuses on three *Sceloporus* species: the eastern fence lizard (*S. undulatus consobrinus*), the striped plateau lizard (*S. virgatus*), and the mountain spiny lizard (*S. jarrovii*). Each represents one of three possible patterns of sexual dichromatism in signaling traits in *Sceloporus* (Wiens 1999). The most common pattern is *sexual dichromatism* in abdominal coloration (Plate 11; *S. undulatus*). Only males exhibit territorial aggression, and only males have the blue abdominal patches. In a second pattern, *monochromatic feminized*, neither sex has the blue abdominal patches, and for at least one species (*S. virgatus*), rates and intensities of territorial aggressive behavior patterns are reduced in males (Vinegar 1975; Quinn and Hews 2000; Hews and Benard, in press), and male-male home range overlap is the highest documented (74%) for any *Sceloporus* species (Abell 1999b). The sister species to the clade that contains only *S. virgatus* and *S. exsul* (both of which are species with loss of blue abdominal patches in males) is *S. undulatus* (Wiens and Reeder 1997). Thus, the monochromatic-feminized species and the sexually dichromatic species on which we focus our research are very closely related. Finally, in the third pattern, *monochromatic masculinized*, both males and females have blue abdominal patches, and for at least one species (*S. jarrovii*), both sexes exhibit territorial aggressive display behavior. Females use the patches during aggressive female-female territorial encounters that involve the same stereotyped display postures used by males (Ruby 1978; Moore 1987; Woodley and Moore 1999a, b).

Sceloporus species vary in other aspects of ventral coloring. In addition to the abdominal patches, there are throat patches that also exhibit sex and species differences. Throat patches vary in color intensity and size (Stebbins 1985). Even within a species, populations of *S. occidentalis* exhibit extensive variation in the intensity and amount of blue color, both in abdominal patches and in throat patches, and in the degree to which sexes differ in these traits (Camp 1916). Thus, many *Sceloporus* species differ interspecifically as well as intraspecifically in other aspects of dorsal coloration and throat patterning (see also Wiens and Reeder [1997] and Wiens, Reeder, and Montes de Oca [1999]). It is important to recognize that our terminology *monochromatic* or *sexually dichromatic* refers only to the presence or absence of the blue abdominal patches. Variation in the size and intensity of coloration of the abdominal patches and of the throat patches (including presence/absence) also deserves careful study.

In addition to this evolutionary diversity in the occurrence of sexual signal dichromatism and the well-corroborated phylogeny, other features recommend *Sceloporus* as a group for integrative comparative studies of sexual signals. In many phrynosomatid lizards, territoriality and courtship are central to male reproductive success (e.g., Hews 1990, 1993; Abell 1997; see also chapters 1, 3, and 9). Lizards are excellent species for field studies of territorial aggression (e.g., Hews 1993; see also chapters 1, 5, 7, 9, and 10). Lizards are also excellent subjects for hormonal studies in the laboratory (reviews in Crews and Greenberg [1981] and Moore and Lindzey [1992]; Hews, Knapp, and Moore 1994) and field (e.g., Marler and Moore 1988; DeNardo and Sinervo 1994; Moore, Hews, and Knapp 1998). Unpublished electroretinograms and microspectrophotometry studies (E. Loew and L. Fleishman, personal communication) suggest that *Sceloporus* lizards have four cone types and good hue discrimination well into the blue end of the wavelength spectra. Finally, there is a wealth of behavioral, ecological and evolutionary research on this genus (Sites et al. 1992). In particular, a variety of functional studies have focused specifically on blue coloration in *Sceloporus* (e.g., Vinegar 1975; Cooper and Burns 1987; Cooper and Greenberg 1992; Dixon 1993; Abell 1997, 1998a, b, c, 1999b; Quinn and Hews 2000; Wiens 2000; Hews and Benard, in press).

In sum, *Sceloporus* offers an excellent array of features for a comparative neuroendocrine study of sexual dimorphism. Some key features are as follows:

1. Loss in males or gain in females of sexually dichromatic signaling traits
2. Reduction of sexual dimorphism in signaling behavior (via increased aggression in females or reduced aggression in males)
3. A well-corroborated phylogeny
4. A substantial body of ecological research providing a rich context for interpreting comparative studies
5. The ability to conduct detailed adult and neonatal endocrine studies including those that manipulate steroid hormone profiles in adults or hatchlings

Endocrine Studies of Species Variation in Sexual Dimorphism of Signaling Morphology and Signaling Behavior

In the following sections, we present an overview of the approaches we are using to study how hormonal controlling mechanisms have evolved in this signaling system as well as an overview of the endocrine regulation of sexual differentiation in vertebrates with a particular focus on reptiles. Then, after providing a brief background on reproduction in *Sceloporus*, we present a series of questions we are asking, as well as some of our findings, about the effects of sex steroid hormones on the development and expression of the signaling trait in *Sceloporus*. We then turn to a discussion of brain regions mediating aggression in vertebrates. We overview our first endocrine studies related to the signaling behavior—aggressive display—focusing on describing the distribution of androgen receptors (ARs) in the brains of males and females. Our endocrine work is only in the beginning stages, but we feel that this overview of our preliminary results provides a useful illustration of a research program that seeks to integrate endocrine approaches in an evolutionary context.

Organization and Activation: The Endocrine Basis of Sexual Differentiation

Sex steroid hormones in vertebrates are central to the development of sexual differences, a process known as sexual differentiation (Becker, Breedlove, and Crews 1992). The organization and activation hypothesis has provided a successful construct in elucidating the role of sex steroid hormones in sexual differentiation (Phoenix et al. 1959; Arnold and Breedlove 1985; Kelley 1988). This hypothesis proposes that sex steroid hormones affect sexual differentiation by *organizational* effects, which are permanent and occur early in life during a discrete critical period, and by *activational* effects, which are temporary and occur in adults. These two modes of hormone action can be thought of as representing extremes on a continuum (Arnold and Breedlove 1985). Traits may require only one or the other type of action, but often traits require both organization and then later activation for complete sexual differentiation. Although the generality of this paradigm is being reassessed (e.g., Crews 1993; Arnold 1996; Kendrick and Schlinger 1996; Wade 1999), it remains a powerful guide for endocrine research on sexual differences.

Sexual differentiation in reptiles appears to follow the basic paradigm for tetrapod vertebrates, involving both organizational and activational effects with resulting dimorphism in brain, behavior, and morphological traits (Crews 1985; Crews and Silver 1985; Adkins-Regan 1987; Moore and Lindzey 1992; Hews and Moore 1995; Godwin and Crews 1997; O'Bryant and Wade 1999; Rhen and Crews 2000). There is a body of endocrine work on reptiles with temperature-dependent sex determination such as the leopard gecko (*Eublepharis macularius*) (Tousignant and Crews 1994, 1995; Rhen and Crews 2000) and on bisexual and unisexual whiptail lizards (*Cnemidophorus* species; reviewed in Godwin and Crews [1997]). However, fewer studies have examined early organizational effects of sex steroid hormones on sexual differentiation for the more typical species with genetic sex determination, such as *Sceloporus*. Nevertheless, studies on tree lizards (*Urosaurus ornatus*) (Hews and Moore 1995) and sexual whiptails (*Cnemidophorus inornatus*) (Wade, Huang, and Crews 1993) indicate that expression of male-typical traits requires androgen exposure during development. Our *Sceloporus* work provides a detailed examination of sexual differentiation of a perhaps more typical bisexual species with genetic sex determination.

Several fundamental endocrine mechanisms that have been found to contribute to sexual differences are likely to contribute also to interspecific variation in sexual differentiation of color and aggression. First, different species could vary in sexual differences in levels of circulating hormone, or in the timing of elevations in plasma hormone concentrations at critical periods in ontogeny, thus affecting organizational actions of the hormone (or hormones). Second, species could vary in sex-specific sensitivity to circulating steroid hormones. Such differences can be owing to tissue-specific differences in the hormone receptors (e.g., distribution, abundance, or regulation) or in the genetic regulation of trait expression (e.g., the trait is no longer under hormonal control). Third, species could vary in sex-specific activity of key metabolic enzymes (aromatase, 5α-reductase) that convert testosterone to other biologically active hormones (17 β-estradiol, 5α-dihydrotestosterone, respectively). Besides mediating differentiation of the color signal, these differences in receptors and metabolic enzymes could themselves result from earlier sexual differences in steroid hormone profiles. Our research examines these endocrine attributes, all of which could contribute to the sex and species differences observed in *Sceloporus* lizards.

Evolutionary Endocrine Studies of Sexual Differences

A growing body of work examines the endocrine mechanisms underlying species differences in sexual signals. Pioneering neuroendocrine work identified anatomical brain dimorphism in vocal control regions that correlated with sexual differences in singing behavior among bird species (Brenowitz and Arnold 1985; Arnold et al. 1987). Brain sexual dimorphism in the distribution of steroid hormone receptors and in the metabolic activities of key brain regions correlates with species differences in pseudocopulation and copulation behavior in the unisexual whiptail lizard and its sexual congener (*Cnemidophorus uniparens* and *C. inornatus*, respectively) (Crews, Wade, and Wilczynski 1990; Wade and Crews 1991a, b, 1992; Godwin and Crews 1997). Seasonal variation among some bird species in sexual dimorphism in sexual behavior correlates with male-female differences in plasma testosterone levels of adults (Wingfield 1994) or differences in other hormones (Kimball and Ligon 1999). Evolutionary alterations in placental endocrine enzymes apparently contribute to the extraordinary masculinization in morphology and behavior of female spotted hyenas (*Crocuta crocuta*) (Glickman et al. 1992). Variation in behavior patterns that define mating systems (e.g., pair-bond behavior; parental behavior) is correlated with species differences in the abundance of receptors for the peptide hormone arginine vasopressin in several vole species (*Microtus* spp.) (reviewed in Young, Wang, and Insel [1997]). Transgenic mice with the vasopressin receptor from a monogamous vole species exhibited increased vasopressin-induced affiliative behavior, which characterizes the monogamous pair-bonded voles (Young et al. 1999). These studies collectively suggest that endocrine mechanisms underlie much of the evolutionary variation in sexual differences (but see also Arnold [1996]).

However, little of this work on naturally occurring variation among species in sex differentiation has used an explicit phylogenetic context, and most other studies have not focused on sexually selected behavior and signaling traits. Emerson and colleagues provide a notable exception (Emerson, Rowsemitt, and Hess 1993; Emerson 1996), examining acoustic signaling, combat behavior, and morphological traits used in fighting within a clade of the frog genus *Rana*. The loss or diminution of these seasonally activated male traits correlates with reduced plasma androgen levels. Similarly, Staub and coworkers (Dempsey, Reilly, and Staub 1996) have initiated a hormonal study of sexual differences in aggression and morphology in plethodontid salamanders (*Aneides*), and measure levels of several hormonal parameters that could mediate these differences, including plasma concentrations of sex steroid hormones and the abundance of hormone receptors in the dimorphic target tissue (jaw musculature). Sexual dimorphism in the distribution of brain receptors and metabolic capabilities may explain sex and species differences in behavior in bisexual and unisexual *Cnemidophorus* (reviewed in Godwin and Crews [1997]). Although it is not known for anurans if color functions in sexual signaling, Hayes (1997) provides an example of how hormonal mechanisms may constrain the evolution of sexual differences. He details specific endocrine mechanisms regulating pigmentation and proposes that they might limit the evolution of sexual dichromatism in anurans (Hayes 1997). Birds, however, are known to recognize and respond to sexual differences in coloration (Andersson 1994), and substantial endocrine work has examined plumage dichromatism. Kimball and Ligon (1999) surveyed studies on the hormonal control of sexually dimorphic plumage in four major avian orders (Galliformes, Anseriformes, Charadriiformes, and Passeriformes). They found that three major endocrine mechanisms (estrogen dependence, testosterone dependence, luteinizing hormone dependence) and one nonendocrine mechanism (strict genetic control) were responsible for the plumage dichromatism, with little species variation in the mechanism within each order (studies on a total of 26 species were analyzed). Note that sexual differences in color signals often are *not* mediated by endocrine mechanisms (e.g., birds, Owens and Short [1995]). Thus, although there are comparative studies of species differences in sexual signals, there is clearly a need for more studies with the explicit aim of examining the evolution of endocrine mechanisms of sexually dimorphic signaling traits in a phylogenetic context.

Androgen Dependence of Aggression in *Sceloporus*

One of the common themes resulting from endocrine work in many vertebrates is that aggressive behavior is often mediated by androgens. Thus, a starting point for the endocrine study of species differences in aggression in *Sceloporus* is an examination of the evidence that such behavior is androgen mediated. Such work includes documenting seasonal expression of the behavior, which suggests a correlation with seasonal elevation in plasma androgen levels that is often associated with spermatogenesis.

There has been considerable work on the neuroendocrine basis of behavior of selected reptiles (reviews in Moore and Lindzey [1992] and Godwin and Crews [1997]), and on *Sceloporus* lizards in particular. Territorial aggressive behavior is seasonally expressed in most, if not all, of the North American *Sceloporus* species. These *Sceloporus* lizards are seasonal breeders, and seasonal variation in intensity of aggression has been well described for males of all three species that we are studying, and for females of the masculinized species, *S. jarrovii*. Here is what we know about *Sceloperus* species:

TYPICAL DICHROMATIC SPECIES

An association between seasonal changes in aggression and plasma androgens occurs in many populations of the "typical" species, *S. undulatus*

(McKinney and Marion 1985; Klukowski and Nelson 1998; Smith and John-Alder 1999).

MONOCHROMATIC SPECIES WITH MASCULINIZED FEMALES

Considerable work in *S. jarrovii* has documented seasonal steroid hormone profiles and correlations with elevated levels of territorial aggression in adult males (Ruby 1978; reviews in Moore and Lindzey [1992]) and females (Woodley and Moore 1999a, b). Experimental work with male *S. jarrovii* has identified androgen-dependent and androgen-independent components of territorial aggression in males (reviewed in Moore and Lindzey [1992]). Correlative and experimental work (Woodley and Moore 1999a, b) suggests a potential role for androgens in aggression of these masculinized females. Because this and several other monochromatic *Sceloporus* species with masculinized females are ovoviviparous, masculinization could result from exposure to sex steroid hormones in the maternal environment (e.g., vom Saal 1979). This hypothesis is currently being explored for *S. jarrovii* (D. Painter and M. C. Moore, personal communication). However, other monochromatic *Sceloporus* species with masculinized females also include oviparous forms, such as *S. occidentalis taylori* (California) and *S. undulatus tristichus* (New Mexico, Colorado). Thus, there may be substantial evolutionary variation in the endocrine mechanisms underlying such masculinization of females within *Sceloporus*.

MONOCHROMATIC SPECIES WITH FEMINIZED MALES

Only one of the monochromatic species with feminized males, *S. virgatus*, has been studied. In this species, seasonal elevations in plasma androgen concentrations in males (Abell 1998c) correlate with the single bout of breeding activity and territoriality in late May and early June (Vinegar 1975; Rose 1981; Abell 1998b).

Species Variation in Actions of Sex Steroid Hormones in *Sceloporus*

For some vertebrates, adult sexual differences in circulating androgen levels are involved in adult differences in trait expression. Often such adult activational actions of hormones involve sexually dimorphic substrates that also require early organizational effects of sex steroid hormones in perinatal stages of development. For example, in many male passerine birds, seasonal increases in plasma androgens in males mediate breeding-season territorial singing, and the brain regions involved in song production require early masculinization (Nelson 2000).

Examination of the relative importance of organizational versus activational effects of sex steroid hormones requires studies that measure plasma concentrations of hormones in adults and in hatchlings, as well as experiments that manipulate hormone levels in both.

DO SEX AND SPECIES DIFFERENCES IN EXPRESSION OF ABDOMINAL COLOR PATCHES CORRELATE WITH ADULT DIFFERENCES IN SEX STEROID HORMONE PROFILES?

One endocrine hypothesis to explain sex and species differences in abdominal patches in *Sceloporus* is that, as adults, individuals with abdominal blue patches have higher androgen levels. That is, the expression of these traits differs because of differences in activational actions of androgens. However, unlike plumage in many passerine birds, the occurrence of abdominal blue patches in *Sceloporus* lizards does not decline in the nonbreeding season; the patches are expressed year-round. There may be some seasonal variation in the intensity of the blue hue and/or the size of the blue patch, but there are few studies directly examining this possibility.

The lack of seasonal change in this male-typical trait suggests that plasma androgen levels might not correlate with occurrence of the color patches, when comparing males of species with and without the trait. Indeed, published studies on hormones of various *Sceloporus* species provide provisional support for this conclusion. For example, during the breeding season, male *S. virgatus,* the species with male loss of the abdominal blue patches, have androgen levels similar to breeding levels of androgens in congeners (Abell 1998c; Abell and Hews 1999). Hormone manipulations in adults are also ineffective in altering the sexually dichromatic expression of the trait. The sexual difference in abdominal blue patches is not altered by castration of adult males or by androgen implants in adult females (Kimball and Erpino 1971; Moore 1987; Rand 1992; John-Alder et al. 1996). Thus, data on circulating hormone concentrations and results of adult hormone manipulations both are consistent with the hypothesis that sexual differences in the patches do not result from differences in activational actions of sex steroid hormones.

IS SEXUAL DICHROMATISM IN THE ABDOMINAL SIGNAL IN *S. UNDULATUS* MEDIATED BY EARLY ORGANIZATIONAL ACTIONS OF ANDROGENS?

The sexual differences in abdominal blue patches could arise from differences not in adult plasma levels of sex steroid hormones, but in levels in the hatchlings. Thus, there could be differences in the organizational actions of hormones. Sexual differentiation of peripheral traits (i.e., outside the central nervous system [CNS]) in male vertebrates is often due to the 5α-reduction of testosterone (T) to dihydrotestosterone (DHT) (McEwen, Luine, and Fischette 1988; Becker, Breedlove, and Crews 1992). DHT is a biologically active metabolite of T in many vertebrates, and the enzyme that converts T to DHT is 5α-reductase. Female-typical plumage in males of the domestic Sebright breed of chickens involves a mutation altering expression of this enzyme (George, Nobel, and Wilson 1981).

This enzyme that converts T to DHT is implicated in the sexual differentiation of blue abdominal patches in male tree lizards, *Urosaurus ornatus*, a species in the sister genus to *Sceloporus*. Castration of hatchling males abolishes expression of the trait, indicating the organizing role of androgens (Hews, Knapp, and Moore 1994). Females will express the male-typical abdominal patches only if given DHT as intact hatchlings but will not express the trait if given T at this age or given either androgen as adults (Hews and Moore 1995). Thus, given the role of DHT in the expression of blue abdominal patches in tree lizards, we focused on the hypothesis that DHT is also necessary for hormonal organization of blue abdominal patches in *Sceloporus*.

The endocrine mechanisms underlying sexual differentiation of the abdominal blue patches may be the same for sexually dichromatic taxa in the sister genera *Urosaurus* and *Sceloporus*. Specifically, our results (Quinn and Hews, unpublished data) indicate that organizational effects of DHT are also central to the expression of this sexually dichromatic trait in *S. undulatus*. Both DHT and T influenced the expression of traits that are activated by androgens in adult lizards (Fig. 8-2A, cloacal gland secretions in the tail base and femoral gland secretions). In addition, as was found for tree lizards, only hatchling *S. undulatus* females given intraperitoneal DHT implants expressed the male-typical abdominal patches (Fig. 8-2B).

Blood levels of androgens in hormone-implanted animals must be assayed to confirm that plasma DHT levels due to the implants were physiological and not pharmacological. Similarly, a study determining the effects of castration on male hatchlings remains to be conducted. However, these data support the hypothesis that the organizational role of DHT in the expression of the male-typical signaling morphology in these two sister genera, *Urosaurus* and *Sceloporus*, are the same.

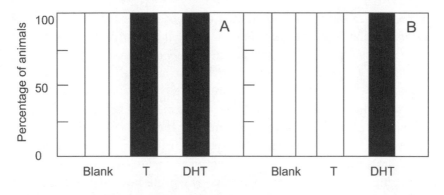

Figure 8-2 Effects of long-lasting androgen implants on expression of male-typical traits in hatchling *Sceloporus undulatus* females when implanted 20 d posthatching. (*A*) Cloacal glands and femoral pores. These are activated by androgens in several lizard species. (*B*) Blue abdominal patches. Open bars = female-typical trait expression; solid bars = male-typical trait expression; blank = empty control implant; T = testosterone implant; DHT = 5α-dihydrotestosterone implant. *n* = 4–6 per treatment group.

ARE DIFFERENCES IN THE SEXUAL DICHROMATISM OF ABDOMINAL COLORATION BETWEEN THE SISTER SPECIES *S. UNDULATUS* AND *S. VIRGATUS* MEDIATED BY SPECIES DIFFERENCES IN THE ORGANIZATIONAL ACTIONS OF ANDROGENS?

One hypothesis for the lack of abdominal patches in *S. virgatus* involves differences in organizational effects of androgens. This hypothesis proposes that blue patches are not organized by DHT in *S. virgatus* because levels of circulating androgens in hatchling *S. virgatus* differ from those in young males of the dichromatic *S. undulatus* during the time period when androgens affect sexual differentiation of the trait. There are currently no data on hormone profiles of hatchlings for any *Sceloporus* species during this stage of ontogeny. However, some data indirectly support the hypothesis that the *timing* of a critical period appears similar, at least when it ends. Implants of T given to intact male *S. virgatus* at 40 d posthatching had no effect on abdominal coloration (Abell 1998a). This posthatching age is when abdominal blue coloration just begins to be expressed in some males in two other dichromatic species (*U. ornatus*, Hews and Moore [1995; 1996]; *S. undulatus*, Hews, unpublished data). This would suggest that, like *U. ornatus*, the critical period for this trait (if it were to occur in *S. virgatus*) is earlier than day 40. Specifically, it is possible to have a tissue's fate (blue, white) determined earlier in ontogeny by a hormonal effect but have the differentiation take place later in ontogeny. For example, the organizational effect of the hormone may be to increase the number or type of hormone receptors in target cells. Then, when higher levels of the hormone (or hormones) arise later, perhaps during puberty as in *U. ornatus*, the hormone (or hormones) act and the skin cells are altered, causing the permanent expression of factors that produce the blue coloration. Thus, if the process of sexual differentiation seen in *U. ornatus* also occurs in *Sceloporus*, then intact male *S. virgatus* given long-lasting T implants early after hatching (during the time when the fate of the abdominal skin patch is determined) should develop the abdominal patches. We are currently testing this hypothesis with such a manipulation. If this manipulation produces blue abdominal patches in *S. virgatus* males, it would suggest that the evolutionary loss is the result of decreased circulating levels of T during the critical period.

Another endocrine hypothesis about the evolutionary patch loss in *S. virgatus* concerns 5α-reductase activity in the abdominal skin. Recall that this enzyme converts T into DHT, and that in *U. ornatus* expression of blue abdominal patches requires organization by DHT. We are currently rearing hatchling *S. virgatus* to test this reductase hypothesis. If hatchlings given DHT implants develop blue abdominal patches, and those given T implants do not, it would suggest that the evolutionary loss of the patches is due to decreased activity of 5α-reductase in the skin during the critical period.

Endocrine manipulations can reveal if hormones affect sexual differentiation. To confirm that this naturally occurs, one must verify that hatchlings of the sexes and species with different abdominal patches naturally have different levels of the endocrine attributes (hormone levels, enzymes that metabolize hormones, hormone receptors). For example, our preliminary results indicate that DHT implants given to female *S. undulatus* hatchlings result in expression of blue abdominal patches, explaining the sex difference in the expression of this trait in this dichromatic species. It is therefore important to establish that there are indeed endogenous endocrine differences. If androgen manipulations alter the expression of abdominal patches, then one would predict that there are naturally occurring differences in circulating androgens at the ontogenetic stage when androgen manipulations were successful. Several major endocrine differences could contribute to endogenous differences in exposure to DHT. For example, hatchlings could differ in circulating levels of T and/or in the activity of 5α-reductase, and thus in levels of DHT.

Species Variation in Target Tissue Sensitivity to Hormones: ARs

Another attribute of the endocrine system that could also contribute to these sex and species differences among *Sceloporus*, both in signal morphology and in signaling behavior (aggression), is sensitivity to androgens. Sensitivity to a hormone could vary because of differences in either the abundance or nature of the AR, or because expression of the trait is no longer hormone dependent. Mutations in ARs are associated with a variety of clinical disorders in humans (McPhaul et al. 1993), and they are also known for strains of mice (e.g., Freeman et al. 1995). Mutations in the ARs affect levels of aggression in rodent strains (e.g., Simon and Whalen 1986).

Similarly, ARs could also vary among species and affect trait expression. For example, ARs could differ in occurrence and abundance in particular targets (e.g., Kelley et al. 1989; Boyd et al. 1999). Alternatively, species that differ in trait expression may have mutations in the AR gene that result in differences in receptor function, altering the binding specificity or affinity. In addition, work on other steroid hormone receptors reveals that, within a species, there can be multiple forms of a steroid hormone receptor. For example, in the rough skinned newt (*Taricha granulosa*, Orchinik, Murray, and Moore [1991]) and in the house sparrow (*Passer domesticus*, Breuner and Orchinik [1999]), there are at least two glucocorticoid receptors. In each species, one form is a fast-acting membrane-associated receptor and one is a slower-acting intracellular receptor. Thus, differences between species in selection acting on such variation in receptor populations could produce differences in trait expression.

Our research group is beginning to describe the distribution of ARs in the brain of *Sceloporus*. In the near future, we will begin work to examine the abundance of ARs in abdominal skin. In the following section, we provide a brief review of brain regions mediating aggression in vertebrates in general, and the evidence supporting this role for these regions in reptiles, specifically. In the next section, we present some of our initial results from a study determining the distribution of ARs in the brain of *S. undulatus*. Describing the brain distribution of ARs in this sexually dichromatic species is the first step in exploring the differences in aggression among *Sceloporus* species.

WHAT BRAIN REGIONS MEDIATE AGGRESSION?

In tetrapod vertebrates, several brain areas appear to be key mediators of aggression, and these areas are being examined in our research. In a variety of birds and mammals, studies using lesions or electrical stimulation have implicated the hypothalamus and parts of the limbic system (hippocampus, septum, amygdala) as important brain regions involved in aggression (Albert and Walsh 1984; Crews and Silver 1985; Albert et al. 1990; McGregor and Herbert 1992). Androgens often act in these particular brain regions and affect aggression. For example, in castrated male ring doves (*Streptopelia risoria,* Barfield [1971]) and quail (*Coturnix c. japonica,* Watson and Adkins-Regan [1989]), androgen implants in the preoptic area of the anterior hypothalamus stimulate aggression.

In reptiles, the basal portion of the dorsoventricular ridge is considered to be homologous to the mammalian amygdala (although the terminology is somewhat contradictory when comparing neuroanatomical studies of various lizards (cf. Peterson 1980; Propper, Jones, and López 1992; Wade 1997). This view has recently been upheld and clarified by a detailed phylogenetic analysis of the connections of amygdalar subnuclei in representative tetrapod taxa (Bruce and Neary 1995). Comparison among representatives of several major vertebrate groups revealed that there are similarities in homeobox genes that are expressed during brain development in specific brain regions (Fernandez et al. 1998). These comparisons clarified relationships among vertebrates in their respective telencephalic subdivisions (Fernandez et al. 1998), and these telencephalic relationships concord with the homologies proposed by Bruce and Neary (1995).

Functional experiments in reptiles support the role of the amygdalar region (basal dorsoventricular ridge) in the expression of aggression (Peterson 1980), a role that is a conserved functional homology with other tetrapod vertebrates. Lesions in the amygdalar area in caimans (*Caiman crocodilus*) decreased aggression (Keating, Korman, and Horel 1970). Bilateral lesions of the amygdalar area in western fence lizards (*S. occidentalis*) abolished the aggressive responses of dominant males to conspecifics (Tarr 1977). In the collared lizard (*Crotaphytus collaris*), aggressive and defensive postures were elicited with electrical stimulation of the anterior dorsoventricular ridge, amygdaloid complex, septal and preoptic areas, hypothalamus, thalamus, regions adjacent to the nucleus profundus mesencephali, and the reticular formation (Sugerman and Demski 1978). Bilateral lesions of the ventromedial nucleus of the amygdalar area of green anoles (*Anolis carolinensis*) resulted in unimpaired assertion and challenge displays, but a reduction in courtship behavior. By contrast, lesions in the paleostriatum reduced assertion and challenge displays but had no effect on courtship behavior (Greenberg, Scott, and Crews 1985). Studies involving lesions in another telencephalic region, the anterior two-thirds of the dorsal ventricular ridge (the striatum) (Peterson 1980), showed no change in social behavior and animals resumed their presurgical positions in the dominance hierarchy (reviewed in Peterson [1980]).

DO DIFFERENCES IN DISTRIBUTION OF ARs CORRELATE WITH SEXUAL DIFFERENCES IN SIGNALING BEHAVIOR?

Using a technique called immunohistochemistry, brain cells containing ARs can be labeled with an antibody that is AR specific. Cells thus labeled by the antibody are considered "AR positive." Visualization of the location of AR-positive cells on histological sections of the brain then allows one to describe the distribution of AR-positive cells in different brain regions, and densitometry allows quantification of the abundance of AR-positive cells in each brain region.

We have documented the distribution of AR-positive cells in the brains of adults of the sexually dichromatic *S. undulatus,* using the polyclonal AR antibody PG21 (Hews, Moga, and Prins 1999) (Fig. 8-3). We found AR-positive cells in many identified brain nuclei (dense clusters of neuronal cell bodies, as distinguished from a cell nucleus). Such nuclei in and near the hypothalamus, for example, play important roles in mediating sex-typical behavior in many vertebrates (Becker, Breedlove, and Crews 1992; Nelson 2000). Specifically, we found AR-positive cell nuclei in males ($n = 6$) in several regions, including the external nucleus of the amygdala; the arcuate, ventromedial, and periventricular nuclei of the hypothalamus; and the basal forebrain. In contrast with males, we did not find AR-positive cells in the amygdala or in hypothalamic regions in females ($n = 3$).

We also found dense concentrations of AR-positive fibers in male *S. undulatus* in the medial cortex, periventricular hypothalamus, and lateral forebrain bundle. AR-positive neuronal fibers were also present in the preoptic area of the hypothalamus, habenula, and deep layers of the optic tectum. In females, AR-positive fibers were relatively sparse but showed a similar distribution to that in males.

What is known about the distribution of ARs in the brains of other vertebrates with sexual differences in aggression? In the well-studied rat

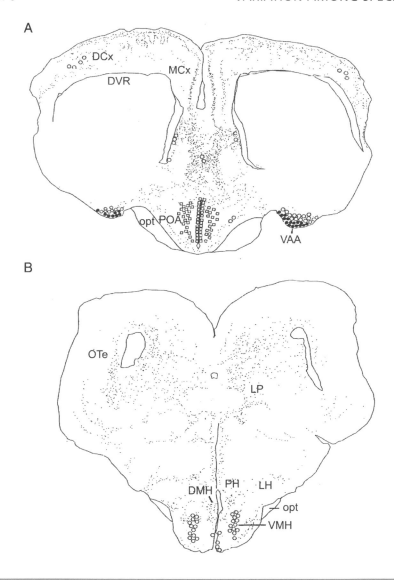

Figure 8-3. Drawings from an immunohistochemical study indicating locations of AR-positive fibers, cell nuclei, and cell soma for (*A*) rostral and (*B*) midlevel sections of brain in a male of sexually dichromatic *Sceloporus undulatus*. Small dots = fibers (possible axonal staining); large solid circles = darkly stained cell nuclei; large open circles or open squares = lightly stained cell nuclei and/or cytoplasmic staining. DCx = dorsal cortex; DMH = dorsomedial hypothalamic nucleus; DVR = dorsal ventricular ridge; LH = lateral hypothalamus; LP = lateral posterior thalamic nucleus; MCx = medial cortex; opt = optic tract; OTe = optic tectum; PH = posterior hypothalamus; POA = preoptic area; VAA = ventral anterior amygdalar nucleus; VMH = ventromedial hypothalamic nucleus (following Bruce and Neary [1995]).

(*Rattus norvegicus*), areas of the brain with the highest density of ARs were the amygdala, septum, hippocampus, medial preoptic area, anterior and lateral hypothalamus, median eminence, and adenohypophysis (Simerly et al. 1990). Similar work involving in situ hybridization in *Cnemidophorus* lizards revealed concentrations of AR mRNA and estrogen receptor (ER) mRNA in septal, amygdaloid, cortical, preopotic, and several hypothalamic brain nuclei (Young et al. 1994). In the amygdalar area, the medial external nucleus was labeled only with the AR probe (and not the ER probe), parallel to AR labeling seen in the medial nucleus of the amygdala of the rat (Simerly et al. 1990). Our preliminary results for the typical sexually dichromatic *S. undulatus* are, in general, consistent with results from other sexually dimorphic vertebrates. Our current studies are now comparing these distributions of AR-positive cells observed in *S. undulatus* with the distributions in males and females of the two monochromatic species (both sexes blue and aggressive, both sexes white and with decreased aggression).

Histological Study of Sexually Dichromatic Color Signals

As part of a study of aggression and color signals, one can also seek to understand the cellular basis of the color signal, and possible roles of sex steroid hormones in the sexual differentiation of these differences in the skin that comprise the signaling trait. In general, very little is known about the endocrine processes involved in sexual differentiation of the cells involved in the production of sexual color signals in vertebrates. There is some endocrine work examining cellular targets, and examples include studies of carotenoid-based signals of adult guppies (reviewed in Houde [1997]), and seasonal avian plumages (see citations in Kimball and Ligon [1999]).

We are examining the cells in the skin that are targets of the sex steroid hormones. The cellular basis of hue and intensity of color patterns has been studied in only a handful of lizard species (Taylor and Hadley 1970; Bagnara and Hadley 1973; Sherbrooke and Frost 1989; Morrison and Frost-Mason 1991; Cooper and Greenberg 1992; Morrison, Sherbrooke, and Frost-Mason 1996). Color patterns in animal skin result from differences in the abundance and relative locations of several types of skin pigment cells, or chromatophores (Cooper and Greenberg 1992; Morrison 1995). Melanophores have melanin-containing organelles called melanosomes. Melanin is a pigment that absorbs all wavelengths, usually imparting a black appearance to the skin (or brown, if chromatophores containing other pigments lie above the melanophore). Melanophores are found deep in the dermal layer but can have finger-like projections extending up into the more superficial layers of the dermis. Another chromatophore type, the iridophore, is located superficial to (above) the deeper melanophores and can (along with other chromatophores) alter which wavelengths are reflected from the skin, and thus the appearance of the skin. Iridophores do not contain pigments but instead have intracellular platelets (often of guanine) that can selectively reflect blue wavelengths via interference. Other chromatophores (e.g., xanthophores, erythrophores) occur in lizard skin.

The model for production of blue versus white skin that we are testing involves only iridophores and melanophores (Cooper and Greenberg 1992). This model proposes that blue wavelengths are reflected off the skin by the iridophore layer. The remaining wavelengths are transmitted through the iridophore and either are absorbed by underlying melanin in the melanophores, yielding a blue appearance to the skin, or are reflected back by underlying layers such as collagen if melanin is absent, yielding a white appearance (Fig. 8-4).

Does the Abundance of Melanin Differ in Blue and White Skin?

We tested the hypothesis that sexual differences in the abundance of dermal melanin correlate with sexual differences in the occurrence of abdominal blue patches in *S. u. consobrinus*. We sampled abdominal skin from blue patches or from the "patch location" in females and prepared the tissues using standard dehydration, paraffin embedding, and staining procedures. We quantified the area of darkly stained melanin in sets of transects across the samples. Male and female *S. undulatus* differed significantly in the density of dermal melanin (Fig. 8-5).

We also prepared histological samples from the two sexually monochromatic species, *S. virgatus* and *S. jarrovii*, and the differences were as predicated by the melanin hypothesis. White abdominal skin (taken from where the blue patch would be located in other *Sceloporus* males) of *S. virgatus* males almost entirely lacked a melanin layer. Conversely, the density of dermal melanin in samples of skin taken from the blue abdominal patches of female *S. jarrovii* was relatively high and did not differ statistically from the melanin density in abdominal patch skin samples from males (Quinn and Hews, unpublished data).

The melanin hypothesis predicts that abundance of dermal melanin should be higher in the blue versus white skin in hormone-manipulated animals that are induced to express blue abdominal patches. Thus, the

A

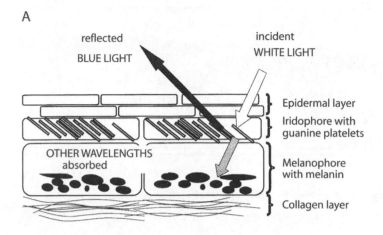

reflected
BLUE LIGHT

incident
WHITE LIGHT

Epidermal layer

Iridophore with
guanine platelets

OTHER WAVELENGTHS
absorbed

Melanophore
with melanin

Collagen layer

B

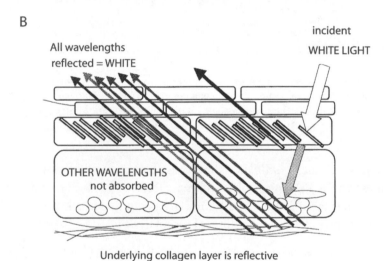

All wavelengths
reflected = WHITE

incident
WHITE LIGHT

OTHER WAVELENGTHS
not absorbed

Underlying collagen layer is reflective

Figure 8-4 Illustration detailing model proposed by Cooper and Greenberg (1992) for production of (*A*) blue versus (*B*) white skin in lizards. For blue skin, iridophore cells have intracellular guanine platelets (open rectangles) that selectively scatter blue wavelengths; other wavelengths are transmitted through the iridophore and absorbed by melanin (solid ovals) in the underlying layer of melanophore cells, resulting in the appearance of blue skin. For white skin, melanin is absent, and the remaining wavelengths are reflected back by an underlying reflective layer (e.g., collagen), along with the blue from the guanine platelets, resulting in the appearance of white skin.

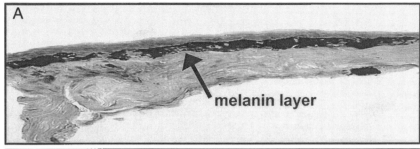

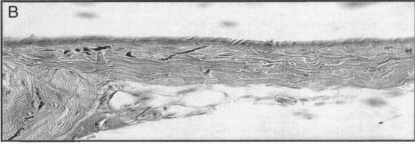

Figure 8-5 Histological section through skin taken from patch location in a male ([A] blue skin) and a female ([B] white skin), in a typical sexually dichromatic species, *Sceloporus undulatus consobrinus.*

hatchling female *S. undulatus* that we have successfully manipulated to express the male-typical blue of this species (Fig. 8-2B) should have greater dermal melanin density compared with that in white abdominal skin of control female hatchlings. This analysis is under way.

Does the Ultrustructure of Iridophores Differ in Blue and White Skin?

Another direction to explore involves the ultrastructure of the dermal chromatophores. In particular, the arrangement of the reflecting guanine platelets in the iridophores (Fig. 8-4) deserves attention. These platelets are present in the skin from both a white-bellied horned lizard (*Phrynosoma modestum*, Sherbrooke and Frost [1989]) and the blue abdominal patches of tree lizards (*U. ornatus*, Morrisson, Sherbrooke, and Frost-Mason 1996) and sagebrush lizards (*S. graciosus*, Morrison and Frost-Mason [1991]). However, the guanine platelets are highly organized in the blue abdominal skin of *S. graciosus*, with a regular brick-like arrangement within the cell. By contrast, in the white abdominal skin of *P. modestum*, the reflecting platelets lack an organized layered arrangement and reflect white light rather than blue wavelengths produced by interference phenomena. Differences in techniques are not a likely explanation for these species differences in platelet organization because these studies were all conducted in the same laboratory.

Differences among *Sceloporus* in occurrence of blue abdominal patches might result, at least in part, from species differences in the arrangement of platelets in the iridophores. Besides the arrangement of the platelets, the individual size of the platelets (Fox 1976) may be key to whether the skin appears blue or not. Thus, transmission electron microscopic (TEM) analysis of the ultrastucture of iridophores will be necessary for a more complete understanding of the variation in the production of this visual signal. Initial TEM work on adult males of the white-bellied *S. virgatus* indicates that iridophores are present (Fig. 8-6) (K. Yanek and D. Hews, unpublished data). We currently are quantifying variables that can affect whether blue wavelengths of light are reflected, including platelet size, interplatelet spacing, and average number of platelet layers.

Conclusion

To understand fully the evolution of communication systems, and the selective forces that act on these systems, one must explore the many components of the communication system. These components include the production and control of the signal, as well as how it is transmitted in the environment, received by other individuals and processed in the CNS, and responded to by the receiver. The functional study of conspicuous male secondary sex characteristics, and conspecific responses to these traits, has led to a better understanding of the process of sexual selection (Andersson 1994), especially when such studies are carried out in a phylogenetic context. For example, analysis of signals and receivers has provided support for the sensory bias mechanisms of sexual selection (Basolo 1990, 1995; Ryan and Rand 1990, 1995; Endler 1992). Our work on functional aspects of the signal, which we do not describe in this chapter, currently involves documenting male and female responses to the presence and absence of abdominal blue patches (e.g., Quinn and Hews 2000).

Studying the roles that hormones play in the development of sexually dimorphic traits leads to a more complete understanding of these traits. In addition, studying these aspects of sexual signals in a phylogenetic context allows comparisons to be made among species, rather than simply between males and females, and the tracking of changes in endocrine mechanisms that underlie phylogenetic character state transi-

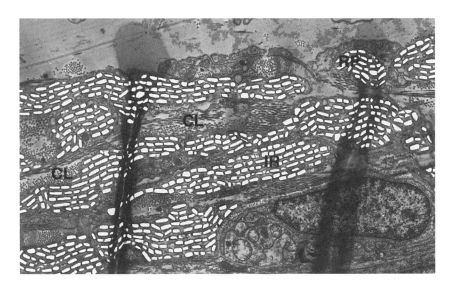

Figure 8-6   Transmission electron micrograph of abdominal skin from a male of white-bellied *Sceloporus virgatus*. The skin was taken from the same location as where the blue patches occur in other species. IR = iridophore; RP = reflecting platelet; CL = collagen layer.

tions in sexual differences in aggression and associated color signals. In this chapter, we focused on (1) aspects of signal production, (2) how hormones are involved in the development of a sexually dichromatic signaling trait, and (3) how hormones may be acting on the target tissues to result in species differences in the expression of the signaling trait. We discussed our work on three species of *Sceloporus* that vary both in the expression of aggression and in the development of an associated color signal. Manipulating pre- or posthatching hormone levels in the context of the organization and activation hypothesis in male and female *S. undulatus* and *S. virgatus* will reveal whether abdominal coloration is controlled by the sex steroid hormone, DHT, as it is in *Urosaurus*, the sister genus to *Sceloporus*. Future studies on the receptors in the skin and the presence and activity of enzymes that convert T into DHT are planned and should reveal the relative contributions of these components to the sex and species differences in signal morphology. Sex and species differences in the occurrence of the signal (abdominal coloration) are likely due, at least in part, to differences in the abundance of melanin in the layers of dermal melanophores that underlie the iridophore layer. Other elements of the dermal chromatophore unit require additional attention.

Intimately related to this work on the signaling trait is the study of the endocrine regulation of the development and expression of signaling behavior. Woodley and Moore (1999a, b) have begun work on the masculinized females of *S. jarrovii*, exploring the activational roles of sex steroid hormones. Our work in progress is determining the distributions of ARs in the brain regions known to mediate aggression in vertebrates, including reptiles, and we will be comparing between the sexes and among the three *Sceloporus* study species. Determining the relative importance of organizational and activational effects of sex steroid hormones in contributing to sex and species differences in aggressive behavior will also need attention, as will assessing the relative importance of endocrine and nonendocrine mechanisms.

The genus *Sceloporus* provides an excellent opportunity for conducting such comparative endocrine and functional studies. Because of the multiple independent events of both loss in males and gain in females of the color patches (Wiens 1999), we can ask, Do similar endocrine mechanisms underlie similar but independent character state transitions?

CHAPTER 9

The Interplay among Environment, Social Behavior, and Morphology

Iguanid Mating Systems

Paul J. Gier

Animal mating systems, the stages on which traits evolve in response to sexual selection, have come to be understood as outcomes of competition among individuals to maximize their reproductive success (Emlen and Oring 1977). The strategy chosen by a given sex to achieve this end will be dependent in part on the strategy employed by members of the other sex. For example, the best mating strategy for a female is partly a function of whether the males contribute care to the offspring (Trivers 1972), whereas male strategies are influenced by the spatiotemporal availability of females. Establishing clear cause and effect strictly within such an interdependent system would clearly be difficult, were it a goal in the first place. Instead, mating systems are best understood in the ecological context within which they occur, because the range of available strategies for a given sex is limited by the distribution of available resources. For instance, in many species in which the male's contribution to female reproductive success is limited to sperm, males defend territories that encompass resources of use to the females, thereby gaining access to the females by default (resource-defense polygyny) (Clutton-Brock 1989). In such systems, the dispersion of food supplies, thermal microhabitats, refuges from predation, or a host of other factors set spatial and temporal boundaries of animal movement, which, in turn, influence the possible set of mating strategies.

Because all environmental resources are inevitably patchy at some scale, determining which resources will most influence movements and social behavior requires tracking variables that may change in time or space. In principle, one would expect the resource most responsible for shaping the evolution of mating systems to be the most narrowly distributed of the subset of crucial resources, not necessarily the one most closely tied to growth, survival, or reproduction. Studies of various taxa show that the degree of patchiness of food (Ims 1987; Davies et al. 1995), shelters or resting sites (Baird and Liley 1989; Baldi et al. 1996), nesting or oviposition sites (Howard 1978a; Alcock 1987; Sato 1994), and thermal patches (Pleszczynska 1978; Carey 1991) play crucial roles in determining the nature of habitat use and the opportunity for polygamy.

Environmental "patchiness" is a scale-dependent phenomenon; a resource that is evenly distributed from the perspective of a small animal may be highly clumped from a larger individual's perspective. For instance, larger animals require more specific microhabitat qualities in providing shelter, because they cannot take refuge in leaf litter or in small crevices or burrows useful to smaller animals. Movements of such species should be constrained to parts of the habitat where adequate predator retreats (e.g., large burrows or caves) are available.

Habitat patchiness also can indirectly influence sexual dimorphism (see chapter 3). Although sexual dimorphism may be influenced by purely ecological factors (Schoener 1967; Hedrick and Temeles 1989; Anderson and Vitt 1990), most studies have emphasized the role of sexual selection operating within the context of a given mating system (e.g., Andersson 1994). Female aggregation around clumped resources, for instance, may set the stage for a high degree of polygyny (i.e., a high environmental potential for polygamy; *sensu* Emlen and Oring 1977). Such a system has a wide disparity in male mating success and may increase sexual dimorphism by favoring the evolution of male morphological traits used in combat and display (Alexander et al. 1979).

Here, I compare the dynamics of resources, movements, social behavior, and sexual dimorphism in two lizard species. *Ctenosaura similis*, the spiny-tailed iguana of Central America, and *Dipsosaurus dorsalis*, the desert iguana of southwestern North America, are both members of the family Iguanidae. This family contains eight genera of herbivorous, medium- to large-bodied species distributed throughout the northern Neotropics and Galápagos Islands but extending north into the deserts of the southwestern United States and with an isolated genus in Fiji (de Queiroz 1987; Frost and Etheridge 1989). For a family of such low diversity and ecological uniformity (large, sedentary herbivores residing in hot climates), the iguanids exhibit a great range of mating systems, including leks, which are not known in any other reptiles (Wikelski, Carbone, and Trillmich 1996).

Iguanids vary in sexual dimorphism, especially in body size and head morphology, and this dimorphism is generally held to be a product of sexual selection, not resource partitioning, since male and female diets or microhabitats do not differ (Carothers 1984). The two focal species of this study represent extremes within the family: *D. dorsalis* is among the least dimorphic in size and shape (Norris 1953), and *C. similis* is among the most (Fitch and Henderson 1978). A second feature of the iguanids that facilitates comparative studies is that studies of environmental patchiness are simplified in these species and in iguanian lizards in general, by the high degree of philopatry among females (see reviews in Rose [1982] and Stamps [1983b]). Female philopatry, in turn, promotes male mate monopolization and the evolution of resource-defense polygyny, the most common mating system among iguanians (Stamps 1983b). Because female home ranges are generally small, it is usually possible to discover which resources most constrain their movements. Past studies have shown individual space use by iguanians to be determined by thermal patches (Christian, Tracy, and Porter 1983; Christian and Tracy 1985; Adolph 1990), availability of basking sites or predator retreats (Gil, Perez-Mellado, and Guerrero 1990), food patches (Simon 1975; M'Closkey, Deslippe, and Szpak 1990; Hews 1993), or social factors (Stamps 1988).

The two species in the present study live in habitats that differ widely in terms of movement constraints. *Ctenosaura similis* inhabits tropical dry forests and other semiopen habitats from southern Mexico through western Panama (Fitch and Hackforth-Jones 1983). Like other members of the genus, it tends to be restricted in movements to open or semi-open areas with variable sun exposure and access to retreats (Evans 1951; Fitch and Henderson 1978; Carothers 1981). Its breeding season coincides with the driest part of the year, when most trees have lost their leaves. However, many trees flower at this time, producing very abundant but highly localized food sources. Thus, the dispersion of burrows and sources of food are a priori considerations in structuring the movements of *C. similis.*

Thermal considerations are expected to be more important in determining the movement patterns of *D. dorsalis* than *C. similis.* The desert iguana ranges throughout the low-elevation deserts of the southwestern United States and adjacent Mexico in habitats that are thermally rigorous, with burrows and patches of shade from shrubs providing relief from extremely high temperatures in the open. Here, microhabitats exposed to the sun represent not just suboptimal patches, but lethal ones, because lizards without shelter would quickly become hyperthermic. Unlike many other desert ectotherms, however, *D. dorsalis* remains active at midday during the hottest times of the year, using shade patches as thermal refuges to maximize above-ground activities such as territorial defense and mating (Norris 1953; DeWitt 1967a, b; Krekorian 1976; 1984). Since *D. dorsalis* is entirely herbivorous (Mautz and Nagy 1987), the sources of shade in the environment double as the main sources of food. However, shrubs differ not only in their quality as food sources (Minnich and Shoemaker 1970; Porter et al. 1973), but also in the quality of shade they provide, owing to species-specific differences in growth form (Hillard 1996). Except in the provident case of a single, common shrub species providing both high-quality food and optimum-temperature microclimates at midday, the maximization of territorial quality in *D. dorsalis* relies on a trade-off, in which the relative importance of the two resources are weighed.

Over the course of several years, I examined how environmental factors shape the mating system for both lizards. I also examined the pattern of male dominance, and how male status influences courtship success. Finally, I related these behavioral traits to patterns of sexual dimorphism. If ecological factors influence the degree of local mate competition, sexual dimorphism should reflect this, and species with more competitive mating systems should have stronger sexual dimorphism. In lizards, these selective pressures are often manifested in differences in head allometry between males and females, because the head is used for display (e.g., crest scales, dewlaps, head coloration) (Carpenter 1982; Cooper and Vitt 1988; Fleishman 1988b) and is the primary weapon used in male-male combat (Carothers 1984; Vitt and Cooper 1985a; Hews 1990; Olsson 1992).

Methods

I conducted fieldwork on *C. similis* during January–March 1992 in Palo Verde National Park, Guanacaste, Costa Rica (10°21′ N, 85°21′ W). This time frame apparently encompassed the bulk of the courtship period, which accelerated during January and declined noticeably in late March. The study population consisted of seven scattered groups of lizards within 1 km of the Organization for Tropical Studies (OTS) field station. In 1994–95 (March–May, plus late June–July 1995), I performed a comparative study on *D. dorsalis* on a 1.7-ha study site near the intersection of Ogilby Road and I-8 in Imperial County, California (32°45.4′N, 114°50.2′ W). Mating activity in this species was most intense on hot days in the spring, indicating that most reproductive behavior fell within the field season; however, mating behavior may have continued, at a lesser frequency, into the summer.

The project's goals required a clear picture of the spatial variability of resources important to each population, coupled with individuals' use of those resources and patterns of male courtship success. In each of the two habitats, I constructed scale maps of the burrows and large plants, staking out a 10 × 10 m grid of flags to allow plotting of lizard movements. I mapped burrow openings (which were permanent for *C. similis* and relatively ephemeral for *D. dorsalis*, owing to the frequency of sandstorms) and sources of food (plants, identified to species, or patches of flowers on the ground in the case of *C. similis*). For *D. dorsalis*, I used two methods to estimate the percentage of cover of different plant species. First, I counted and measured (canopy diameter to the nearest centimeter) all living plants with a diameter >5 cm on nine 100-m² subplots, evenly spaced across the site. Although offering exact measurements and including all plant species within the subplots, this technique is biased by subplot location because plant distribution was not uniform on the site and some plants were underrepresented in this sample. Second, I mapped and counted all permanent shrubs on the entire study area with a canopy diameter >50 cm, which neglects small plants but equally represents all portions of the study area and is a useful measure of the most important sources of shade. To calculate cover area for each of six species of permanent shrubs, I used the averaged results of these two methods to assign a relative abundance to each shrub species.

I also noted the thermal qualities of both species' habitats; the forested environment of *C. similis* offered a relatively uniform thermal environment in all microhabitats, in contrast with the thermally heterogeneous desert habitat of *D. dorsalis*. The methods that allowed me to draw a "thermal map" of *D. dorsalis*'s habitat are described in the results for that species (given later).

I attempted to capture all adults in the study areas. In each case, I captured and marked the majority of them (58 adult *C. similis* and 67 adult *D. dorsalis*). I recorded individual size (mass and snout-vent length) and sex, as well as head width at the widest point on the head, head height at the highest point on the head, and head length from the anterior edge of the ear opening to the tip of the snout. To examine allometric differences in morphology among groups, I analyzed the residuals of the overall regression between the variable of interest and body length. This allowed me to examine morphological differences among groups independent of body size.

To discern any possible pattern between habitat use and dominance status of males, I took detailed focal samples during which I recorded movements, assertion displays (equivalent to Carpenter and Ferguson's [1977] "signature bobs"), foraging behavior (location and source of food noted), and interactions with other individuals. I also continuously recorded the focal animal's exposure to sun or shade. To compare these patterns with the space use of females, I supplemented the male focal samples with "spot censuses" of females at regular intervals throughout the field season, in which I recorded the locations, sun exposure, and behavior (feeding, moving, nonmoving) of all visible females. To quantify female philopatry, I designated "female basking home ranges" as that space encompassing the set of locations in which females were observed motionless during focal censuses. A basking home range is smaller than a general home range, because it omits areas through which a female moves only briefly; however, it emphasizes the areas in which a female spends the most time and, thus, is a preferred method of mapping overall female availability to males. I used the minimum polygon method to delineate basking areas for each female.

I designated as courtship any behavior involving a male directing vibratory head nodding (Distel and Veazey 1982; also equivalent to Dugan's [1982] "shudder-bob") at a nearby female. This head movement is easily distinguishable from the slower and higher-amplitude assertion display employed by iguanids. Vibratory head nodding was never observed outside the context of male-female interactions and was virtually identical in basic structure for both species. For each courtship sequence, I timed the events and categorized the behavior that led to the cessation of courtship as follows:

1. *Female leaves:* The female moves >3 m from the male or makes herself unavailable (e.g., running down a burrow).
2. *Tail slap:* The female strikes the male with her tail, usually on his snout.
3. *Male stops:* The male ceases head nodding even though the female has moved <2 m and has not used the tail slap.
4. *Male chase:* The male is chased by another male, or leaves to chase another male.

These four categories are mutually exclusive. I did not include in the analyses any ambiguous outcomes (e.g., female moving between 2 and 3 m from the male). Once the male grasped the female's neck in his jaws (the "neck-grab" stage of the courtship sequence), I classified as "female resistance" any sequence in which the female struggled. The difficulty of implying female intent in these struggles is somewhat mitigated by the fact that in all cases, the struggles were unambiguous, even bordering on the violent; in one case a female lost the tip of her tail during the thrashings that accompanied resistance. As testament to the earnestness of these struggles, more than half of those females that struggled did escape the neck grab of the male and immediately ran down a burrow.

I placed all males into one of three dominance categories:

1. *Dominant males*: those that chased all other males within their range but were never observed to be chased or supplanted by any other male
2. *Subordinate males:* those that ranked lowest and never were seen to chase or supplant any other male
3. *Midrank males*: those that were chased by dominant males, but chased subordinate males

These rankings were based on 43 natural altercations among male *C. similis* in which a winner could be designated, and 42 for *D. dorsalis*. No reversals of dominance were indicated with either species, and the ease of observation of dominant and midrank males ensured the correct rank assignment of those males. Smaller males, however, were more furtive and difficult to observe, and some "subordinate" males may have engaged in chases that I did not detect. For some analyses (i.e., those in which sample sizes for subordinate males were limited), I lumped midrank and subordinate males into one "low-rank" category.

Physical Environment and the Social Scene
C. similis: Burrows, Food, and Female Movements

The deciduous tropical forest habitat of Palo Verde National Park is typical of that inhabited by *C. similis* throughout its range, and the species breeds there, as elsewhere, during the height of the dry season (Fitch and Hackforth-Jones 1983). Most of the forest habitat at Palo Verde was not used by adult *C. similis*; rather, groups of 6–16 lizards were aggregated in isolated areas ranging in size from approximately 0.4 to 1 ha. I designated these isolated sites "burrow areas" owing to the invariable presence of burrows or other retreats near where the females were seen basking. Four of the seven burrow areas studied were centered on large hollow trees with numerous openings; two included patches of burrows dug into sandy soil (e.g., Fig. 9-1), and one was centered on a limestone outcrop with deep crevices. These localized populations were rarely in sight of one another, typically being separated by at least 200–300 m. The burrows apparently are dug by the females and are permanent

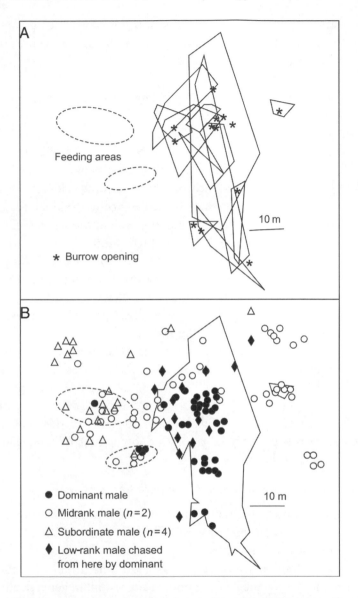

Figure 9-1 (*A*) Female basking home ranges (polygons) associated with one of the burrow areas at the study site, and nearby feeding areas (dashed ellipses). Each female home range is based on sightings of basking individuals (mean number of sightings per female = 10.6). The feeding areas indicated include two *Mangifera* trees within the smaller ellipse and a *Tabebuia* tree within the larger one. (*B*) Sightings of dominant, midrank, and subordinate males at the same locality, superimposed on combined female basking home ranges (polygon) and feeding areas (dashed ellipses). Data are from spot censuses and focal animal observations.

(Burger and Gochfeld 1991). The distribution of females was closely tied to these retreats; of 320 female sightings during spot censuses, 206 (64.4%) were stationary and within 2 m of a retreat (burrow, hollow tree, or rock crevice).

Food was a secondary factor affecting female distribution and movements. Many trees in this forest flower during the dry season despite their leafless condition, producing locally abundant, large flowers that are an important food source for herbivores such as *C. similis*. Of 74 observed feeding bouts, 52 (70%) were on the flowers of trees, particularly *Bombacopsis quinatum* (Bombacaceae) and *Tabebuia ochracea* (Bignoniaceae). Observed feeding events on the only other food sources—the foliage of unidentified herbaceous plants—were more widely scattered. I designated as "feeding sites" all sites surrounding flowering trees in which more than 10 feedings were observed. Since these feeding sites were spatially separated from the burrow areas, the distribution of females was bimodal: of female sightings 84.4% were within the boundaries of the local burrow area, 9.1% were at feeding sites, and most long-range movements were between the two.

The thermal environment in the seven areas studied was a mix of open sun and dappled shade. Although lizards were frequently observed to bask in the sun and shuttle between sun and shade, the locations of burrow areas did not seem based on sun exposure—one was exposed to full sun; four were placed amid the dappled shade of the understory; and two, which were centered on the hollow trunks of nondeciduous trees, were deeply shaded.

C. similis: Male Dominance and Movement Patterns

Like the distribution of females, the dominance hierarchy among male *C. similis* was based on the locations of burrow areas. Each of the seven burrow areas was occupied by a single dominant male and a variable number of midrank and low-rank males. Dominant males spent most of the time in the female basking home ranges and, in general, enjoyed greater physical proximity to females than did males of lower rank. Dominant males were sighted within female basking home ranges significantly more often than either midrank or subordinate males ($F_{2,25} = 18.9$, $P < 0.0001$; Scheffe post hoc comparisons: $P = 0.0023$ and $P < 0.0001$, respectively). The mean distance between a male and the nearest female (taken from spot censuses) was lower for the 7 dominant males than for the 15 most observable low-rank males ($t_{20} = 5.57$, $P < 0.0001$). In addition, the number of females within 5 m of a male during the censuses was greater for the dominant males than for the 15 low-rank males ($t_{20} = 9.72$, $P < 0.0001$).

By contrast, the midrank and subordinate males arranged themselves on the periphery of the burrow areas. Their occasional incursions into the burrow area were almost invariably countered by the resident dominant male, indicating that the burrow area approximated the territory of those males (Fig. 9-1B). The females themselves were not defended, however. For instance, when the females foraged away from the burrow areas, they were often courted by low-rank males, and dominant males did not challenge or give chase under those circumstances. Furthermore, dominant males and low-rank males were often observed feeding in close proximity to one another without altercation, as long as the feeding areas were far from the burrow areas.

D. dorsalis: Distribution of Burrows, Temperature, and Food

The study site for *D. dorsalis* consisted of a series of stabilized sand dunes interspersed with gravel-sand flats, on which most of the vegetation grew. *Larrea tridentata, Ambrosia dumosa, Stephanomeria pauciflora, Eriogonum deserticolum,* and *Psorothamnus emoryi* were the dominant permanent shrubs, with smaller numbers of large, widely scattered *Chilopsis linearis* and *Olneya tesota*. Like most sites in the lower Colorado Desert, the majority of the study site was barren of vegetation, with only 5.0% plant cover. More than half of the plant cover on the site was occupied by permanent shrubs, with *E. deserticolum* dominating. The rarest shrub was *C. linearis,* of which a single large individual was present. The largest plants on the site were some individual *L. tridentata* and the *C. linearis,* which exceeded 3 m in crown diameter; however, all species of permanent shrubs were represented by several individuals with crown diameters exceeding 2 m.

In contrast with the clumped burrow distribution of *C. similis,* burrows were more evenly distributed in the sandy habitat of *D. dorsalis.* The spatial distribution of burrows was somewhat ephemeral; openings would disappear during sandstorms, and new openings would appear. The lizards also made occasional use of the relatively widely dispersed burrows dug by kangaroo rats (*Dipodomys deserti*). Although many of the burrows were aggregated around the bases of shrubs, others were in the open.

Because of the thermal complexity of this environment and the presence of thermally lethal microhabitats, it was necessary to analyze *D. dorsalis* movements against a thermal map of the habitat. Throughout 25 d during the 1995 season, I sampled microhabitat temperatures beneath all permanent shrub species with the exception of *O. tesota*, as well as in the open. Except for unusually large individuals of the herbaceous *Palafoxia linearis* and *Asclepias subulata*, these six shrubs were the only plants capable of providing full shade to an adult lizard. I used bare thermocouples shallowly submerged beneath the sand to record substrate temperature (T_{sub}), and elevated 2 cm to record air temperature (T_{air}). I used a five-channel OMEGA® datalogger (Omega Engineering Inc., Stamford, Connecticut) to sample five microhabitats at 15-min intervals simultaneously, moving thermocouples daily to represent as many individual shrubs as possible. During the June–July 1995 sample, I took equivalent temperatures at hourly intervals with bare-bulb Miller-Weber® thermometers (shaded when in the open). To examine the effects of shrub species on shade temperature, I used repeated measures analysis of variance (ANOVA) on T_{sub} and T_{air} of each shrub species to factor out the effect of time; the microhabitat "open" was excluded from these analyses. Separate analyses including open microhabitats revealed that all plants were significantly cooler than open microhabitats (Scheffe multiple comparisons: $P < 0.0001$).

Temperature profiles for spring (mid-March to late May) and summer (late June to early July) revealed that substrate temperature varied more among microhabitats than air temperature (Fig. 9-2). In the spring, the effect of shrub species was significant on both T_{air} (repeated measures ANOVA: $F_{1118,16,3} = 51.97, P < 0.0001$) and T_{sub} ($F_{563,16,3} = 102.53, P < 0.0001$). There was no significant shrub × time interaction for either T_{air} or T_{sub} ($P = 0.76$ and 0.34, respectively), indicating that shrub species did not differ in their temporal pattern of warming or cooling. In the summer, there was a significant shrub effect on T_{sub} ($F_{40,10,4} = 22.5, P = 0.005$) but no shrub effect on T_{air} ($F_{40,10,4} = 3.22$, $P = 0.18$) and no shrub × time interaction on either T_{sub} ($P = 0.09$) or T_{air} ($P = 0.24$). In general, *E. deserticolum* and *P. emoryi* had the highest temperatures and *L. tridentata* and *C. linearis* had the lowest.

Clearly, different shrubs produce different thermal environments, which can impact a basking lizard's body temperature (T_b). To quantify the influence of shrub temperature on lizard T_b, in July 1995, I conducted an experiment in which I tethered 10 adults with a 1-m long string beneath the shade of each of the five most widely available shade sources (*E. deserticolum*, *L. tridentata*, *S. pauciflora*, *P. emoryi*, and *A. dumosa*; the order of shrubs was randomized for each lizard). These experiments took place between 0830 and 1200 h, during the normal time of lizard activity.

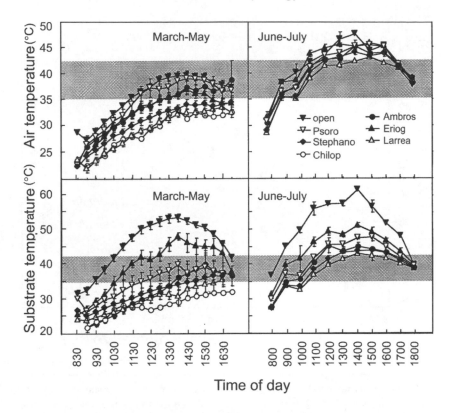

Figure 9-2 Air and substrate temperatures beneath six shrub species and in the open (means ± 1 SE) in spring (mid-March to late May) and summer (late June to early July) in habitat of *Dipsosaurus dorsalis*. Ambros = *A. dumosa*; Eriog = *E. deserticolum*; Larrea = *L. tridentata*; Psoro = *P. emoryi*; Stephano = *S. pauciflora*; Chilop = *C. linearis*. Shaded area represents range of body temperatures of noosed *D. dorsalis*.

For each trial, I allowed 10 min for the lizard's T_b to respond to the microclimate, then immediately measured T_b (cloacally), T_{sub}, and T_{air}. Although the tethered lizards were free to move within a 1-m radius beneath the shrub and into sun exposure, most lizards moved little during the trials; typically, they briefly explored the environment before remaining motionless. I terminated the trial immediately if the lizard began panting, indicating a T_b approaching 44°C (DeWitt 1967b); this occurred in 5 of the 10 *E. deserticolum* samples and 4 of the 10 *P. emoryi* samples, because these shrubs provide relatively poor shade cover. Thus, end T_b values for these two shrub species are underestimates.

The results of the tether experiment revealed a significant shrub effect on T_b (ANOVA: $F_{45,4} = 6.33$, $P = 0.0004$) (Fig. 9-3). The fol-

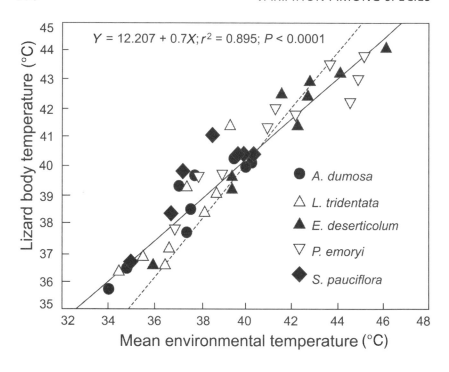

Figure 9-3 Body temperature of *Dipsosaurus dorsalis* as function of mean environmental temperature $[(T_a + T_b)/2]$ taken after being tethered for 10 min beneath each of five shrub species. Dashed line represents isotherm.

lowing shrub pairs differed significantly in resultant lizard T_b (in each case, the plant associated with the cooler temperature is listed first; Scheffe multiple comparisons: $P < 0.05$): *L. tridentata–E. deserticolum*, *L. tridentata–P. emoryi*, and *S. pauciflora–P. emoryi*. Comparisons of the mean environmental temperatures $[(T_{air} + T_{sub})/2]$ of the shrubs revealed the same significant differences that the T_b tests showed, in addition to which the following shrub pairs differed significantly: *A. dumosa–E. deserticolum*, *A. dumosa–P. emoryi*, and *S. pauciflora–E. deserticolum* ($F_{45,4} = 8.64$, $P < 0.0001$; Scheffe multiple comparisons: $P < 0.05$).

In summary, shrubs do differ in the thermal environments they provide, with the coolest temperatures found beneath *L. tridentata* and *C. linearis* (hereafter referred to as cool-shade shrubs). By contrast, *E. deserticolum* and *P. emoryi* (referred to as warm-shade shrubs) provide the warmest temperatures. The warm-shade shrubs' temperatures at midday exceed the range of T_b taken from free-ranging lizards (based on cloacal temperatures from 90 noosed lizards; mean ± 1 SE = 39.47°C + 0.15; range: 35–42.5°C) (Fig. 9-2).

Dipsosaurus dorsalis in this population was extremely catholic in its diet, feeding at least occasionally on almost all the plant species on the study site (a total of 21 species; unpublished information). Of 118 observed feeding events, 50 (42.4%) were on the six permanent shrub species. The most important single food plant was *S. pauciflora* (flowers), which accounted for 27.1% of observed feedings. Only one other species, the herbaceous *Tiquilia plicata* (leaves; 13.6% of feedings), accounted for >10% of the diet. I focus here on permanent shrubs rather than herbaceous plants because they tend to dominate the landscape and diet, especially during the latter half of the lizards' breeding season, when most herbaceous plants have dried and withered. Lizards did not select shrubs as food sources or resting sites in proportion to their abundance (Fig. 9-4). The two most abundant permanent shrubs (*E. deserticolum* and *L. tridentata*) were among the least-used food sources (one and zero feedings observed, respectively). Shrubs disproportionately selected as food sources included *S. pauciflora, P. emoryi,* and *C. linearis* (32, 11, and 4 observations, respectively). By contrast, those disproportionately selected as basking sites included *L. tridentata; C. linearis;* and, to a lesser extent, *A. dumosa* (Fig. 9-4).

D. dorsalis: Individual Movement Patterns and Male Dominance

Most *D. dorsalis* have small home ranges, typically with a cluster of sightings around a particular shrub; others have larger, more dispersed home ranges with no clear center of activity. I designated a particular shrub as an individual's home range center (HRC) if 50% or more of the sightings were beneath the shrub or within 2 m of its edge. Female *D. dorsalis* were not as clumped in space as female *C. similis* (cf. Figs. 9-1A and 9-5A), and basking home ranges tended to be larger. The overall wider dispersion of females results in part from the dispersed distribution of the chief food plant (*S. pauciflorum*). The highest levels of overlap between female home ranges occurred beneath large, cool-shade shrubs (*L. tridentata* and *C. linearis*), where up to three females occasionally shared overlapping basking home ranges. These females were less tied to these shrubs, however, than female *C. similis* were to their burrow areas; female *D. dorsalis* occasionally dispersed among *S. pauciflorum* to feed and bask. In addition, some individual females had widely dispersed home ranges with no definable HRC (Fig. 9-5A). This

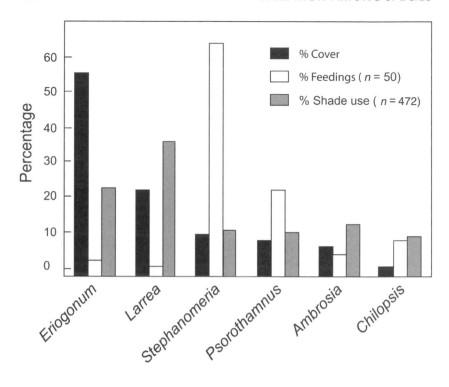

Figure 9-4 Six shrub species arranged in descending order of percentage of cover (calculated as percentage of total for only these six species), with percentage of feeding events and shade use by *Dipsosaurus dorsalis* (shade use calculated from spot census observations of lizards using plants as shady basking sites).

pattern was generally mirrored by the distribution of male home ranges. The cool-shade shrubs with the highest levels of female home range overlap formed the nuclei of dominant male territories (Fig. 9-5B). Only 1 of 10 territories based on either *L. tridentata* or *C. linearis* was defended by a low-rank male, and low-rank males were more likely than dominant males to have wide-ranging activity areas, with no HRC (Table 9-1). As in *C. similis*, dominant males defended the centers of their activity areas from intruding males of lower status.

In summary, both *C. similis* and *D. dorsalis* exhibit resource-defense polygyny, although they differ in the proximate object of male territoriality. Female *C. similis* are most restricted in movements to areas near burrows, and these areas are invariably defended by dominant males. Female *D. dorsalis* often (though not always) center their activity on cool-shade shrubs, which tend to be defended by dominant males. Home range overlap among females was higher for *C. similis* (up to 12

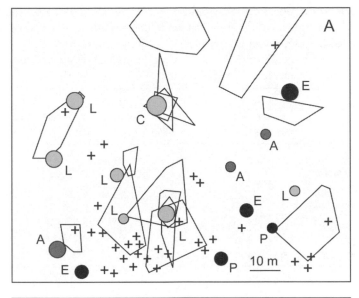

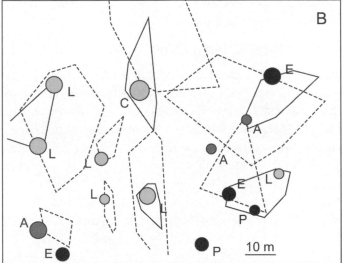

Figure 9-5 (A) Distribution of female *Dipsosaurus dorsalis* basking home ranges (minimum polygon method) in western third of the study site, with all shrubs serving as home range centers (HRCs) and all shrubs >2 m in diameter shown. Light gray circles represent cool-shade shrubs (L = *L. tridentata*; C = *C. linearis*), medium gray circles represent *A. deltoideum* (A), and black circles represent warm-shade shrubs (E = *E. deserticolum*; P = *P. emoryi*). Plus signs represent *S. pauciflora* used as food plants. (B) Distribution of dominant male (solid-line polygons) and low-rank male (dashed-line polygons) home ranges over same area as (A).

Table 9-1 *Dipsosaurus dorsalis* home range centers (HRCs) as a function of sex and male status. Shrubs are arranged in descending order of abundance (see Fig. 9–4). Designation of a shrub as the HRC indicates that >50% of the sightings were beneath or within 2 m of the shrub; None - a dispersed site use pattern, with no clear center of activity.

Plant	No. of Female HRCs	No. of Dominant Male HRCs	No. of Low-Rank Male HRCs
E. deserticolum	1	0	2
L. tridentata	9	7	1
S. pauciflorum	0	0	0
P. emoryi	1	1	0
A. dumosa	3	1	2
C. linearis	4	2	0
None	3	1	7

female home ranges contained within a dominant male's territory, compared with a maximum of 3 for *D. dorsalis*).

Male-Female Interactions: Effect of Male Status

I observed 90 complete courtships for *C. similis* and 101 for *D. dorsalis*. In the case of *C. similis*, males of different status differed in the locations in which they courted. Forty-five of 56 courtships (80.4%) by dominant males ($n = 6$) took place within the burrow areas, supporting the notion that this area represents the territory of those males. Among 43 courtships by 14 low-rank males, however, 24 (55.8%) took place in the burrow areas, 8 (18.6%) in feeding areas, and 11 (25.6%) between the burrow areas and feeding areas. Most of the courtships by low-rank males within the burrow areas consisted of attempted "sneak" copulations and differed from the courtships of dominant males (discussed subsequently). In *D. dorsalis*, there was no clear spatial segregation of courtship behavior among males of different status. In this species, courtships tended to be drawn out and often took place over a large area. It was common for a female to be courted continuously by a single male as she moved from the shade of an *L. tridentata* shrub to open sun and then to another shrub to feed. Thus, it was not possible to assign specific localities to individual courtships.

There was a striking similarity between the patterns of courtship outcomes for both species (Figs. 9-6 and 9-7). In each case, a minority of courtships reached the neck-grab stage (although somewhat more

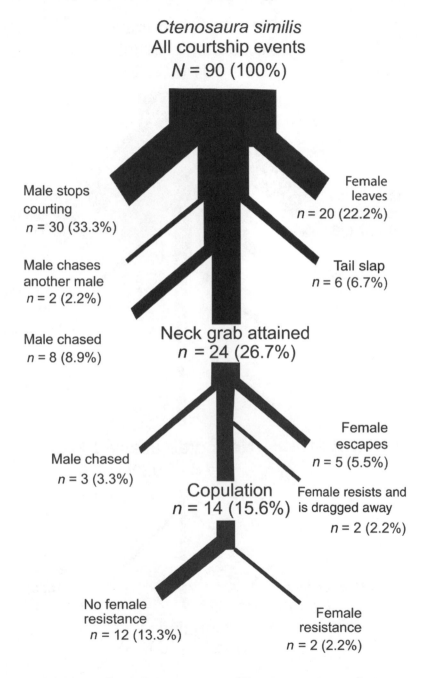

Figure 9-6 Sequence of courtship and frequencies of events that end courtship for *Ctenosaura similis*. Widths of lines are proportional to the number of observations. See text for descriptions of behavioral categories.

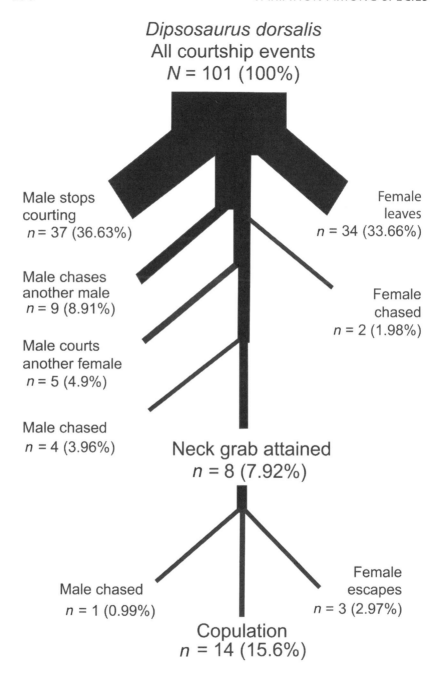

Figure 9-7 Sequence of courtship events and frequencies of outcomes for *Dipsosaurus dorsalis*. Widths of lines are proportional to the number of observations. See text for descriptions of behavioral categories.

for *C. similis*; 26.7 versus 7.9%, respectively), and only about half of those terminated in copulation. Overall rates of copulation were somewhat higher for *C. similis*. Prior to the neck grab, many courtships terminated because of male-based behavior. About one-third of courtships terminated when the courting male simply stopped vibratory head nodding (the index behavior for courtship). In these cases, the male moved away from the female or remained in proximity to her in a basking posture. In *C. similis*, it was not uncommon for a dominant male to court a single female multiple times during a single day. Other male-based behavior that led to cessation of courtship included courting males leaving one female to court another, or male-male chases. In each species, courting males were occasionally chased away from a female, or they left a female to chase another male.

The second single most important cause of courtship termination (after cessation by the male, as just discussed) was the act of a female leaving the courting male. Females typically moved quickly away from the male, making a short sprint >2 m away or down a burrow. Female *D. dorsalis* did not employ the tail slap, as female *C. similis* did, but two females were chased away from courting males by other females (Fig. 9-7).

Females continued to resist some courtships beyond the neck-grab stage. Five female *C. similis* and three *D. dorsalis* struggled free of neck grabs and escaped. Two *C. similis* females were dragged, struggling, out of sight of the observer by low-rank males; the outcomes of these "courtships" are unknown. Two female *C. similis* continued to resist during the copulation (Fig. 9-6).

The nature of courtship by male *C. similis* varied as a function of male status. Although dominant and low-rank males utilized the same suite of movements to court females, courtships by subordinate and midrank males tended to be quicker than those of dominant males. The normal range of times spent in the pre-neck-grab portion of courtship was 0–513 s for low-rank males, and 5-686 s for dominant males (means = 91 and 170 s, respectively). Dominant males spent more time in the pre-neck-grab portion of courtship than did low-rank males (*t*-test: $t_{77} = 3.17, P = 0.002$). The courtships of dominant males also included more separate bouts of vibratory head nodding than did those of low-rank males (4.9 and 2.0, respectively; $t_{35} = 2.27, P = 0.03$). However, there was no difference between dominant and low-rank males in the total portion of the daily time budget spent courting ($t_{13} = 0.91$, $P = 0.38$). By contrast, male *D. dorsalis* of different status did not court females differently. Dominant males and low-rank males courted for the same duration, whether measured as the time spent head nodding (*t*-test: $t_{34} = 1.69$, $P = 0.10$); the number of head-nodding bouts involved in the courtship ($t_{100} = 0.52$, $P = 0.61$); or the total duration of courtship, including head-nodding time and pauses between bouts ($t_{27} = 1.25$, $P = 0.22$).

Figure 9-8 depicts the frequencies of the nature of courtship termination on males of different status. Female *C. similis* were significantly less likely to leave when courted by a dominant male (χ^2 on courtship outcome against male status; contingency coefficient $= 0.574$, $P = 0.0002$). Dominant males were also more likely than low-status males to stop courting on their own. Females also resisted the neck grab of dominant males less than those of low-rank males. There was an overall skew in the number of copulations as well; dominant males achieved 12 of the 14 observed copulations.

In *D. dorsalis*, male status did not significantly affect the nature of courtship termination; all males were equally likely to stop courting on their own, and female response did not differ as a function of male status (χ^2 on courtship outcome against male status; contingency coefficient $= 0.262$, $P = 0.26$). Although the trends are suggestive of some female discrimination among males (more low-rank males than dominant males appeared to meet resistance during the neck grab), the rarity of courtship by low-rank males limits the analyses.

These results suggest that females do respond differently to the courtship of males of different status, at least for *C. similis*. However, it is difficult to determine whether females are responding to male status per se or to courtship style, because in *C. similis*, courtship style varies as a function of male status—low-rank males court more aggressively and are quicker to move to the neck-grab stage of courtship. This perhaps explains the greater frequency with which the neck grabs of low-status males are met with female resistance. For both species, courtships by low-rank males are generally rare. For example, for *D. dorsalis*, I observed only a single individual of each status in the "neck-grab, struggle" category. More long-term studies, including more observations of these rare events, will be required to examine trends among males of different status with more statistical rigor.

The frequency with which males stop courting on their own brings into question the measurement and function of courtship. From a methodological viewpoint, it may be difficult in practice to pin down the stop point for a single courtship. Extensive periods of close proximity between male and female, with isolated bouts of head-nodding

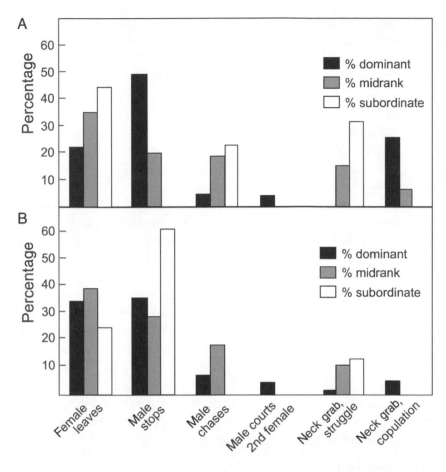

Figure 9-8 Outcomes of courtship for males of different status for (*A*) *Ctenosaura similis* and (*B*) *Dipsosaurus dorsalis*. Height of bars represents the percentage of total courtships for males of a given status. See text for descriptions of behavior terminating courtship. "Male chases" includes males leaving to chase another male and males being chased by other males.

behavior by the male, may function as courtship and may be required to bring a female into reproductive condition ("mate priming"; Bull 1994). Male displays have been shown to stimulate ovulation in teiid lizards (Crews 1975), and a female's receptiveness to the neck grab is also hormone dependent, at least in anolines (Tokarz and Crews 1981). Thus, what appear to be isolated, aborted courtships, especially by dominant males, may be needed to synchronize the reproductive efforts of the pair behaviorally and physiologically. If so, then the advantage of dominance for males defending female basking areas becomes clear: they gain extended access to females during the majority of their activity time. Low-rank males, by contrast, are courting "moving targets," because the females typically walk extensively during their foraging bouts beneath flowering trees and (at least in the case of *C. similis*) quickly return to their basking home ranges.

What are the opportunities for female choice in the mating systems of these two species? Females were not observed to initiate courtship with males, which on the surface seems to indicate that female mate choice may be limited. However, the universal low receptivity toward males (especially in *D. dorsalis,* in which females left males of all status with equal frequency) (Fig. 9-8) still may be considered a form of mate choice, because it clearly contributes to variance in mating success between males of different status (Tokarz 1995b). Low receptivity toward males regardless of status ensures that the only males that will achieve copulations are those able to invest the necessary time in courtship and/or to fend off other males (Cox and Le Boeuf 1977). Female rejection of male courtship is common in lizards (Stamps 1983b), but in few species do males force copulations with unwilling females (but see Hews [1990], Rodda [1992], and Olsson [1995]). Such behavior is most likely when a given social system restricts access to females to a small number of males, and when sexual size dimorphism is pronounced, making it difficult for a small female to resist the attempts of a larger male. Both criteria are met in the case of *C. similis.*

Differences in Male Courtship Success

The distribution of courtship success among males (defined as the number of females courted in a season) was more skewed for *C. similis* than for *D. dorsalis* (Fig. 9-9). Those *C. similis* males that defended burrow areas (invariably the dominants) courted significantly more females than did peripheral males (midrank and subordinate combined; $t_{19} =$ 2.4, $P = 0.02$). Among *D. dorsalis,* males who defended cool-shade shrubs (*L. tridentata* and *C. linearis* combined) courted more females than those that defended other shrubs or that lacked an HRC ($t_{27} =$ 2.79, $P = 0.01$). Here, the effect of shrub type is complicated by the variable of male status: one *L. tridentata* was defended by a midrank male and three dominant males defended sites other than cool-shade shrubs. A separate analysis examining the influence of male dominance on courtship success, independent of shrub type, showed a significant effect of male status ($t_{27} = 2.47$, $P = 0.02$).

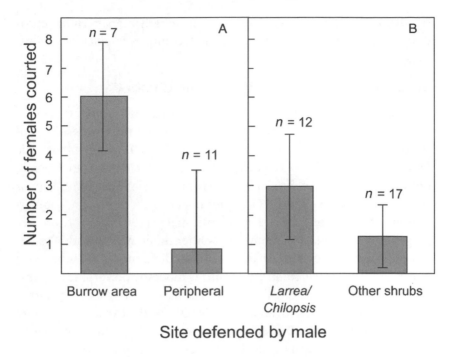

Figure 9-9 Courtship success as a function of male site defense for (*A*) *Ctenosaura similis* and (*B*) *Dipsosaurus dorsalis*. For *C. similis,* "burrow area" males were those that defended clumps of burrows and attendant females, and "peripheral" males were all others. For *D. dorsalis,* males that defended the shrubs *Larrea tridentata* and *Chilopsis linearis* (which provide the coolest shade; see Figs. 9-2 and 9-3) are grouped; "other shrubs" includes males that defended other shrubs, as well as those with no defined home range center.

Although male courtship success is a function of site defense in both species, the overall disparity in mating success appears greater in *C. similis* (Fig. 9-9). This general comparison suggests that the more extreme movement restrictions of *C. similis*—which seem to result primarily from a limitation of burrows—lead to a greater degree of polygyny than for *D. dorsalis.* Cool-shade shrubs, although arguably advantageous to the animals' above-ground survival, are simply too widely available to lead to a highly skewed pattern of mating success; more male *D. dorsalis* have access to females than do males of *C. similis.*

A limitation of this comparison is that both courtship success and mating success are imperfect measures of reproductive success. Studies on many taxa, especially birds (e.g., Westneat, Sherman, and Morton 1990), have revealed that unseen copulations by furtive males may be a significant source of fertilizations in polygynous systems. Thus, the

trends in this study are suggestions of an underlying pattern of reproductive success that will require genetic techniques to elucidate definitively.

Sexual Dimorphism and Male-Male Combat

Both species exhibit sexual dimorphism, with males proportionately
larger than females in all measured dimensions (*t*-tests on residuals of
the overall regression with log body length; $P < 0.01$), except that male
D. dorsalis did not differ proportionately in mass from females. Overall,
however, the degree of dimorphism was more pronounced in *C. similis*
(Fig. 9-10). Males of this species exhibit a head allometry that is qualitatively different from that of females: the male's head is proportionately
longer and more massively muscled. Other differences include the development of the dorsal crest and a twofold difference in mass between
males and females. *Dipsosaurus dorsalis* is modestly dimorphic by comparison; although males of this species have proportionately larger
heads, the growth trajectories of head length for the two sexes are more
nearly parallel (Fig. 9-10). Males and females are also more similar in
size and in development of the dorsal crest than *C. similis*.

In *C. similis*, dominant males did not differ significantly from
midrank males in body length or mass (ANOVA: $F_{2,18} = 15.60$, $P =
0.0001$; Scheffe post-hoc comparison: $P = 0.96$; and ANOVA: $F_{2,18} =
18.23$, $P < 0.0001$; Scheffe post-hoc comparison: $P = 0.98$, respectively),
but both dominant and midrank males had longer bodies and greater
mass than subordinate males (Scheffe post-hoc comparisons: domi-
nant-subordinate, $P = 0.006$ and $P = 0.0002$; midrank-subordinate, $P
= 0.002$ and $P = 0.001$). Dominant males also differed from low-rank
males (midrank and subordinate status lumped) in having proportionately longer heads (Fig. 9-11). However, dominant males did not differ
from low-rank males in proportional head height, head width, or body
mass (*t*-tests on residuals of the regressions of these variables on log
body length; $t_{19} = 1.5$, $P = 0.15$; $t_{19} = 0.67$, $P = 0.51$; and $t_{19} = 0.75$, $P
= 0.46$, respectively).

Unlike *C. similis*, dominant male *D. dorsalis* did not differ from low-
status males in relative head length (ANOVA: $F_{2,27} = 2.10$, $P = 0.42$) or
in any other relative body proportion. Dominant male *D. dorsalis* did
have longer heads than subordinate males, but not midrank males
(ANOVA: $F_{2,27} = 6.10$, $P = 0.007$ and $P > 0.05$, respectively). Male size
varied as a function of site defense as well; males whose HRC was *L. tridentata* or *C. linearis* had significantly longer bodies than those whose

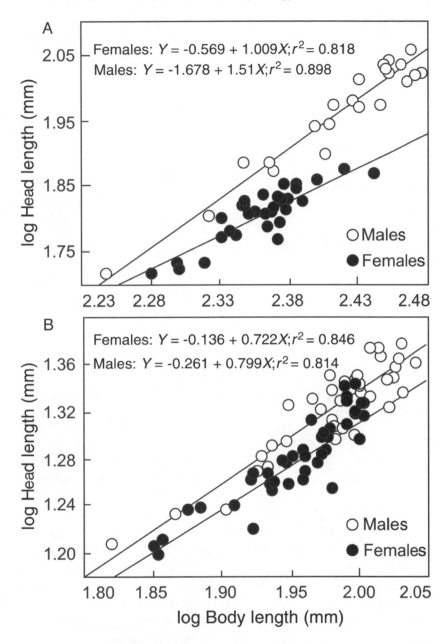

Figure 9-10 Regression of log head length versus log body length for males and females of (A) *Ctenosaura similis* and (B) *Dipsosaurus dorsalis*.

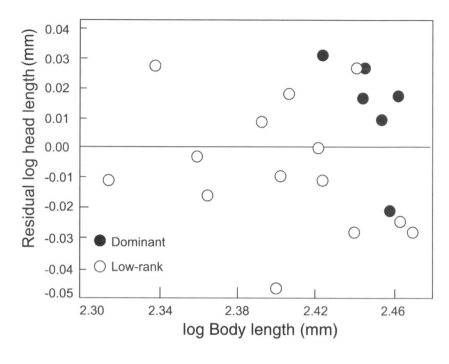

Figure 9-11 Residuals of log head length versus log body length regression for adult male *Ctenosaura similis*, plotted against log body length. Dominant males had proportionately longer heads than low-rank males ($t_{19} = 2.12$, $P = 0.0467$).

territories centered on *E. deserticolum* or those that had no HRC (ANOVA: $F_{26,6} = 5.86$, $P = 0.001$; Scheffe multiple comparisons: $P < 0.05$). In addition, males whose HRC was beneath *L. tridentata* and *C. linearis* had longer heads than those who had no HRC ($F_{26,6} = 5.62$, $P = 0.001$; Scheffe multiple comparisons: $P < 0.05$). Males whose HRC was beneath *L. tridentata* were also significantly heavier than males with no HRC (ANOVA: $F_{26,6} = 5.34$, $P = 0.001$). However, residual analyses indicated no differences among these males in head-body proportions, not in head length, width, or height ($P > 0.05$). Thus, males with territories centered on cool-shade shrubs had longer bodies and absolutely longer heads but were not proportionately larger headed.

The difference in proportional head length for male *C. similis* of different status should be accepted as a preliminary result, given the marginal *P* value of the test ($P = 0.047$; Fig. 9-11). Even if interpreted as only a trend, however, it suggests intense selection on head length for this species, since it is less easily explained in terms of differences in male age. One possible explanation for the differing degrees of head length dimorphism in these two species emerges when the trait is viewed in light of the mode of combat. Fights between males were rare (I observed six for *D. dorsalis,* and three for *C. similis*) and followed extended lateral displays in which both males circled and bobbed their heads. The initial phases of aggression and combat were remarkably similar in these two species; however, they differed in the mode of physical contact between males. A pair of *D. dorsalis* would, after extended escalation of display, lunge and bite one another's forelimbs or thighs and, while maintaining this mutual jaw lock, would slap each other with their tails (also described in Norris [1953]). Fights rarely lasted more than 3 min, during which there were usually several bouts of biting and tail slapping separated by brief pauses. *C. similis* males engaged in a more ritualized contest in which, after escalating, the two faced each other with foreparts overlapping, and each laid his head against the flank of his opponent with the mouth gaping to its maximum extent (Fig. 9-12A, B). They would alternate remaining motionless in this gaping posture, jaws in contact with but not biting down on their opponent, with bouts of lunging and pushing against one another (Fig. 9-12C). At no time was actual biting observed.

Why such a singular and distinctive mode of combat in *C. similis?* Possibly, it serves as an assessment contest; with gaping jaws laid across one another's flanks, each lizard could conceivably judge his opponent's head length, which would be a function of the distance between the open jaw tips. If this interpretation is correct, it is reminiscent of assessment matches between male ungulates, which are believed to have evolved as an indirect result of large size and potentially lethal weaponry; it is safer to assess rather than to fight (Maynard Smith 1974; West-Eberhard 1979). Male *C. similis* are certainly equipped for fighting, with elongated front teeth that make savage weapons when used against human captors, and the ritualized display may have evolved as an evolutionary stable strategy to minimize unnecessary combat (e.g., Parker 1974). However, scars do occur on the thighs and flanks of large males, so more observations are needed to ascertain how often males actually abstain from biting.

Conclusion

The study of animal mating systems and sexual dimorphism is inherently complex because the interactions of ecology, behavior, and morphology are myriad. Parsing out cause-and-effect relationships in

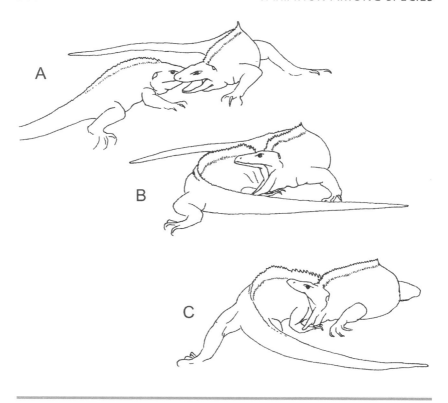

Figure 9-12 Sequence of behavior in male-male "combat" in *Ctenosaura similis*, drawn from photographs. Following extensive lateral displays and head bobbing, two males approach one another (*A*) and align their heads alongside their opponent's flank, mouths gaping (*B*); they may remain in contact for 1 min or more before lunging and circling one another (*C*). This sequence was repeated multiple times in long combat/assessment matches, which lasted more than 20 min.

a single study relies on a holistic approach that sometimes may run roughshod over details of ethology and ecology that are arguably important. The strength of such broad-brush approaches, however, is that they portray animal social systems as inseparable from the ecological setting in which they are embedded. For *C. similis* and *D. dorsalis,* a complex relationship emerges from a set of resources important to the animals. In each case, the one that most closely determines the mating system is the one most narrowly distributed in space.

The close association among the location of burrows, female home ranges, and male courtship success in *C. similis* suggests that access to retreats, either for escape from predation or for other purposes, is crucial to the social system. Similar ties between movements and burrows have been shown for other large-bodied iguanids (e.g., *Ctenosaura pectinata:* Evans [1951]; *Ctenosaura hemilopha:* Carothers [1981]; *Sauromalus:* Berry [1974]; Case [1982]; *Cyclura:* Iverson [1979]; Dugan and Wiewandt [1982]). Herbivorous lizards in general tend to be large, and herbivory itself may require ideal thermoregulatory conditions (Pough 1973). Therefore, iguanid social life often centers around shelters, basking sites, or both. This situation may lead to resource defense polygyny if these activity centers are defensible, in which case smaller or younger males may adopt alternative mating strategies, becoming satellite or "sneaker" males, as has been observed in many taxa (Waltz and Wolf 1984). However, if males cannot monopolize resources important to females, such as occurs in some iguanids (*Amblyrhynchus:* Wikelski, Carbone, and Trillmich [1996]; *Iguana:* Dugan [1982]; Rodda [1992]), males may defend display territories, which may be clumped into leks.

Dipsosaurus dorsalis, however, is not as constrained by large size. Instead, the social system and patterns of male mating success seem to be most tightly correlated with thermally superior home ranges. This is not surprising given this species' unusual habit of maintaining activity in a habitat comprising mostly space that is functionally uninhabitable. *D. dorsalis* has been hypothesized to tolerate excessively high environmental temperatures as a cost of maintaining above-ground activity (Porter et al. 1973), with individuals exhibiting "voluntary hyperthermia" of T_b values of 44–46°C (Norris 1953; Porter et al. 1973). However, while this range of tolerance may allow lizards to move through a mosaic of thermal patches, including open sun and high-temperature shrubs such as *E. deserticolum* and *P. emoryi,* many of these microhabitats would be lethal to the animals after extended exposure times. Hence, although *D. dorsalis* can briefly tolerate high temperatures, a home range centered on a cooler retreat is crucial.

Much of the previous discussion focuses on the male's perspective, but some of the most interesting unanswered questions involve benefits and costs to the females of the dispersion of resources and of the female-directed behavior of males. For example, the precise function of female philopatry has not been demonstrated, at least for *C. similis.* Do the burrow areas function solely in predator escape, or are thermoregulatory benefits of equal or greater importance? In addition, since these females may oviposit communally in these burrows (Mora 1989), female philopatry may be related to a particular egg deposition strategy, and not to predator avoidance. Intrasexual aggression among females is common in both species (unpublished observations; in the case of *C.*

similis I observed more female-female altercations than male-male ones). However, lizard ethologists currently lack a conceptual framework on which to hang such behavior, and investigations into female dominance hierarchies in the Iguanidae would seem profitable.

Finally, can we say something about the mating patterns of these two species, based on studies of a single population of each? When there appears to be a link between morphology and a social system (such as for the large-bodied *C. similis* and its absence of widespread useable burrows), it is tempting to designate a specieswide pattern, arguing from the details of my studies that since all populations of *C. similis* contain large individuals, all populations should encounter the same habitat constraints. One risks being too simplistic, however, by assuming that local resource limitations will be unvarying. Movement patterns, and the social systems to which they contribute, are expected to be fluid, responding to local resource patterns and predation pressures that may vary from population to population (see chapters 5-7). For example, the avoidance of *L. tridentata* as a food source in this population of *D. dorsalis* is crucial in reaching the conclusion that the value of this shrub to the lizard is not in its leaves and flowers, but in the shade that it provides. If *L. tridentata* had been a valuable food source, the effects of food and shade would be confounded. In fact, several studies have reported that flowers or leaves of this plant do serve as a food source for other populations of *D. dorsalis* (e.g., Norris 1953; Minnich and Shoemaker 1970). This suggests possible differences between the mating systems in those populations and the one in the present study. Intraspecific variation in mating systems has been documented in species ranging from guppies (Endler 1980, 1983) to iguanids (*Sauromalus*: Ryan [1982]) to artiodactyl mammals (Langbein and Thirgood 1989; Nefdt and Thirgood 1997). Such shifts seem tied to local resource dispersion and predation pressure, supporting the notion of a direct ecological influence on mating patterns and serving as a warning against characterizing an entire species as having a single mating system.

Acknowledgments

Earlier incarnations of parts of this chapter benefited from the comments of Michael Bull, John Carothers, Stanley Fox, Victor Hutchison, Charles Peterson, Caryn Vaughn, Laurie Vitt, Linda Wallace, and Martin Wikelski. In addition, Stanley Rand and Patricia Schwagmeyer provided suggestions during the early phase of the work. Special thanks go to Karla Feist for encouragement and assistance. My interest in this project was sparked by a pilot project in Costa Rica during my participation in a field course of the Organization for Tropical Studies. Further work was supported by a Smithsonian Tropical Research Institute Short-Term Fellowship. The University of Oklahoma Department of Zoology and Graduate College provided financial assistance and the invaluable loan of a truck for fieldwork in California.

CHAPTER 10

Social Behavior at High and Low Elevations

Environmental Release and Phylogenetic Effects in *Liolaemus*

Stanley F. Fox and Paul A. Shipman

One of the axioms of evolutionary ecology is that animals are modified to the specifics of their environment through the process of natural selection (Darwin 1859). In fact, it was the observation of distinct differences among species in the radiation of geospizine finches on the various islands of the Galápagos, later documented by Lack (1947), that stimulated Darwin to formulate the theory of natural selection. This adaptation to the environment is obviously true for morphological traits that allow certain phenotypes, compared with alternative phenotypes, to better survive and reproduce against challenges of their environment, both biotic and abiotic, but it is also true for behavioral traits. Indeed, on arrival to a novel environment (or following some change in environmental conditions), animals are thought to change behaviorally first, then morphologically (MacArthur and Wilson 1967).

Interspecific Examples

One of the first to recognize such adaptational change in suites of traits related to social behavior was John Crook (1964), in his comparative studies of some 68 species of African weaver birds in different environments. Species inhabiting the forest were monogamous, sexually monomorphic, solitary insectivores that defended large territories and built cryptic, solitary nests. By contrast, savannah species were polygynous, colonial, sexually dimorphic granivores that nested in colonies and did not defend territories. Crook argued that these divergent patterns of social behavior were best understood by considering food and predation. In the forest, food is spread out and not overly clumped, which promotes territoriality. Both parents are needed to feed the nestlings, so monogamy is favored. Spaced-out cryptic nests serve to minimize predation of nests (Tinbergen et al. 1967; Andersson and Wiklund 1978). Because females are not clumped, males are monogamous and sexual selection is not strong (Emlen and Oring 1977). Thus, sexual dimorphism does not develop. In the savannah, however, the distribution of seeds is patchy and in places food is very abundant. A single male cannot economically defend such a rich patch due to potentially high intrusion pressure from conspecifics, so the birds feed as a group. Their group living also aids in finding these superabundant clumps of food (Krebs, MacRoberts, and Cullen 1972; De Groot 1980). Suitable nest sites are rare, so nests are constructed in colonies in thorny acacia trees as protection against predators and a colony of birds can show group defense. Because birds are clumped and females can care for nestlings by themselves, opportunity for sexual selection is high and, thus, polygyny and sexual dimorphism develop (Emlen and Oring 1977). A third, smaller group was grassland granivores. Group living was favored for finding rich patches of seeds, but nests were on the ground and therefore subject to predation. Consequently, these grassland species showed a social organization intermediate to the forest and savannah species; they fed in flocks but spaced out their nests in loose colonies.

Jarman (1974) analyzed 74 species of African ungulates in a similar manner. Smaller species, because of their higher mass-adjusted metabolic demands and smaller mouths capable of selective foraging, lived in the forest and fed on well-dispersed, highly nutritive shoots and berries. Consequently, male and female pairs mostly defended small territories, and sexual selection was minimal, as was sexual dimorphism. The animals escaped from predation by escaping detection. At the other extreme of the continuum of body size were the large ungulates, which subsisted on lower-quality forage and lived in large herds on the plains. Because animals were congregated, there was an elevated opportunity for sexual selection, and sexual dimorphism, defense of female harems, and dominance hierarchies among males were more prevalent. The largest herds showed group defense against predators, and those in smaller herds relied on flight or safety in numbers.

A comparative approach to the study of social organization among primates likewise categorized species into groups according to aspects of their ecology, particularly aspects relating to food and predation (Crook and Gartlan 1966). Similar to Crook (1964) and Jarman (1974), categories ranged from small, solitary, insectivorous, only slightly sexually dimorphic, forest-dwelling primates such as lemurs, to larger baboons that live in large groups, browse vegetation in the plains, and show marked sexual selection and sexual dimorphism.

Intraspecific Examples

Intraspecific variation in social behavior has been described and related to food and predation, as well. Franklin (1982, 1983) conducted long-term field studies of the socioecology of the South American camelid, the guanaco (*Lama guanicöe*), principally at two sites in southern Chile. At both sites, the social organization during the breeding season was characterized by family groups of a territorial male with his group of females and young, and males without females that were either congregated into bachelor herds or solitary. Animals from the colder, harsher site of Torres del Paine were seasonally migratory, grouping together in large mixed-sex bands in the winter and territorial only in the summer, whereas territories at the more benign site on Tierra del Fuego were defended year-round. Forage was better at Tierra del Fuego and, consequently, territory size (and family size) was much smaller than at Torres del Paine. Females were a scarcer commodity at the harsh site, and most males were either aggregated with other males or defending barren territories waiting to attract females. The puma (*Felis concolor*) is a principal predator of the guanaco (Franklin et al. 1999). Because pumas are common at Torres del Paine (Franklin et al. 1999), but absent from Tierra del Fuego, larger family group size, large, mixed-sex groups in the winter, and relative lack of females at the mainland site may partially be a consequence of puma predation.

In fish, Farr (1975) showed how predation can mold the social behavior of the guppy (*Poecilia reticulata*) in different habitats of Trinidad and Tobago. In populations with no predators or only large characid and cichlid ones, guppies live at the edges of streams in dense aggregations, sexual selection is strong, and the very brightly colored males court the dull-colored females with prolonged, elaborate displays. The large groups and shallow waters offer protection from large predators. In those populations with only a small predator (*Rivulus hartii*), guppies are dispersed throughout the stream, and the somewhat drab males do not engage in prolonged, flashy courtship. The method of ambush hunting of this small predator that can enter shallow water selects against the bright coloration and eye-catching courtship of groups of guppies in shallow water seen in populations with no predators or only large ones.

However, Endler (1980, 1983), who studied color dichromatism and polymorphism in guppies and not primarily behavior, found a

somewhat different pattern related to predation. Through direct observation and his classic greenhouse and field experiments, Endler
(1980) showed that the color pattern of males was a balance between
sexual selection for bright colors to attract mates, on the one hand,
and natural selection for crypsis for avoiding predators, on the other.
In the presence of dangerous predators (characids and cichlids), natural selection evolved less colorful males that matched their background well, at odds with Farr (1975). In the absence of these predators (or presence of only less dangerous ones such as *Rivulus*, as
evaluated by Endler), epigamic sexual selection evolved colorful males
that diverged from a good background match. Stoner and Breden
(1988) further demonstrated a strong preference among females for
bright, actively courting males in populations with light predation and
the reverse in populations with heavy predation. Although the conclusions of Farr and Endler differ (apparently Farr [1975] confounded
the effects of predation pressure and female density [Stoner and Breden 1988]; Luyten and Liley [1985] later showed that when densities
were controlled, active courtship frequency by males was inversely correlated to predation pressure), it is clear that predation pressures influence dichromatism and the strength of sexual selection in the
guppy, and related concomitant differences in social behavior likely
coevolve with the relative degree of dichromatism (Houde 1997).

Lizards, just as other animals, adapt their social organization to the
environment in which they live. Tinkle (1967a) compared the autecology of *Uta stansburiana* from Texas and Colorado. Whereas those from
Texas were strongly sexually dimorphic, aggressive, and territorial, those
from Colorado were barely dimorphic and only weakly aggressive. Tinkle (1967a) concluded that males in Colorado did not defend territories; instead, they formed dominance hierarchies. The habitat in Colorado was vegetatively more closed and visibility was limited, thus
precluding efficient territorial defense. Because lizards lived longer in
Colorado and males had more opportunities for mating, the premium
for heightened aggression was less there than in Texas, where males had
but one chance within their short lifetime to acquire a territory through
intrasexual competition, and mate.

Lizard Social Organization
Effect of Food Availability

Abundance and distribution of food can influence spacing patterns
and behavior of lizards. Simon (1975) and Krekorian (1976) showed

an inverse relationship between food abundance and territory size. Food supplementally added to treatment plots sometimes can provoke a decrease in home range size (Simon 1975; Ferguson, Hughes, and Brown 1983; but see Stamps and Tanaka [1981b], Waldschmidt [1983], and Guyer [1988b]); an increase in home range overlap (Stamps and Tanaka 1981b; Ferguson, Hughes, and Brown 1983; Guyer 1988b); an increase in lizard density through enhanced survival and heightened immigration from nearby, unaltered sites (Guyer 1988a; Hews 1993); and elevated mating opportunities for adult males (Hews 1993; Guyer 1994). Among species of lizards, diet and variability of food supply can influence social organization. Perhaps because insectivores experience less variation in food supply than herbivores (Stamps 1983b), their social organization is less variable. In general, insectivorous lizards are territorial and polygynous (especially those with small home ranges of females), whereas herbivorous lizards exhibit a wide range of social organization from territoriality to leks, to dominance hierarchies, to absence of territoriality (Stamps 1983b; Carothers 1984; Wikelski, Carbone, and Trillmich 1996).

Effect of Predation

Predation also can influence the behavior of lizards as well as their social organization. Even Darwin (1845) commented on the astonishing tameness of the large marine iguanas on the Galápagos, noting the inherited lack of fear to humans after thousands of years of existence with no natural terrestrial predators (Galápagos hawks do, however, prey on gravid females [Boersma 1983]). Male lizards displaying with visual signals from exposed, elevated perches may run an elevated risk of predation compared with nondisplaying males (Jakobsson, Brick, and Kullberg 1995; see also chapter 4), and the choice to carry on courtship (or intraspecific aggression) or not in the presence of a predator is influenced by the cost of breaking off the social behavior for the benefit of evading the predator (Cooper 1999b; see also chapter 4).

One important defense against predation that many lizards have is tail autotomy (Arnold 1988), but individuals that have lost their tails in such an encounter may be socially impaired (Fox and Rostker 1982; Martín and Salvador 1993a; Salvador, Martín, and López 1995; Fox, Conder, and Smith 1998). Nevertheless, in species of lizards with heavy predation, tail loss was more frequent and tail autotomy was induced more easily than in species with lighter predation, even when correcting for phylogeny (Fox, Perea-Fox, and Castro Franco 1994). The social behavior of *Ctenosaura similis* is molded around the fact that females are grouped at burrows (see chapter 9), probably for the benefit of escape from predators (but perhaps also for thermoregulatory reasons). Brattstrom (1982) mentions an interesting relation among predation, vegetative density, probable food density, and social organization in two endemic species of tree lizards of the Revillagigedo Islands off the tip of Baja California. Only one species of lizard lives on Clarion Island and also only one species of predator, the endemic racer, *Masticophis anthonyi*. On Socorro Island, a different species of tree lizard and its sole predator, an endemic red-tailed hawk, are found. On Socorro Island, the lizards are not territorial and are very wary, supposedly because of the predation threat of the hawk, which keeps lizards away from open areas, where they could defend territories from conspecifics. Consequently, the lizards are congregated in the more densely vegetated parts of the island, and because of the high lizard density there, presumably greater abundance of food, and reduced visibility, they appear to have adopted a dominance hierarchy. Territoriality would not be favored in such a situation (Tinkle 1967a; Stamps 1977b; Alberts 1994). By contrast, on Clarion Island, which is less vegetated and presumably poorer in food abundance, the lizards are territorial and not wary; males defend territories from elevated perches in the open, where they can readily detect the terrestrial, diurnal snake and conspecific territorial interlopers.

Stamps (1983a) concluded that the social organization of *Anolis aeneus* is molded by the habitat affinities of the predatory *Anolis richardi*, which inhabits shady habitats of Grenada. Juveniles are aggressive and fight for territories in scarce, sunny clearings devoid of *A. richardi*. As adults, they are less aggressive and live in shadier habitats along with their former predator, having grown too large to be suitable prey. Chapters 5–7 also describe how predation pressures can influence morphological and behavioral traits of populations of lizards, primarily in the relative degree to which the strength of sexual selection for obtaining mates may be compromised by the strength of natural selection for escaping predators, just like color pattern in populations of a lizard on the island of Tenerife (Thorpe and Brown 1989), and just like the studies of the guppy we have discussed.

Effect of Thermal Environment

Finally, in contrast to endothermic birds and mammals, lizards, as ectotherms, often are found to be more seriously constrained by thermal aspects of their environment (Huey 1982), with consequences to their social behavior. *Uta stansburiana* in western Colorado confines its social behavior to spring and early summer; later in the summer, thermal properties of microhabitats (biophysical predictions of lizard body temperatures from micrometeorological measurements) are too hot for such activity (Waldschmidt and Tracy 1983). In June, when predicted body temperatures in more microhabitats were closer to those temperatures at which the lizards were able to sprint the fastest, males had significantly larger home ranges than females; after June, most microhabitats were simply too hot for lizards to thermoregulate adequately, and home ranges of males dropped to a size comparable with that of females.

Spatial locations and microhabitat use (and presumably social interactions) of Galápagos land iguanas (*Conolophus pallidus*) were strongly affected by thermal aspects of their environment over different parts of the year (Christian, Tracy, and Porter 1983). Home ranges of these lizards were smallest during the cool, foggy Garua season when thermal conditions were most stressful (Christian and Tracy 1985). At one locality in Big Bend National Park, Texas, operative (environmental) temperatures of *Sceloporus merriami* (as measured by internal temperatures of copper lizard models) were too high during most of the day, except for about 2 h beginning at local sunrise and a brief period in the late afternoon (Grant and Dunham 1988). That is the time lizards there also showed the most movement and social behavior. Even in the late morning when some microhabitats offered suitable operative temperatures, lizard activity was subdued because these small patches were surrounded by thermally unsuitable microhabitat, and it proved too costly for lizards to use these "insular" patches.

Even though thermal conditions normally constrain lizard activity in pursuits such as feeding, territorial defense, courtship, and mating, sometimes the benefits of these activities outweigh the costs of being active at warmer operative temperatures. At a lower-elevation site at Big Bend National Park, *S. merriami* adjusted body temperatures upward in the evening so as to be able to be active under the warmer operative temperatures then, despite the elevated cost of compromised locomotion, elevated water loss, and faster metabolism (Grant 1990). Apparently, the need for social activities such as territorial defense and mate acquisition under the high lizard densities at this site made up for the cost of activity at a higher body temperature. Moreover, operative temperatures in the evening were only somewhat higher than body temperatures observed in the morning, whereas operative temperatures by late morning were largely much too high for use of those microhabitats by lizards. Thus, an upward adjustment of body temperature in the evening, but not in the morning, opened up considerably more microhabitat to the lizards and constrained their activity less in the evening than it would have in the morning. Schäuble and Grigg (1998) observed the same phenomenon in the Australian agamid *Pogona barbata*, but over seasons. More careful thermoregulation and higher body temperatures in summer versus autumn was not a consequence of limiting optimal thermal habitat in autumn but, instead, to the demand in summer for territorial defense and social interactions. Lizards in the summer maintained their body temperatures close to (or briefly exceeding) the maximal limits in order to be able to continue social behavior. As such, they had to thermoregulate carefully so that they would not overheat. In the autumn, with lower body temperatures, they were not subject to the same need for careful thermoregulation.

Phylogenetic Effects

One very important contribution of recent years is the rediscovery of the importance of phylogeny in the interpretation of biological traits, and the presentation of a multiplicity of statistical methods to gauge the importance of phylogenetic history in comparative studies (Felsenstein 1985; Brooks and McLennan 1991; Harvey and Pagel 1991; Miles and Dunham 1993; Martins and Hansen 1996, 1997; Martins 2000). These new phylogenetic comparative methods (PCMs) have been applied especially in animal behavior (Brooks and McLennan 1991; Martins and Hansen 1996), and every modern comparative study of behavioral traits among species now uses one or another of these new approaches. Phylogenetic comparative methods are applied not only to address the statistical problem of inflated degrees of freedom when related species are considered as independent data points (Felsenstein 1985), but also to infer the form and direction of phenotypic evolutionary change along a phylogeny (Brooks and McLennan 1991; Martins and Hansen 1996). We developed a new PCM that uses partial, sequential canonical correspondence analysis to identify and test for phylogenetic correlation in a set of multiple behavioral characters of a small group of species of lizards from central Chile.

Social Organization of Chilean Lizards: An Integrative Hypothesis

Previous studies have shown or suggested that two important influences on lizard social behavior—predation (Brown and Ruby 1977; Van

Damme et al. 1989; Medel et al. 1990; Fox, Perea-Fox, and Castro Franco 1994; Lemos-Espinal and Ballinger 1995) and thermal rigor (Ruby and Dunham 1987; Marquet et al. 1989; Van Damme et al. 1989; Grant and Dunham 1990; Lemos-Espinal and Ballinger 1995; Sorci, Colbert, and Belichon 1996; Bashey and Dunham 1997)—vary with elevation. We studied the social behavior of seven closely related *Liolaemus* species of lizards (Schulte et al. 2000) along a steep elevational gradient in central Chile that varied in thermal rigor and predation pressure. We hypothesized that social organization would be more complex and developed at those sites with reduced thermal constraints and less predation. We further speculated that this adaptation of social organization would develop locally despite any phylogenetic constraints.

Methods

Study Sites and Species

We studied populations of *Liolaemus* species in their natural habitat at three sites in the Andean cordillera of central Chile during the austral summer/fall of 1986 and spring/summer of 1986–87. The sites were El Colorado (33°14' S, 70°16' W), located 75 km northeast of Santiago at an elevation of 2,900 m; Curva 20 (33°15' S, 70°19' W), located 45 km northeast of Santiago at an elevation of 2,300 m; and San Carlos de Apoquindo (33°23' S, 70°30' W), located 20 km east of Santiago at an elevation of 1,200 m. The area of each study site was 0.24, 1.10, and 0.82 ha, respectively. Central Chile is characterized by a Mediterranean climate of warm, dry summers and cold, rainy winters (with snow at the higher elevations) (di Castri and Hajek 1976). At El Colorado, the hottest month is January, with a mean maximal temperature of about 21°C; annual precipitation is 270 mm, mostly between May and October (Rozzi, Molina, and Miranda 1989). Lizards there were active from November to early March. At Curva 20, summer temperatures are somewhat warmer than at El Colorado, but clear skies are less common (Fox, unpublished data). Lizards were active at Curva 20 from October through March. At San Carlos, January is the hottest month, with a mean maximal temperature of about 26°C; annual precipitation is 425 mm, concentrated between May and September (del Pozo et al. 1989). Lizards at this site were active mostly from October to early April, but they may be seen on warm days even throughout the mild winter (J. Jiménez, personal communication).

Lizard species encountered at the three sites (in order of decreasing relative abundance) are as follows: El Colorado: *L. bellii* (=*altissimus*),

L. leopardinus, L. nigroviridis; Curva 20: *L. lemniscatus, L. nitidus, L. monticola, L. schroederi, L. fuscus;* San Carlos: *L. fuscus, L. lemniscatus, L. monticola, Callopistes maculatus. Liolaemus nigroviridis* and *C. maculatus* were rare on our study sites and were not part of our behavioral study. Table 10-1 provides the elevations of each site and the species we studied at each site. Table 10-2 gives the average sizes of adults and some natural history information about the seven species included herein; Plates 12–16 illustrate some representative species.

Field Methods

We placed numbered flags at prominent landmarks and drew a scale map of each site using a theodolite/plane table. During weather conditions permitting lizard activity, we censused for lizards, spending all day at one site for 2 to 3 d, then moving to a different site. When each lizard was first captured, we individually marked it with a combination of clipped toes and dorsal paint spots that allowed continuous recognition (Fox 1978). We determined sex from body coloration and presence of precloacal pores in males. When a marked subject was sighted, we recorded the date, hour, location, microhabitat, shaded air temperature 1 cm above its perch, sky conditions, and relative wind speed. If a sighted lizard lacked paint codes (and also on first capture), we captured it by noose and also measured its snout-vent length (SVL), tail length, head width, head length, tail loss status, length of tail stub if previously autotomized, and body temperature (T_b) with a rapid-reading cloacal thermometer. We did not measure T_b if a lizard was chased or if there was a delay of more than 20 s between capture and attempt to read its T_b. We computed relative density of each species at each site as the number of marked adults per hectare. Additionally, we maintained quantitative records of all potential predators (saurophagous snakes and raptors; the predatory *C. maculatus* was never observed within the boundaries of a site) seen on or flying over each study site during our lizard censuses.

Thermal Rigor

In the first days of each month, we measured thermal conditions at each site at each hour of a clear day from sunrise to sunset. We placed paired copper lizard models of three sizes (SVL of 44, 73, and 100 mm) in each of six microhabitats (sun/rock, sun/soil, filtered sun/rock, filtered sun/soil, shade/rock, and shade/soil) in two body positions (perpendicular to the sun, adpressed to the substrate; and parallel to the

Table 10-1   Study sites, their elevations above sea level, and lizard species from central Chile that were subjects of a behavioral study during 1986–87.

Study Site	Elevation (m)	Species
El Colorado	2,900	*L. bellii*
		L. leopardinus
Curva 20	2,300	*L. schroederi*
		L. nitidus
		L. monticola
		L. lemniscatus
		L. fuscus
San Carlos de Apoquindo	1,200	*L. monticola*
		L. lemniscatus
		L. fuscus

sun, upright off the substrate) each hour of these measurement days. The models were made of thin 40-gauge copper sheeting rolled and fashioned by hand into tubular shapes of lizard heads, torsos, and tails to which smaller tubular legs were glued. Each model was spray painted neutral gray and fitted with an internal copper/constantan thermocouple read by an Omega thermocouple reader Model 450 ATT to the nearest 1°C. Models equilibrated within 3 min to ambient conditions and were read 4 min after placement in each microhabitat. With this paradigm, the models integrated prevailing conditions of effective thermal radiation area of a lizard, air temperature, substrate temperature, solar and thermal absorbed radiation, convective and radiation conductance, and wind, physically modeling the operative (environmental) temperature (T_e) (Bakken and Gates 1975; Bakken 1992), that is, the equilibrial T_b of a live lizard of that size and posture in that microhabitat at that hour as per similar studies (e.g., Grant and Dunham 1988; Grant 1990; Schäuble and Grigg 1998; Bauwens, Castilla, and Mouton 1999). By using pairs of models of each size in divergent postures and orientations to the sun, we were able to record the minimal and maximal T_e of each size under the given meteorological conditions (Waldschmidt and Tracy 1983). A live lizard of the same size in the same microhabitat under the same meteorological conditions was assumed to fall within the range of T_e expressed by the pair of copper models.

On the same days in which the models were utilized, we also determined hourly availability of the same six microhabitats at each site. Each hour we measured the proportion of sun, filtered sun, and shade intersecting four randomly placed 15-m transects. Previously we had

Table 10-2 Average adult SVLs (mean SVL ± 1 SE) by sex and natural history information of seven *Liolaemus* species of central Chile.

Species	Males		Females		Diet[a]	Reproductive Mode[a]	Elevational Distribution (m)[b]
	SVL (mm)	n	SVL (mm)	n			
L. bellii	74.4 ± 0.7	48	70.2 ± 0.9	30	Omnivorous	Viviparous	2,100–3,000
L. leopardinus	85.8 ± 1.1	15	81.0 ± 1.8	13	Omnivorous	Viviparous	2,100–3,000
L. schroederi	52.8 ± 1.4	19	59.9 ± 1.1	9	Insectivorous	Viviparous	1,800–2,590
L. nitidus	86.9 ± 3.0	14	82.1 ± 2.0	16	Insectivorous	Oviparous	1,500–2,450
L. lemniscatus	48.2 ± 0.4	84	44.7 ± 0.5	56	Insectivorous	Oviparous	1,500–2,400
L. monticola	61.6 ± 1.2	34	56.9 ± 1.1	26	Insectivorous	Oviparous	1,500–2,450
L. fuscus	46.5 ± 0.4	98	43.3 ± 0.4	67	Insectivorous	Oviparous	1,500–2,100

[a]Data from Donoso-Barros (1966).

[b]Data from Carothers et al. (1997).

measured the overall proportion of rock and soil using 12 randomly placed 15-m transects at each site. By multiplying these fixed rock and soil proportions against the hourly proportions of sun, filtered sun, and shade, we estimated the hourly proportion of the availability of the six microhabitats in which the models were placed.

We compared the models' response against live lizards of matching sizes held in the shade and against freshly killed lizards of matching sizes held in the sun and shade at appropriate orientations at different hours over a clear day. (Live lizards would not maintain fixed orientations in the sun long enough to reach equilibrial T_b under all conditions.) We read model temperatures using the thermocouple/reader combination described earlier, and lizard temperatures using a rapid-registering cloacal thermometer. The two methods gave identical readings when the thermocouple and thermometer were immersed together in a water bath regulated over a range of temperatures experienced by lizards in the field. The copper models simulated T_b of live lizards held in the shade and of freshly killed lizards in the sun and shade over an appreciable range of ambient temperatures very precisely ($r = 0.886, n = 18$, $P < 0.001$, and $r = 0.994, n = 44, P < 0.001$, respectively) and accurately (regression of model temperature versus lizard temperature: slopes of 1.19 and 0.99, respectively). Thus, we have confidence that this method adequately estimated T_b of lizards in the field under various thermal conditions.

Agonistic Behavior

In both the austral spring/summer and summer/fall, we collected lizards from all sites (summer/fall: March 10–24, 1986; spring/summer: November 24–28, and December 15–21, 1986) and conducted encounters of size- and sex-matched pairs of adults of each species in a laboratory room held at optimal temperatures of 30–35°C (Fuentes and Jaksíc 1979). We introduced pairs of lizards into a circular neutral arena (1-m diameter) with a sand substrate and a 150-W incandescent lightbulb suspended over the center of the arena above a single flat rock on which the subjects could bask. After a 5-min acclimation period, we tallied frequencies of all agonistic behavior for 30 min from behind a blind. We distinguished the following aggressive behavior patterns: chase, fight, bite, superimposition (subject lays head, front leg, hand, or most of anterior body over opponent's body), supplant, lateral display, circle (subject moves in semicircle in front of opponent, presenting flank), gape, push-up, approach, and lick (subject licks opponent). We also distinguished the submissive behavior patterns retreat, flee, and flatten and the uncategorized behavior patterns tail switch and head scrape (side of the jaw is scraped along the substrate) (Carpenter 1978b; Baird, Fox, and McCoy 1997). In the summer/fall, we used each lizard only once, but in the spring/summer, we used some lizards also (on a different day) in heterosexual encounters (discussed next), then returned to the point of capture.

Courtship Behavior

In the austral spring/summer, we collected lizards from all sites (November 24–28 and December 15–21, 1986) and conducted encounters of pairs of adult males and females of each species in a laboratory room held at optimal temperatures of 30–35°C in the same neutral arenas and using the same protocol as for agonistic behavior (as already discussed). We tallied the same behavior patterns as listed in the previous section, plus shudder-bob and stotting (rapid short jumps on tiptoe).

We observed courtship behavior in the spring/summer in both the field and the laboratory in *L. fuscus, L. lemniscatus,* and *L. monticola;* thus, our laboratory encounters were conducted at the appropriate time of year. We did not directly observe courtship in the other species (although males and females interacted with one another), and information as to the season of courtship for them is not as certain and somewhat contradictory. *Liolaemus bellii, L. leopardinus,* and *L. schroederi* are viviparous and medium- to high-elevation lizards (Table 10-2). Many species of high-elevation, viviparous, temperate lizards (including *Liolaemus*) breed in the fall, females overwinter pregnant, and parturition occurs the following spring (Pearson 1954; Ramírez Pinilla 1991; Flemming and van Wyk 1992). Some viviparous species have an extended period of vitellogenesis (spring, summer, and winter) and ovulate the next spring, but not every year (Ibargüengoytía and Cussac 1996, 1998). *Liolaemus bellii* and *L. leopardinus* (from the same localities where we worked) undergo vitellogenesis in the fall and winter and ovulate and breed in the spring (Leyton and Valencia 1992; H. Núñez, personal communication). There is, however, some evidence for fall breeding in these two species (A. Labra, personal communication). Since *L. nitidus* lays its eggs in spring (Donoso-Barros 1966), *L. schroederi* gives birth in spring through summer, and males of *L. nitidus* and *L. schroederi* exhibit peak testicular sizes in spring, it appears that courtship and breeding occur in the spring/summer in these two species, as well (Ramírez Pinilla 1991; A. Labra, personal communication; H. Núñez, personal communication). Thus, on balance, the majority of evidence points to spring/

summer courtship and breeding for all seven species we studied, and that is when we conducted male-female laboratory encounters.

Results

Thermal Rigor

We used the monthly data from the models and transects to assess thermal rigor at each site. To compare the three sites, we first defined a common optimal range of body temperatures for all sites and species pooled. We defined the upper bound as the maximal T_b (to the nearest whole degree) observed in the field and the lower bound (to the nearest whole degree) as that temperature such that the interval from the upper to lower bound would encompass approximately 90% of the observed T_b in the field. This interval from 39 down to 28°C included 93% of all observations.

It is now standard to define the optimal, or selected, temperature range as that range of temperatures selected by lizards in a laboratory temperature gradient free from environmental constraints (Van Damme et al. 1989; Bauwens, Hertz, and Castilla 1996; Díaz 1997; Andrews et al. 1999), but we conducted our study in 1986 before such methodology was sufficiently appreciated (Hertz, Huey, and Stevenson 1993). Instead, we defined the common optimal range as just described (see also Grant and Dunham [1990], Bashey and Dunham [1997], and Bauwens, Castilla, and Mouton [1999] for similar methodology), which we acknowledge may be an imperfect substitute. Nevertheless, we believe that our data can still shed light on the relative thermal rigor of our study sites. We reasoned that a lizard in the field would not voluntarily exceed its ecologically maximal T_b (Huey 1982), but some fraction of observations (arbitrarily set here to ~10%) would be of lizards in the process of warming up, that is, still below their optimal range. Although lizards of the different species at the different sites of this study exhibited significantly different field T_b (Fox and Jaksíc, unpublished data), we elected to define a common optimal interval for all species over all sites rather than species- or site-specific intervals. Because a particular site might offer generally lower T_e than the other sites, the lizards there might be active under suboptimal T_b only because opportunity for optimal thermoregulation is limited. Defining species- or site-specific optimal intervals might obscure such a situation and characterize a site as thermally optimal when it may not be. (This is why current methodology measures the selected temperature range in a laboratory gradient.) Additionally, even though *Liolaemus* is found in very different thermal environments, its various species are active at very similar T_bs (Pearson 1954; Pearson and Bradford 1976; Fuentes and Jaksíc 1979; but see Marquet et al. [1989]). Carothers, Marquet, and Jaksíc (1998) found no significant relationship between mean T_b and elevation among the same set of *Liolaemus* species, and even though there was a significant difference in T_b among species, mean T_b among species spanned a range of <2.5°C, well within the optimal temperature range defined here. In a related laboratory study of thermal minima and maxima of the same set of species, Carothers et al. (1997) found no relation between elevation and these thermal end points, and although mean experimental voluntary maxima significantly differed among species, it varied <1.8°C, and all maxima were slightly less than the upper bound of 39°C reported here.

For each site for each month, we estimated the hourly proportion of the site available for optimal thermoregulation. Size of the copper models did not significantly affect these estimates (Fox and Jaksíc, unpublished data); therefore, we averaged results gathered for the three size classes of models. For each hour, we summed the areal proportion of microhabitats in which the interval of T_e defined by the lower and higher T_e of each pair of sizes of copper models intercepted the optimal range of 28–39°C. Figure 10-1 gives those average hourly availabilities for the three sites, respectively.

The highest site, El Colorado, offered the greatest opportunity for optimal thermoregulation over the day for all months (Fig. 10-1). In early to mid-summer (December–January), 100% of the site at midday was composed of microhabitats with optimal T_e. We observed no bimodal pattern of availability. By contrast, the lowest site, San Carlos, showed much reduced opportunity for optimal regulation over the day for all months (Fig. 10-1). We saw distinct bimodal availability in December–February, when midday operative temperatures in almost all microhabitats were too hot for the lizards. Curva 20, the site of intermediate elevation, was intermediate in its proportion of availability and bimodality. Mean percentage of availability (averaged over months) from the highest to lowest site was 64, 42, and 40%, respectively. The highest site differed the most from the other two.

Predation

Based on records of time in the field each day at each site, we calculated the number of potential predators seen per hour at each site (Table 10-3). The principal lizard predators in central Chile are a snake (*Philodryas*

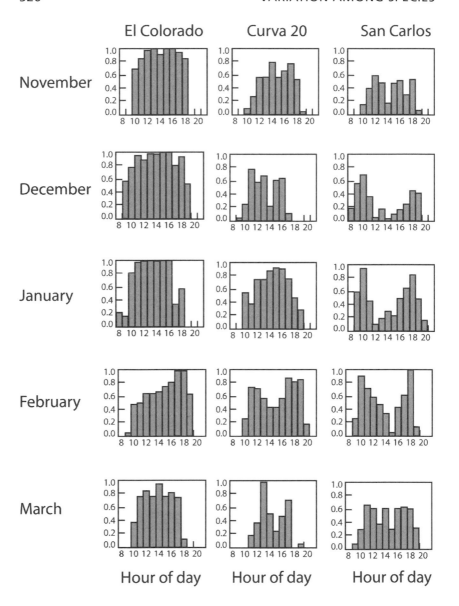

Figure 10-1   Hourly availability of optimal T_es over active season at El Colorado (2,900 m), Curva 20 (2,300 m), and San Carlos de Apoquindo (1,200 m), using microhabitat availability and copper lizard models.

Table 10-3   Hourly rates of predator sightings at three study sites in central Chile.

	Study Site (cumulative observation hours)		
	El Colorado	Curva 20	San Carlos
Predator Species	(177)	(204)	(220)
Philodryas chamissonis (snake)	0.000	0.024	0.072
Falco sparverius (kestral)	0.000	0.142	0.000
Total	0.000	0.166	0.072
Buteo polyosoma (hawk)	0.033	0.005	0.050
Parabuteo unicinctus (hawk)	0.000	0.000	0.122
Geranoaetus melanoleucus (hawk)	0.006	0.328	0.100
Grand Total	0.039	0.499	0.344

chamissonis) and the American kestrel (*Falco sparverius*) (Jaksíc et al. 1982). *Philodryas* is not found at the elevation of El Colorado and was quite rare even at the elevation of Curva 20. Considering just the two principal predators, as well as all potential ones, El Colorado presented less predation risk than San Carlos, which in turn presented less than Curva 20 (Table 10-3).

As a second measure of predation pressure, we calculated the rate of tail loss over the interval of the study for each marked subject (Rand 1954; Pianka 1967, 1970; Schoener 1979; Fox, Perea-Fox, and Castro Franco 1994, but see also Jaksíc and Greene [1984] and Medel et al. [1988]). For each site, we calculated the number of tail losses occurring during our study divided by the total days each subject was monitored times the number of subjects (no. of losses $\times$ lizards^{-1} $\times$ d^{-1}), i.e., a dynamic rate, rather than the static occurrence of tail loss used in past studies. This statistic thus avoids the confounding effect of elevated opportunity for tail loss among longer-lived species or those with longer activity seasons (Tinkle and Ballinger 1972). When the numbers were converted to annual rates of tail loss per lizard, San Carlos showed the highest rate of tail loss (0.55 $\times$ lizard^{-1} $\times$ yr^{-1}; $n = 70$), Curva 20 intermediate (0.36 $\times$ lizard^{-1} $\times$ yr^{-1}; $n = 80$), and El Colorado the least (0.15 $\times$ lizard^{-1} $\times$ yr^{-1}; $n = 80$). If these dynamic rates of tail loss relate directly to predation attempts, then predation risk was inversely related to elevation among sites.

As a third indicator of predation risk, for each species at each site, we calculated Schoener's index of predation intensity (Schoener 1979), which takes into account the relative inefficiency of predation as it gauges predation pressure. In general, San Carlos showed the highest predation intensity and Curva 20 and El Colorado showed the lowest (Table 10-4). Predation intensity was negatively correlated with elevation (Fig. 10-2), but because of interspecific differences in susceptibility to predation and our limited set of species, this was not a significant correlation ($r = -0.46$, $df = 7$, $P > 0.05$). Nevertheless, on balance, our data suggest that predation was lax at El Colorado and more intense at San Carlos, with Curva 20 somewhere in between.

Density

Figure 10-3 shows the relative density of the seven species at the three sites. The two species from the high-elevation site, *L. bellii* and *L. leopardinus*, were generally more dense than species at the other two sites, especially for *L. bellii*. Adult densities of *L. bellii* reached impressive values for lizard populations, although still not as high as have been reported in other studies (Beuchat 1982; Boersma 1983; Bauwens, Hertz, and Castilla 1996; see also chapters 6 and 7).

Space Use

From size and color pattern, we distinguished adults from juveniles; space use by adults only is presented here. We plotted sightings of all adults on a scale map for each site, separately for spring/summer and summer/fall seasons. For those individuals seen at least three times per season, we plotted their home ranges using the convex polygon technique. Most lizards were seen many more times than this minimal number of sightings (males: $\bar{x} = 6.5$, range $= 3–22$, SD $= 4.1$, $n = 139$; females: $\bar{x} = 5.9$, range $= 3–23$, SD $= 3.6$, $n = 102$). Because a home range is defined as that space within which an individual conducts most of its activities (Burt 1943), we eliminated any sightings that were beyond a distance of 1.96 SDs from the geometric center of sightings (Tinkle 1967a), although, in reality, this procedure eliminated very few sightings. Thus, we objectively defined the home range area as that convex polygon that encompassed all sightings within 95% of the distance from the geometric center of activity. Because little is known about space use in most of these species and because our sample size was limited, we pooled home range estimates drawn separately from both seasons for a single analysis for each sex. Furthermore, we reasoned that whereas social use of space may change somewhat from spring/summer to summer/fall (e.g., territoriality that wanes over the season), it is unlikely to change completely (e.g., from territoriality to nonterritoriality). To wit,

Table 10-4   Annual survivorship, tail-break frequency, and Schoener's index of predation intensity for each lizard species by study site.

Site	Species	Annual Survivorship	Tail-break Frequency	Predation Intensity
El Colorado	L. bellii	0.40	0.51	1.86
	L. leopardinus	0.28	0.47	2.39
Curva 20	L. schroederi	0.25	0.40	2.31
	L. nitidus	0.44	0.37	1.30
	L. monticola	0.43	0.33	1.25
	L. lemniscatus	0.30	0.53	2.53
	L. fuscus	0.00	0.50	—
San Carlos	L. monticola	0.37	0.52	2.07
	L. lemniscatus	0.19	0.57	3.87
	L. fuscus	0.32	0.53	2.43

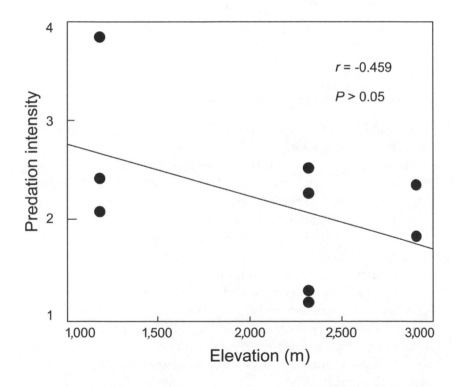

Figure 10-2   Negative correlation of Schoener's index of species- and site-specific predation intensity and elevation.

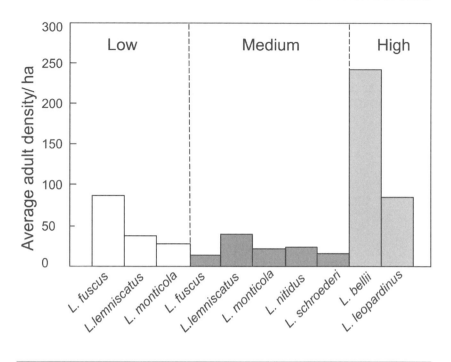

Figure 10-3   Relative adult lizard density by species by site.

we did not observe any such drastic change at any site during our field work (but see Discussion, *L. bellii*).

From our site maps with plotted locations of lizards and software provided by A. E. Dunham, for each species we computed (1) home range area, (2) percentage of overlap by same-sex individuals, (3) percentage of overlap by opposite-sex individuals, (4) number of same-sex individuals overlapping, and (5) number of opposite-sex individuals overlapping. Categories 4 and 5 include adults seen fewer than three times in addition to those seen more than three times. Table 10-5 reports average home range areas and intrasexual percentage of overlaps for males and females of the seven species we studied. We pooled data for species inhabiting more than one site, then conducted discriminant functions analysis for each sex separately in an attempt to use the five aforementioned independent variables to separate the seven species in multivariate space.

For males, species were significantly separated (Wilks' $\lambda = 0.301$; Rao's *F*-statistic $= 5.436$; $df = 30, 466$; $P < 0.001$). Discriminant factor 1 (canonical correlation $= 0.724$) was loaded heavily by the number and percentage of overlap of both males and females (Table 10-6), whereas

Table 10-5 Mean home range size (± 1 SE), mean intrasexual percentage of overlap (± 1 SE), and male:female ratio of mean home range size for seven species of *Liolaemus* from central Chile.

Species		Males			Females			Male:Female
								Home Range Size
	n	Home Range Size (m²)	Male-Male Overlap (%)	n	Home Range Size (m²)	Female-Female Overlap (%)		
L. bellii	39	53.5 ± 9.3	47.3 ± 5.8	25	36.3 ± 7.6	25.3 ± 6.2		1.50
L. leopardinus	7	112.4 ± 38.3	58.1 ± 11.9	10	106.8 ± 39.0	41.0 ± 9.2		1.05
L. schroederi	6	88.1 ± 21.6	8.6 ± 5.3	2	68.1 ± 53.8	—		1.29
L. nitidus	7	74.9 ± 18.8	0.0 ± 0.0	13	70.3 ± 19.7	11.6 ± 6.2		1.07
L. lemniscatus	28	70.3 ± 15.9	19.8 ± 5.8	14	25.2 ± 5.7	5.6 ± 4.4		2.80
L. monticola	33	242.2 ± 47.8	36.3 ± 6.5	24	110.9 ± 28.0	15.1 ± 5.7		2.18
L. fuscus	19	32.2 ± 14.5	10.7 ± 5.0	14	9.4 ± 3.2	17.3 ± 8.0		3.56

Table 10-6 Loadings of variables of home range area and overlap by males and females (number and percentage) on first two axes of discriminant functions analysis for males (*n* = 127) and females (*n* = 96).

Sex	Independent Variable	Axis I	Axis II
Male	Home range area	−0.129	0.755
	Number of overlapping males	0.613	−0.017
	Percentage of overlap by males	0.444	0.303
	Number of overlapping females	0.521	0.381
	Percentage of overlap by females	0.581	0.442
Female	Home range area	−0.080	0.796
	Number of overlapping males	0.765	0.298
	Percentage of overlap by males	0.365	0.257
	Number of overlapping females	0.477	0.086
	Percentage of overlap by females	0.357	0.207

home range area loaded heavily on discriminant factor 2 (canonical correlation = 0.568). Plotted 90% confidence ellipses of each species on axes of discriminant factor 2 against discriminant factor 1 showed that the high-elevation species (*L. leopardinus* and *L. bellii*) had more overlapping individuals of both sexes, and that more of their home ranges were overlapped by both sexes, than the species of middle and low elevations (Fig. 10-4A). *Liolaemus leopardinus* and *L. monticola* had larger home ranges than the other species.

Likewise for females, species were significantly separated (Wilks' λ= 0.384; Rao's *F*-statistic = 3.770; *df* = 25, 320; *P* < 0.001; *L. schroederi* was excluded owing to lack of sufficient sample size). Interpretation of discriminant factors 1 and 2 was as for the males, although the loading of the number of overlapping males on factor 1 was much higher than the other variables that loaded appreciably on this factor (Table 10-6). Plots of 90% confidence ellipses for each species did not show as great a separation of species for females compared with males, but the higher scores on factor 1 (high overlap in number and percentage) of *L. leopardinus* and *L. bellii* were evident (Fig. 10-4B), just as in the males. Likewise, home ranges of female *L. leopardinus* and *L. monticola* were larger than those of the other species.

Although aspects of space use by lizards are known to be plastic within a species and change with differences in density, food availability, or habitat complexity (Stamps and Tanaka 1981b; Guyer 1988b; Eason and Stamps 1992; Stamps and Krishnan 1998), there do exist generalized species differences in social organization and space use (Stamps

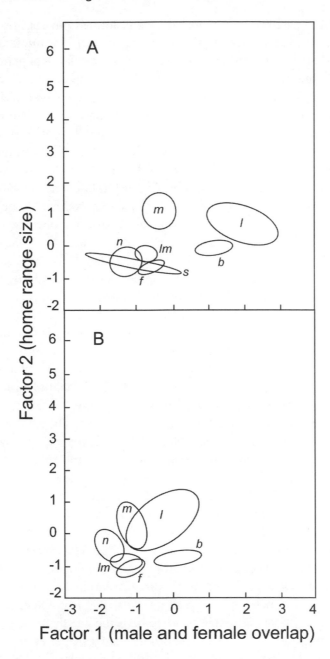

Figure 10-4   Ninety percent confidence ellipses of discriminant factor two versus discriminant factor one from a discriminant functions analysis of home range parameters for (*A*)
males and (*B*) females. *m* = *L. monticola*, *n* = *L. nitidus*, *f* = *L. fuscus*, *lm* = *L. lemniscatus*,
l = *L. leopardinus*, *b* = *L. bellii*, *s* = *L. schroederi*.

1977b; Martins 1994). Consequently, we examined our specific measurements of space use among the seven species of *Liolaemus* in light of their phylogenetic relatedness. Owing to the concordance of patterns of space use among species for both sexes, the species' behavioral traits of home range overlap by both sexes (number and percentage), and home range size can be mapped onto the known phylogeny as per Brooks and McLennan (1991). The topological phylogeny of the set of seven species is taken from Schulte et al. (2000), plus Schulte (personal communication) for placement of *L. schroederi,* and is based on 785 phylogenetically informative aligned base positions of mitochondrial DNA sequences from a set of 60 species of *Liolaemus.* Neither home range overlap nor home range size appears to be strongly phylogenetically constrained (Fig. 10-5). *Liolaemus bellii* and *L. leopardinus,* both with high overlap, are not members of the same clade, and since *L. schroederi* (character state known only in the males) and the remaining monophyletic four species all show low overlap, it would appear that the evolution of high home range overlap minimally occurred twice in this set of species, if we assume that the plesiomorphic state is low overlap. Without an outgroup, the polarization of this character cannot be fixed. However, the sister taxon of the group of seven species we studied is a clade of mostly Argentine *Liolaemus* (Schulte et al. 2000). Not much is known about space use and social behavior for most of the species of this clade, but at least in some of them males appear to be territorial and rather intolerant of other males (Halloy 1996; Halloy and Halloy 1997; Rocha 1999; Frutos, Belver, and Avila 2000; Robles and Halloy 2000). At least one other *Liolaemus* species (*L. tenuis* of the Chilean clade; Manzur and Fuentes 1979) also is territorial and polygynous.

From these rather meager data from a large set of species, we speculate that the ancestral state of this trait for both the Chilean and Argentine clades might be low home range overlap. The two species with high home range overlap, *L. bellii* and *L. leopardinus,* are the two high-elevation species. Large home range size is found in *L. leopardinus* and *L. monticola,* not members of the same clade (Fig. 10-5). Species of the sister taxon to the set of species studied here, the mostly Argentine *Liolaemus,* appear to have mostly small home ranges (Robles and Halloy 2000; M. Halloy, personal communication; but see Frutos, Belver, and Avila [2000]), as does *L. tenuis* of the Chilean clade (Manzur and Fuentes 1979); therefore, we suggest (acknowledging the paucity of data) that small home range size might be the ancestral state of the

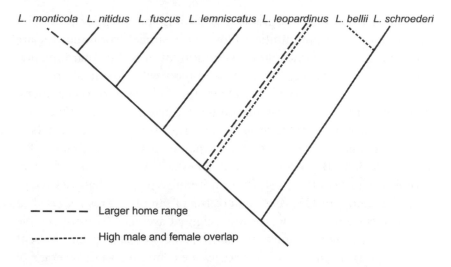

Figure 10-5 Character states of home range parameters of males mapped onto topological phylogeny of seven *Liolaemus* species. Phylogeny taken from Schulte et al. (2000).

lizards we studied. In that case, the state of large home ranges appears to have evolved separately twice, once for *L. leopardinus* and once for *L. monticola*. There is no correlation of this trait with elevation.

Without the ability to polarize either character of space use, we cannot know the evolutionary history of character changes. We have presented a reasonable scenario, but the actual sequence of character changes is less important than the observation that neither home range overlap nor home range size relates strongly to the phylogeny of the seven species we studied. Within that clade of species, these space use traits do not apear to be phylogenetically constrained, and that of high home range overlap appears to correlate with high elevation.

Agonistic Behavior

Using our data on agonistic behavior of the seven species, we propose a new way to analyze data sets composed of counts of behavioral characters of individuals of a set of species. Furthermore, we propose an extension of this method to separate the effect of phylogenetic constraint from local adaptation, that is, a new PCM. Our method is taken after community ecology and is a form of ordination analysis, in which commonly the raw data are a set of plots with measured abundance of species on each plot. Plots are then "ordered" along a hypothetical or known environmental gradient according to similarity of species composition, or communities. Plots with similar communities are grouped together at one end of a continuum of some environmental gradient (e.g., moisture, elevation), and plots with dissimilar communities from these are grouped together at the other end of the gradient, with plots of intermediate communities located in between. One popular current way to ordinate such data is a "direct gradient analysis" known as canonical correspondence analysis (CCA) (ter Braak 1986; Palmer 1993; ter Braak and Šmilauer 1998). CCA is a variant of correspondence analysis (CA), which is an iterative process that uses reciprocal averaging. In CA, (initially arbitrary) sample scores are used to compute species scores, which are weighted averages (sum of sample scores of each plot weighted by the frequency of each species present on each plot). Then new sample scores are computed as the average of the species scores, again weighted by the abundance of each species in each sample. Scores are standardized at each step to prevent their approach to zero, and the process is repeated until scores stabilize. The result is the first CA axis solution. Subsequent ordination along further axes is performed in the same way after the effects of the first axis are factored out. Thus, axes are orthogonal. In CCA, measured environmental variables that describe ecological gradients are included in the algorithm. Consequently, CCA is a form of "direct gradient" analysis. At each iteration, the environmental variables are used as the independent variables in a multivariate linear least-squares regression to predict the new sample scores. Iteration is continued as before until scores stabilize. In CCA, one can plot species scores, sample scores, and independent variables on the same triplot scatter diagram to see how plots with similar communities are related to the measured environmental variables.

The novelty of our method as applied to behavior is the redefinition of samples and species (Shipman and Palmer, unpublished data). In our case, each individual lizard is a "sample" or "plot," and each distinct behavior pattern is a "species." The environmental or independent variables are the actual species of each individual lizard plus attributes of the site where each lizard was collected (e.g., population density, elevation). Just as ecological plots with similar species compositions are expected to cluster together and relate to measured environmental variables along a gradient in CCA, individual lizards with similar behavior are expected to cluster together and relate to the independent variables of species and other variables such as density or elevation. In other words, individuals of the same species should show the same kinds and frequencies of behavior patterns, and if the environment in which they live influences their behavior, species found in similar environments should likewise show the same kinds and frequencies of behavior.

Furthermore, membership in hierarchical clades of a phylogeny from the most basal common ancestor toward the distal species tips can be included as covariables to factor out the effect of phylogenetic correlation. If species from similar environments still cluster together after removing the effects of phylogeny, then the common set of behavior patterns observed in these species is interpreted as an adaptation to the common environment (convergent evolution) and not phylogenetic history. If the pattern of behavior among species is disrupted with the inclusion of phylogeny, then species are not locally adapted to their current environment, and phylogenetic inertia is in evidence.

We performed this CCA first on the agonistic behavior of males. Because so little is known of the social behavior of most of these species and because our sample size was limited, we pooled data drawn from both seasons for a single analysis for each sex. Without phylogenetic covariables, the spatial pattern of behavior, species, density, and elevation across canonical axes one and two appears as in Fig. 10-6. The pattern is significantly different from random across all axes (Monte Carlo permutations test: $F = 4.43$, $P < 0.001$). In this plot, the clustering of behavior patterns (points) represents lizards that tended to show the same relative frequencies of these categories of behavior, and species are depicted as directional vectors whose lengths are proportional to the importance of that variable in establishing the observed pattern of points. Behavior patterns that fall close to a species vector are highly correlated with that species, that is, are characteristic of that species. Finally, site elevation and density are also plotted as directional vectors (with length proportional to the influence of that variable in explaining the pattern), and behavior patterns positioned close to a particular vector correlate highly with it. Density is interpreted as a double-ended arrow; behavior patterns of lizards from high-density populations fall out near the plotted vector, and those of lizards from low-density populations are found diametrically opposed on the opposite side of the origin. Short vectors and points near the origin offer little explanatory value.

In Fig. 10-6, individuals with high frequencies of the more passive agonistic patterns of approach, retreat, circle, gape, head scrape, lick, and superimposition clustered to the right of the ordination diagram and tended to be the species *L. bellii*, *L. leopardinus*, and *L. nitidus*,

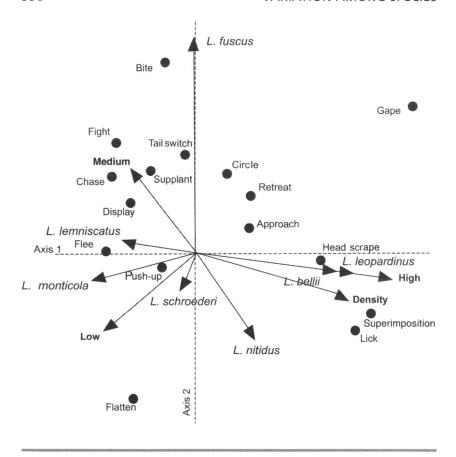

Figure 10-6   CCA of male agonistic behavior patterns with inclusion of no phylogenetic covariables.

associated with high densities and elevations. Individuals with high frequencies of more active agonistic patterns of bite, fight, chase, supplant, display, push-up, flatten, and flee clustered to the left of the figure and tended to be *L. fuscus, L. lemniscatus, L. monticola,* and *L. schroederi,* associated with generally low densities and low to medium elevations.

We called the first set of behavior patterns *chemical signaling/passive* and the second set *visual signaling/aggressive* and plotted these character states onto the known phylogeny of our set of species (Fig. 10-7). We do not mean to imply that in general lizards that mostly use chemical signaling for social communication necessarily are passive in their agonistic interactions, only that in the set of species we studied those that showed the more passive agonistic behavior patterns also displayed behavior patterns associated with chemical signaling more frequently, and

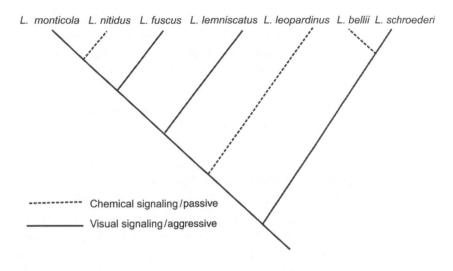

Figure 10-7 Character states of male agonistic behavior as derived from Fig. 10-6 mapped onto topological phylogeny of seven *Liolaemus* species.

vice versa. Lacking an outgroup, we cannot definitively polarize the character sequence. If the sister clade of Argentine *Liolaemus* generally shows more active aggression and visual displays (Halloy 1996; Rocha 1996; Castillo and Halloy 2000), and is thus the ancestral state for both the Argentine and Chilean clades of *Liolaemus,* then the most parsimonious interpretation is that chemical-signaling/passive behavior evolved independently at least three times: in *L. bellii, L. leopardinus,* and *L. nitidus.* If the Argentine clade is instead characterized as mostly chemical signaling/passive (as suggested from data of visual display frequency [Martins 2000]), then that is the ancestral state and apparently *L. schroederi* evolved the derived state of visual signaling/aggressive, *L. bellii* and *L. leopardinus* retained chemical signaling/passive, the ancestor of the remaining species independently evolved visual signaling/aggressive, and, finally, *L. nitidus* reevolved chemical signaling/passive. Both scenarios involve three independent state transitions, and both generally associate high-elevation, high-density species with chemical-signaling/passive behavior, and low-elevation, low-density species with visual-signaling/ aggressive behavior. The behavioral states do not show strong phylogenetic correlations.

In the next step, we included the deepest clade membership (*L. schroederi-L. bellii* versus the others) as a covariable and repeated the CCA. This variable removed the variation due to phylogenetic differences originating from this node. In this case, if most of the initial observed pattern of behavior, species, and independent variables were a function of the basal phylogenetic split into these two clades, then the pattern would disappear. However, the same basic pattern was still present in this second CCA (Fig. 10-8), and it was significantly different from random across all axes (Monte Carlo permutations test: $F = 4.40$, $P < 0.001$). Within each clade, chemical-signaling/passive behavior was generally associated with high-elevation, high-density species, and visual-signaling/aggressive behavior was associated with low- to mid-elevation, generally low-density species (Fig. 10-9).

In the final step, we excluded *L. bellii* and *L. schroederi* and coded for membership in the sister clades of *L. leopardinus* versus the remaining species, thus controlling for the phylogenetic split at that level. The subsequent CCA (Fig. 10-10) retained the pattern of behavior, species, and independent variables (Monte Carlo permutations test: $F = 2.81$, $P < 0.001$), with *L. nitidus* and its characteristic chemical-signaling/passive behavior separated from the remaining species and their visual-signaling/aggressive behavior. (*L. leopardinus* is not plotted because all its variation is absorbed in the covariable that contains only that species.) Thus, within the four-species clade, *L. nitidus* retained its behavioral distinctiveness from the other three species (Fig. 10-11).

We conducted the same analysis of agonistic behavior of females. Without phylogenetic covariables, the spatial pattern of behavior, species, density, and elevation across canonical axes 1 and 2 appears as in Fig. 10-12. The pattern is significantly different from random across all axes (Monte Carlo permutations test: $F = 6.95$, $P < 0.001$). The observed pattern is very much like that of the males: individuals delivering chemical-signaling/passive behavior traits such as approach, retreat, gape, head scrape, lick, and superimposition were separated from the other individuals, and these less overtly aggressive individuals were high-density, high-elevation *L. bellii* and *L. leopardinus*. Different from the males, however, *L. nitidus* did not cluster with *L. bellii* and *L. leopardinus*. Instead of exhibiting the chemical-signaling/passive behavior typical of those latter two species, *L. nitidus* individuals tended to show bites and supplants, and their behavior was more correlated with that of *L. schroederi* and *L. fuscus*, which, in turn, was correlated with that of the low-elevation, low-density species (i.e., visual-signaling/aggressive behavior). When the character states were mapped onto the phylogeny (Fig. 10-13), the character state of chemical signaling/passive appears not to be phylogenetically correlated and associates with the high-ele-

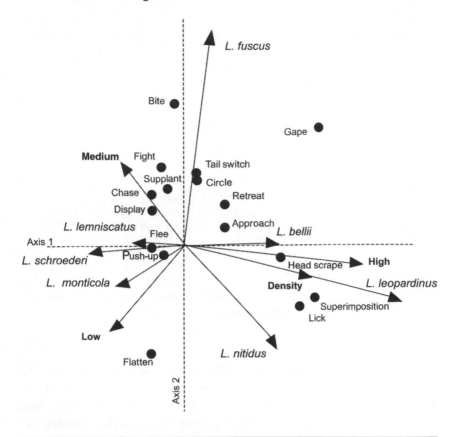

Figure 10-8 Partial CCA of male agonistic behavior patterns with inclusion of clade membership of *L. bellii-schroederi* versus *L. leopardinus-lemniscatus-fuscus-nitidus-monticola* as a covariable.

vation, high-density species. The monophyletic sister clade to *L. leopardinus,* on the other hand, is composed of species all showing the same visual-signaling/aggressive behavior.

Courtship Behavior

We performed the same type of CCA on courtship behavioral data from the male-female pairs as we did for the agonistic behavior of same-sex pairs of lizards. The observed pattern (with no phylogenetic covariables) bears similarities to that derived from agonistic behavior but also shows some differences (Fig. 10-14). The high-elevation species, *L. leopardinus* and *L. bellii,* again group together and are aligned with chemical-signaling/passive behavior patterns and high densities. As in the analysis of agonistic behavior among males (Fig. 10-6), *L. nitidus* also

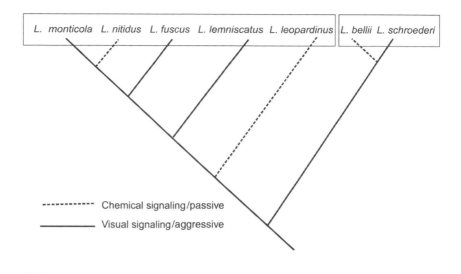

Figure 10-9 Character states of male agonistic behavior as derived from Fig. 10-8 mapped onto topological phylogeny of seven *Liolaemus* species.

clusters with these two species (Fig. 10-14). However, *L. fuscus*, which grouped with other visually signaling, actively aggressive species in the previous analyses (Fig. 10-6), in this analysis falls with *L. leopardinus, L. bellii,* and *L. nitidus. Liolaemus schroederi* also groups with these more passive species, although its relationship to the chemical-signaling behavior traits such as lick, head scrape, and gape (see Discussion) is less strong. *Liolaemus monticola* falls diametrically opposed to *L. leopardinus* and *L. bellii* and can thus be considered to be the behavioral antithesis of a chemical signaler during heterosexual encounters. *Liolaemus lemniscatus* falls diametrically opposed to *L. schroederi* and correlates strongly with active aggressive behavior traits such as chase and fight, as well as with the distinctive courtship patterns of shudder-bob and stotting.

The main differences between the pattern derived from agonistic behavior of males and that from courtship behavior were (1) that *L. fuscus* adopted chemical-signaling/passive behavior and allied with the high-elevation species, and (2) that *L. schroederi* displayed behavior different from all the other species. Regarding the first difference, *L. fuscus* was found at high densities (Fig. 10-3), which may help explain why courtship patterns were like other chemical signalers with high density. On the other hand, *L. nitidus* was not found at high densities and *L. fuscus* did not group with *L. leopardinus* and *L. bellii* in the analyses of agonistic

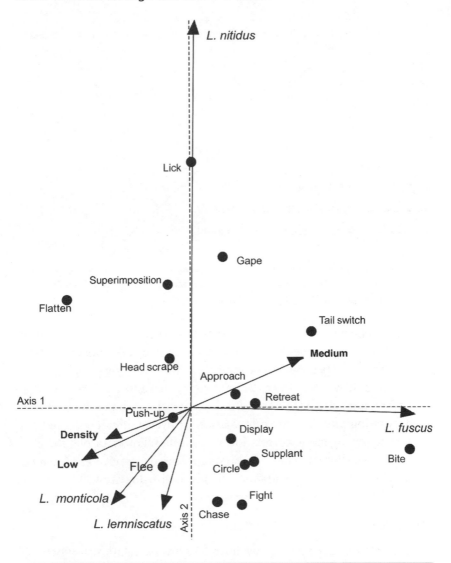

Figure 10-10 Partial CCA of male agonistic behavior patterns with exclusion of *L. bellii* and *L. schroederi* and inclusion of clade membership of *L. leopardinus* versus *L. lemniscatus-fuscus-nitidus-monticola* as a covariable.

behavior. Regarding the second difference, inspection of the raw data for courtship revealed that *L. schroederi* showed some of the same behavior as *L. monticola* and some of the same behavior as the species clustering to the right in Fig. 10-14, while differing drastically in behavior from *L. lemniscatus,* the species that showed the most definitive courtship behavior traits such as shudder-bob and stotting. *Liolaemus schroederi* showed no

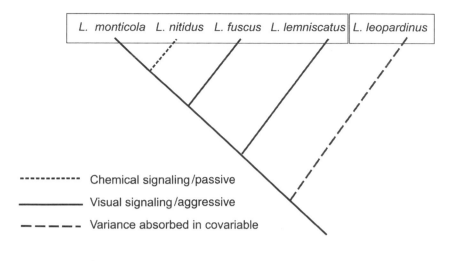

Figure 10-11 Character states of male agonistic behavior as derived from Fig. 10-10 mapped onto topological phylogeny of five *Liolaemus* species.

definitive courtship behavior; perhaps the artificial laboratory conditions or time of year were unsuitable. Consequently, the position on the CCA plot of this species must be interpreted with caution.

Because the CCA pattern derived from courtship was less clear than that from agonistic behavior, we did not continue the sequential CCA incorporating phylogenetic information. However, it can be noted that no strong phylogenetic correlation is indicated. For example, the cluster of species showing chemical-signaling/passive courtship (*L. leopardinus, L. fuscus, L. bellii,* and *L. nitidus*) is not phylogenetically compact within the set of seven species we studied.

Discussion

Contrary to our expectation, we found an inverse relationship between elevation and thermal rigor. Our high-elevation site was more thermally benign for the lizards than our low-elevation site. In other studies in which thermal conditions at disparate elevations were compared, higher elevations were harsher (Crowley 1985; Marquet et al. 1989; Van Damme et al. 1989; Hertz, Huey, and Stevenson 1993; Lemos-Espinal and Ballinger 1995; Díaz 1997; Andrews et al. 1999), usually in that T_es were lower than at lower elevations. However, most microhabitats over most of the day during the principal months of lizard activity at our high-elevation site (2,900 m) offered suitable T_es. By contrast, T_es at the lower-elevation site (1,200 m) at midday were simply too hot for lizards

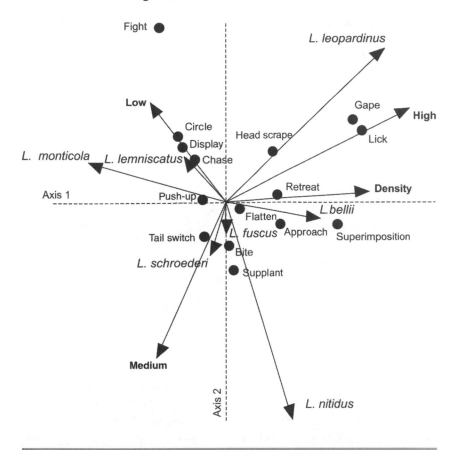

Figure 10-12 CCA of female agonistic behavior patterns with inclusion of no phylogenetic covariables.

in most of the summer months, thus severely limiting thermoregulatory opportunities. Patterns of lizard activity among sites mirrored the patterns of T_es (Fox and Jaksíc, unpublished data). The intermediate-elevation site (2,300 m) offered intermediate opportunities for optimal thermoregulation, and intermediate levels of activity were observed, as well. In quite a number of other studies in which thermal rigor has been compared at different elevations, low-elevation sites were likewise too hot for lizards, especially during the middle part of the day (Burns 1970; Beuchat 1986; Adolph 1990; Grant and Dunham 1990; Bashey and Dunham 1997). Often these extreme midday T_es provoked reduced lizard activity at this time (Beuchat 1986; Grant and Dunham 1990). High T_es at one middle elevation site in west Texas limited the activity of *Sceloporus merriami* to a mere 2 h at the beginning of the day and a

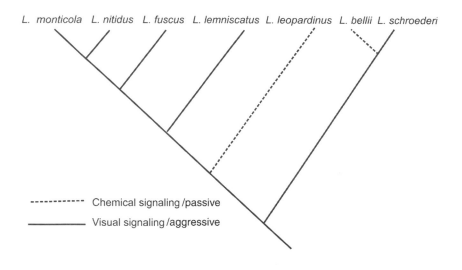

Figure 10-13 Character states of female agonistic behavior as derived from Fig. 10-12 mapped onto topological phylogeny of seven *Liolaemus* species.

brief period in late afternoon (Grant and Dunham 1988). At the same locality, but relative to the thermoregulatory behavior of sympatric *Cophosaurus texanus,* 70% of a high site averaged within a range of optimal T_es, but <20% of a low-elevation site averaged thermally adequate (Bashey and Dunham 1997). In some studies, high elevations were frequently cloudy (Pearson 1954; Beuchat 1986; Van Damme et al. 1989; Adolph 1990; Hertz 1992), thus diminishing the availability of optimal T_es, but we did not observe such weather at El Colorado. Grant (1990) documented an increase in afternoon field T_bs of *S. merriami* at low, but not high, elevations in west Texas, apparently as an adaptation for extended activity by lizards exposed to higher T_es in the afternoon at this thermally rigorous site.

Consistent with previous studies (Brown and Ruby 1977; Crowley 1985; Van Damme et al. 1989; Medel et al. 1990; Fox, Perea-Fox, and Castro Franco 1994; Lemos-Espinal and Ballinger 1995), predation pressure decreased with elevation. Lizards at our high-elevation site enjoyed high survivorship and reached high densities. It seems to be rather general that lizard populations at high elevations reach higher densities than those at lower elevations (Beuchat 1982; Van Damme et al. 1989; Ruby and Baird 1994; Bashey and Dunham 1997; Díaz 1997), likely because of reduced predation, abundant food, and oftentimes low cost of thermoregulation.

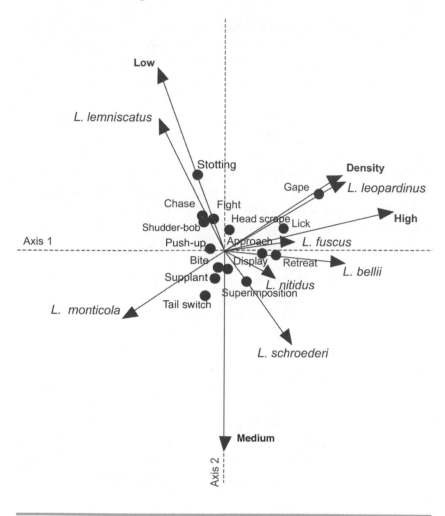

Figure 10-14 CCA of male-female courtship behavior patterns with inclusion of no phylogenetic covariables.

Lizards forage more and carry out more social activities during the hours at which T_es are optimal (Waldschmidt and Tracy 1983; Grant and Dunham 1988; but see Bauwens, Hertz, and Castilla [1996]). When thermal conditions are more limiting, lizards must thermoregulate more carefully (Bashey and Dunham 1997; Díaz 1997; Schäuble and Grigg 1998), taking care to select appropriate thermal microhabitats. On the other hand, when much of a site is thermally optimal, lizards exercise less care in selecting a microhabitat (Grant and Dunham 1990; Lemos-Espinal and Ballinger 1995) and can devote more time and effort to other activities. In a like fashion, when predation is lax, lizards can remain active even when thermal conditions are suboptimal because selection pressure for optimal escape performance is weak (Crowley 1985; Grant and Dunham 1988; Van Damme et al. 1989; Díaz 1997).

The high-elevation lizards of our study were nonaggressive—tolerant of high home range overlap of conspecifics, both males with males and females with females. Visual social displays were infrequent, as were escalated agonistic displays. We found no evidence of territorial social organization. On the other hand, low- and mid-elevation lizards generally were aggressive; males were intolerant of home range overlap by other males; females also showed low spatial overlap; visual, escalated agonistic displays were frequent; and these species showed territorial social organization.

Lizards at high elevations experienced only minor constraints on their activity due to thermal limitations or predation compared with those living at lower elevations. Apparently food was not limiting either, because density and survivorship were high. Life was easy for these lizards. Because of the high densities, we reason that conspecific intrusion pressure preempted defense of territorial borders (Brattstrom 1974; Stamps 1977b); furthermore, thermally optimal microhabitats at high elevations were uniformly distributed and in abundance all through the day, so intraspecific competition for superior space/time elements of the thermal environment (Tracy and Christian 1986) can be presumed to be weak. Because of the lax thermal and predation constraints, the species at high elevations have available to them unfettered time and opportunity for rich social interactions. The high conspecific densities also heighten the frequency of social interchange. Thus, these lizards are not territorial and practice some other kind of social organization. Unfortunately, we do not know what that is, but we suspect that the social organization of the high-elevation *L. leopardinus* and *L. bellii* is rather complex and substantially different from that commonly observed in most lizard species.

Because of the high densities and high levels of activity of *L. bellii,* we suggest that this species may form neighborhood dominance hierarchies. Other species of lizards that live in dense colonies or are aggregated due to patchiness of their habitat form dominance hierarchies (Stamps 1977b; Alberts 1994). *Liolaemus bellii* is not very sexually dimorphic in coloration nor in body size (the ratio of average adult male SVL to average adult female SVL is only 1.06). During and outside of the breeding season, there is considerable male-male home range overlap (Fig. 10-4; Table 10-5).

Stamps (1983b) presented an elegant model relating sexual selection, sexual dimorphism, space use, and territoriality in lizards and predicted that territoriality would be favored in species (1) with small home ranges of females, and (2) wherein males could expand their home ranges in order to overlap home ranges of more females and prevent other males' access to them. In other words, that territoriality would be more likely when home ranges of females were small and the ratio of home range size, male to female, was large. Considering all seven species in our study, home range size of female *L. bellii* (averaging 36.3 m^2) falls rather in the middle, ranking third, small to large (Table 10-5). In the ordered series of male:female ratios of home range size, *L. bellii* falls exactly in the middle (rank 4; Table 10-5). However, the magnitude of this ratio (1.50) is very close to the average male:female home range size ratio reported by Stamps (1983b) for nonterritorial species, and far below the ratio for the low-elevation, strongly territorial species of our study, *L. fuscus, L. lemniscatus,* and *L. monticola* (Table 10-5). Thus, the various predictors of type of social organization tend to suggest that *L. bellii* should be nonterritorial, and that is what we found, at least at the site where we studied its behavior. Male-male spatial overlap was high despite home ranges being small for the size of this lizard (Table 10-5) and being located in habitat with excellent visibility. But the high conspecific density probably precluded effective defense of home range boundaries. Alternative to territoriality, and especially common when lizards are aggregated, is the dominance hierarchy. Unfortunately, we lack the data sufficient to test for this social organization, but we suspect that is how this high-elevation species is socially organized.

If *L. bellii* has a hierarchical social organization, it is maintained with very little agonistic behavior (Fig. 10-6). We observed few instances of agonism in the field, and in our laboratory arenas we likewise observed only weak aggression toward conspecifics. *Lioaemus bellii* never performed lateral display, chase, fight, flatten, or flee. Push-ups and bites were rare. Instead, chemical signals seemed to predominate during social interactions. *L. bellii* delivered more licks and mouth gapes than any other species, and head scrape and superimposition were also quite common. We initially had the bias that the behavior pattern of mouth gape was a visual display, as it seems to be in other species (Carpenter 1978b; Fox and Baird 1982), but its association with the behavior patterns of chemical marking and detection (head scrape and lick) (Simon 1983) suggested that it might play some role in chemical signaling (in addition to a possible visual role). We propose that mouth gape might be used to promote the delivery of some pheromone to the exterior of the body (eyes, jaw, cephalic organs?) or, more likely, to improve the reception of an airborne pheromone into the vomeronasal system. This latter conclusion is supported by observations and experiments with snakes (Graves and Duvall 1983, 1985) and two other species of lizards (Cooper and Alberts 1991; Zuri and Bull 2000). More and more herpetologists are realizing the importance of lizard chemosense in recognizing individuals or classes of conspecifics (Duvall 1982; Cooper and Vitt 1986a; Bull 1994; Cooper 1996; Olsson and Shine 1998; Hanley, Elliott, and Stamps 1999; Bull et al. 2000). Chemosensory social signaling has been suggested or documented even in the genus *Liolaemus* (Halloy and Halloy 1997; Labra and Niemeyer 1999).

In addition to the lack of escalated aggressive behavior and the importance of chemical signaling, *L. bellii* performed a high frequency of the superimposition behavior pattern, especially in same-sex encounters. Torr and Shine (1993) have described the same behavior pattern, which they called lie-on, in same-sex laboratory encounters in a nonterritorial skink (*Lampropholis guichenoti*). In the field, this behavior is observed when skinks are aggregated around favored basking sites, and whereas it may have a thermoregulatory function, it appears also to play a role in signaling status in dominance hierarchies. Interestingly, a similar behavior pattern, called hand placing, has been described in two high-elevation *Liolaemus* species: *L. quilmes* and *L. alticolor* (Halloy 1996). Even though these species are aggressive and territorial, they seemed to use hand placing in the laboratory to signal social rank in a dominance hierarchy. It is possible that *L. bellii* in our study used the same behavior pattern to signal social rank in noncombative social interactions; it certainly did not use escalated agonistic behavior to do so.

A final curious note can be made about the social behavior of *L. bellii*. Although males overlapped space use with other males considerably during the spring/summer breeding season, they overlapped even more in the summer/fall. In the spring/summer, each male lived with an average of 2.3 other adult males in his home range. By summer/fall, cooccupancy had increased to an average of 4.7 other males. This seasonal change in number of overlapping adult males was statistically significant ($t = 3.09$, $df = 37$, $P = 0.004$). In the summer/fall, overlapping males outnumbered overlapping females nearly three to one. No such elevated aggregated behavior was seen among the females ($t = 0.66$, $df = 23$, $P > 0.50$). The number of overlapping males on home ranges of females was less in the summer/fall than in the spring/summer, but not significantly ($t = 1.71$, $df = 23$, $P = 0.10$). Thus, the males were grouped together with other males after the breeding season, even though they (or females) did not reduce the sizes of their home ranges between seasons (both $P > 0.50$). This may be one of a few cases in lizards of *bachelor herds*, aggregations of males mostly separated from females (but also see Boersma [1983]).

The other high-elevation species, *L. leopardinus*, also showed details of unusual lizard social organization. This species was one of the largest of the seven species we studied (Table 10-2) and lacked sexual dimorphism in coloration or size. Sexual size dimorphism (as average adult male SVL:average adult female SVL) was only 1.2. Like *L. bellii*, *L. leopardinus* was nonterritorial, passive, socially tolerant of conspecifics, and used chemical signaling in place of escalated visual communication. Home range overlap among males was the highest of all seven species, averaging 58.1% (Table 10-5). Home range size of females was large and nearly equal to that of males; thus, the male:female ratio of home range size was low (1.05)—the lowest of all the species (Table 10-5). Stamps (1983b) predicted that a species with large home ranges of females and substantial home range overlap among males would not be territorial, and *L. leopardinus* appears to be that species.

Bull (1988, 1994) discovered the same conditions of space use in the large Australian skink, the sleepy lizard (*Tiliqua rugosa*), and found that this species also failed to show territoriality. Instead of defending the space where female mates are located, male sleepy lizards defend mates and move around the home range in constant attendance. *Tiliqua rugosa* forms stable pair bonds that last not just a single mating season, but in some cases up to 10 yr. Both sexes are faithful to their partners and use chemical signals to sometimes find their mates on temporary separations. Females are viviparous and give birth after males have separated (males attend their mates only during the spring mating season). There is no parental care (Bull 1988), although females can recognize their own offspring via pheromones (Bull et al. 1994) and tolerate them within their home ranges during the next spring after the neonates have overwintered (Bull and Baghurst 1998). Males of another long-lived skink (*Niveoscincus microlepidotus*) apparently use chemical signals to recognize specific females and to maintain close spatial proximity to them (mate guarding) over a substantial part of the breeding season, but not between years (Olsson and Shine 1998).

During our field censuses, we noticed what appeared to be family groups of *L. leopardinus*. It was quite common to see an adult male and female plus one to three juveniles in close proximity, often with one or more superimposed on the other. In the laboratory agonistic encounters, superimposition in *L. leopardinus* was observed more often than in all the other species. If *L. leopardinus* is like *T. rugosa* and adults maintain stable pair bonds, we would not expect those pair bonds to last more than a single season since annual adult mortality (as estimated from survival from just a single year to the next) was too high (75%). Unfortunately, we lack any information on the genetic relatedness of these small groups in the field, but we suggest that they may be mother, father, and offspring and that such associations might represent some form of parental care. Furthermore, *L. leopardinus* has a very small litter size of two to three young despite its relatively large size (Leyton and Valencia 1992), which is consistent with the premise of postnatal parental care. Such family groups are not common in lizards, but they have been documented through DNA analysis in an Australian skink, *Egernia stokesii* (Gardner et al. 2001), and have been suggested in other species along with the idea that the young might be receiving some sort of protection (Panov and Zykova 1993; Zykova and Panov 1993; Lemos-Espinal, Smith, and Ballinger 1997; Mouton, Flemming, and Kanga 1999). Presumed family groups and indirect parental care have been suggested in another high-elevation, viviparous *Liolaemus* lizard, *L. huacahuasicus* (Halloy and Halloy 1997).

Stamps (1983b) closed her chapter on lizard territoriality in the oft-cited book *Lizard Ecology: Studies of a Model Organism* with an emphasis on the variety and novelty of patterns of social organization among the herbivorous Iguanidae. Here, we emphasize the variety and novelty of social organization of high-elevation, viviparous lizards. These lizards are often found at high densities and because of often relaxed thermal rigor and predation, develop rich, complex social behavior that is not frequently observed in lizards. Often these high-elevation, viviparous species are also herbivorous (*L. bellii* and *L. leopardinus* are herbivores), so that dietary specialization may join with the characteristics of high elevation and viviparity—which are also highly intercorrelated (Shine 1985; Schulte et al. 2000)—in eliciting some rather complicated social behavior.

However, we do not mean to single out just high-elevation, vivipa-rous *Liolaemus*. For example, Beuchat (1982) observed high-elevation, viviparous *Sceloporus jarrovi* at very high densities and reported them to be nonterritorial, although Ruby and Baird (1994), in a later study at the same site, could not find such elevated population densities and did find territorial behavior among males. However, females at the high site were less aggressive than those at a low site and were much less territo-rial (Ruby and Baird 1994). Females had larger home ranges and more overlap with other females at the high site compared with the low one. In addition, at least at one small colony at the high elevation, two adult males substantially overlapped each other and all six resident adult fe-males, but only the larger male courted females—what appears to be a very simple dominance hierarchy. Lemos-Espinal et al. (1997) reported breeding-season, mixed-sex aggregations of two to nine individuals of viviparous *Sceloporus mucronatus* at 3,400 m in México. Lemos-Espinal, Smith, and Ballinger (1997) described mother-offspring associations in viviparous *Xenosaurus newmanorum* of Mexican cloud forests and sug-gested possible parental care. There seem to be some interesting varia-tions in lizard social behavior among high-elevation, viviparous lizards, and we suggest that studies of their behavior and social organization will bring to light even more surprising aspects.

Conclusion

We did not observe a phylogenetic correlation of space use, agonistic behavior, or courtship behavior in our analyses. It would appear that local conditions have been more important than phylogenetic con-straints in the evolution of the patterns of social behavior that we ob-served. Species at high elevations enjoy relaxed predation and thermal constraints compared with those at low elevations. Under these favor-able environmental conditions, lizard densities are high, territorial so-cial organization is not evident, and rich forms of social behavior have developed. Elements of complex social behavior not often observed in lizards are suggested in these high-elevation species; however, more de-tailed studies of these species are needed to confirm such unusual and intriguing social savoir faire. Although we did not find much correla-tion of phylogeny and social organization in the seven *Liolaemus* species we studied, when the evolution of territoriality is explored among the entire taxon of all lizards, a distinct phylogenetic pattern ap-pears (Martins 1994). Even when just the monophyletic clade of *Scelo-porus* is considered, there is an obvious phylogenetic component to the evolution of territorial behavior (Martins 1994). To test the ideas presented in this chapter—to test more fully for a phylogenetic effect—more data on more *Liolaemus* species are needed; *Sceloporus* and *Anolis* have had their turn, and now it is *Liolaemus*'s turn.

Our new PCM is both graphical and statistically quantitative. Using partial CCA and coding for evolutionary bifurcations at each node, one can test to determine if the split at that node (as an explicit covariable) explains the majority of the pattern of behavior of the species distal to it, or if those species retain their behavioral distinctiveness and fall out in a pattern statistically different from random. As such, our PCM is somewhat like that of Stearns (1983) as calculated by Harvey and Pagel (1991) on data from Millar and Zammuto (1983), except that we controlled for variation in behavior at each node of the phylogeny to ask if further microevolution occurred distal to that node in each sister clade. Thus, our method uses sequential subtraction of variation due to phylogeny at each bifurcation, not all at once as per Stearns (1983). We used the dummy variables of clade membership as covariables in CCA. Most PCMs cannot handle within-species variation (but see Martins and Hansen [1997]), but our proposed method can. Nevertheless, a deficiency in our method is that we must assume that resultant descendants from each node have equivalent phylogenies with equal branch lengths. Obviously this may not be so, even though the same time has passed from a given node to each extant terminal taxon. Some lineages have more speciation than others, which, according to some models of evolutionary change, may induce more trait variation (Martins and Hansen 1997).

In closing, we acknowledge that we know nothing of the genetic basis of the social behavior of these species (nor is much known about the genetic basis of behavior in general in lizards). Behavioral traits, like life history traits, are probably quite labile and lizard populations may be able to modify their social behavior to local conditions even without genetic differentiation, without natural selection. Indeed, just as the thermal environment might affect lizard life history traits in a proximate, nongenetic way through the intermediary of activity times (Adolph and Porter 1993; Sorci, Colbert, and Belichon 1996), so also might thermal (and predation) aspects of the environment proximately mold social behavior. One must exercise caution in attaching a wholly adaptationist interpretation to the observed elevational patterns of social organization. Most likely, the observed pattern is a combination of ultimate and proximate influences (Alcock 1998). Moreover, our conclusions are reached from the study of social behavior of only a small set of species. As more information on spacing and social organization of more lizard species is gathered, it will be possible to incorporate those data into phylogenetic analyses and understand better the variation in lizard social behavior. We hope our chapter and the others in this book will stimulate more such studies.

Acknowledgments

We thank Susana Perea, Rodrigo Medel, Antonieta Labra, Anna Rosa Young-Downey, Pablo Marquet, Karen Folk, Jaime Jiménez, Herman Núñez, and Eduardo Pavez for help in the field. We are indebted to Fabian Jaksíc for logistic and intellectual support and advice. John Carothers lent valuable advice and field expertise to the project. Susana Perea and J. Kelly McCoy helped substantially with data entry and checking, initial analyses, and preliminary data organization. We thank Nora Ibargüengoytía, Monique Halloy, Jerry Husak, and Chuck Peterson for critically reading the manuscript. The Universidad Católica de Chile generously gave office and laboratory space as well as logistical support, which Emilia García and Juan Domingo Molina helped to expedite. This study was supported by National Science Foundation grant INT-8515418, the Fulbright Scholar Program, and Dirección de Investigación de la Universidad Católica de Chile grant 086/86.

CHAPTER 11

Sexual Dimorphism in Body Size and Shape in Relation to Habitat Use among Species of Caribbean *Anolis* Lizards

Jonathan B. Losos, Marguerite Butler, and Thomas W. Schoener

Evolutionary biologists have long used morphological differences between the sexes as indicators of differences in sexual or natural selection operating on the sexes. Within a sufficiently narrow taxonomic group such that other life history factors are more or less uniform, this has been a very fruitful approach, producing voluminous literature. We now know that sexual dimorphism is associated with a variety of factors, including mating system, diet, body size, and habitat (see Andersson [1994] and references therein).

Although males and females differ in a variety of aspects, the vast majority of this work has considered only dimorphism in body size (for a recent review, see Fairbairn [1997]). Shape dimorphism can also indicate evolutionary responses to different selective pressures, but patterns of size and shape dimorphism need not be congruent (Cooper and Vitt 1989; Gittleman and Van Valkenburgh 1997; Lappin and Swinney 1999). Thus, studies of both size and shape dimorphism may be particularly informative.

The traditional explanations for the evolution of sexual dimorphism can be grouped into three general classes (reviewed in Andersson [1994]):

1. Sexual Selection or Competition for Reproductive Success—If one sex can increase reproductive success by obtaining more matings, then any trait that confers a mating advantage will experience strong selection. This is commonly seen in territorial species in which one sex attempts to attract mates by acquiring the best possible territories, and through this intrasexual selection, often

results in larger body size or better developed weaponry in the more territorial sex. Alternatively, sexual selection may operate by mate choice (intersexual [epigamic] selection; reviewed in Bateson [1983]). If one sex controls mating opportunities by exercising choice among mating partners, then any trait that makes the opposite sex more attractive is likely to become exaggerated. This may include overall size, color, or ornamentation.

2. Intersexual Resource Differences—To a degree increasing with the scarcity of resources relative to demand, competition may be lessened if the sexes specialize to use different resources. Thus, intersexual resource partitioning can lead to sexual dimorphism in any trait that is directly linked to resource use, such as body size, trophic structures, or other morphological specializations (Selander 1966, 1972). This intersexual divergence is similar to the classic scenario of character displacement, with the added complication that members of the two sexes share a common gene pool (Slatkin 1984). Differences between the sexes in the use of resources such as food can also evolve, perhaps with difficulty (Slatkin 1984), if the appropriate fitness function is bimodal (i.e., if multiple adaptive optima exist), regardless of whether intersexual competition occurs (Schoener 1969b).

3. Different Reproductive Roles—In most species, the sexes perform different functions in reproduction that require differences in morphology (Arnold 1983). For example, females may require relatively greater abdominal volumes to accommodate eggs or developing embryos. Similarly, females may need larger pelvic openings for the passage of eggs or live young. In other species, females may care for the young while males provide food, which may lead to selection for different structures related to foraging.

An aspect of sexual dimorphism that has been little studied is the impact of variation in habitat. Most studies incorporating environmental variation have examined only the direct impact of intersexual resource competition. However, the habitat may potentially influence several types of selective pressures for dimorphism by controlling the opportunity for sexual selection to operate or by imposing particular habitat-specific functional constraints on the morphology of males and females (Selander 1966; Schoener 1969b; Post et al. 1999).

Sexual Dimorphism in Lizards

Lizard taxa exhibit a diversity of forms of sexual differences. Dimorphism in body size has been detected in almost all lizard families, with male-biased dimorphism more common but not universal (Fitch 1981; Stamps 1983b; Carothers 1984; Zamudio 1998; see also chapters 1, 5, 6, and 9). In addition to differences in size, dimorphism in various size-independent morphological attributes—such as relative tail length, limb length, dewlap size, wing size (in the gliding dragon, *Draco*), horn presence or size, and head size as well as coloration—is common (e.g., Vitt 1983; Cooper and Vitt 1989; Powell and Russell 1992; Mouton and van Wyk 1993; Barbadillo et al. 1995; McCoy et al. 1997; Shine et al. 1998; Herrel et al. 1999; Mouton, Flemming, and Kanga 1999; see also chapters 5, 6, and 9).

Studies of lizards have contributed a great deal to the understanding of sexual dimorphism, providing examples of each of the aforementioned evolutionary causes. For example, competition for mates is probably common, both among the many territorial species of lizards and in species that, although not territorial, engage in physical competition for access to females (Tokarz 1995b). Mate choice has been less studied in lizards, but it has been documented (reviewed in Tokarz [1995b]; see also Baird, Fox, and McCoy [1997]). In addition, intersexual differences in resource use among lizards have been widely reported and are often related to differences in body size and other aspects of morphology (e.g., Schoener 1967, 1968; Schoener and Gorman 1968; Hebrard and Madsen 1984; Vitt, Zani, and Durtsche 1995; Perry 1996).

Differences relating to reproductive roles have also been detected. For example, gravid females of many species have reduced sprinting abilities. Perhaps as a result, relative to other individuals, these gravid females rely more on crypsis than on rapid locomotion to escape potential predators (Bauwens and Thoen 1981; Cooper et al. 1990; Formanowicz, Brodie, and Bradley 1990; Schwarzkopf and Shine 1992; see also chapter 4).

Dimorphism and Habitat Use in *Anolis* Lizards

The evolutionary radiation of *Anolis* lizards in the Caribbean presents a particularly good opportunity to study sexual dimorphism. Anoles have radiated for the most part independently on each island in the Greater Antilles (Jackman et al. 1999), producing, with several exceptions, the same set of habitat specialists, termed *ecomorphs*, on each island

(Williams 1983; Losos et al. 1998). Members of the same ecomorph class, although not closely related, are similar in morphology, habitat use, and behavior (Williams 1983; Losos 1990a).

The general trend among Caribbean *Anolis* species is that males are the larger sex (Schoener 1970, 1977; Stamps, Losos, and Andrews 1997; Butler, Schoener, and Losos 2000). Dimorphism in morphological traits other than overall size has been studied less but has been documented in a number of anole species (e.g., Collette 1961; Powell and Russell 1992; Vitt, Zani, and Durtsche 1995; Glossip and Losos 1997). In addition, the sexes of many species differ in habitat use, behavior, and diet (e.g., Schoener 1967, 1968; Schoener and Gorman 1968; Perry 1996).

Anoles are generally highly territorial, although the degree of territoriality appears to differ among species. Where it has been studied, both sexes are usually territorial, but males may have substantially larger territories than females (Stamps 1977a, 1983b; Schoener and Schoener 1982b; Ruby 1984). Within a species, body size in males is often correlated with territory size, which, in turn, is often related to the number of females residing within the territory (Rand 1967a; Trivers 1976; Stamps 1977b; Schoener and Schoener 1982b; Hicks and Trivers 1983; Stamps 1983b; Ruby 1984). In one species, a correlation between male body size and number of copulations has been reported (Trivers 1976).

Given these characteristics, anoles would seem to present a good opportunity to study the role of habitat in producing sexual dimorphism. Previously, we established a relationship between size dimorphism and habitat use (Butler, Schoener, and Losos 2000). Here, we review those results on size dimorphism and then ask whether a similar relationship exists for shape and, if so, whether patterns of shape and size dimorphism differ, potentially indicating different selective forces at work.

Methods
Size Dimorphism

To investigate whether species occupying different habitats displayed consistent differences in size dimorphism, we used data on body size for Caribbean anoles originally compiled by one of us for another set of studies (Schoener 1969b, 1970; Table 11-1). We used snout-vent length (SVL) as a proxy for body size, taking the mean of the largest one-third of each sex of each species based on large samples of museum specimens. We calculated size dimorphism as the natural logarithm of the ratio of this mean male size divided by the counterpart mean female size. Six habitat categories (ecomorphs) were included: trunk-ground,

Table 11-1 Species of *Anolis* included in study.

Island	Species	Ecomorph	Size Analysis	Shape Analysis
Cuba	A. ahli	Trunk-ground		X
	A. allisoni	Trunk-crown	X	
	A. allogus	Trunk-ground	X	X
	A. alutaceus	Grass-bush	X	X
	A. angusticeps	Twig	X	X
	A. equestris	Crown-giant	X	X
	A. guazumae	Twig		X
	A. homolechis	Trunk-ground	X	X
	A. loysiana	Trunk	X	X
	A. luteogularis	Crown-giant		X
	A. mestrei	Trunk-ground		X
	A. ophiolepis	Grass-bush	X	X
	A. paternus	Twig		X
	A. porcatus	Trunk-crown	X	X
	A. sagrei	Trunk-ground	X	X
	A. vanidicus	Grass-bush		X
Hispaniola	A. aliniger	Trunk-crown	X	
	A. brevirostris	Trunk	X	
	A. chlorocyanus	Trunk-crown	X	
	A. coelestinus	Trunk-crown	X	
	A. cybotes	Trunk-ground	X	
	A. distichus	Trunk	X	
	A. hendersoni	Grass-bush	X	
	A. insolitus	Twig	X	
	A. olssoni	Grass-bush	X	
	A. ricordii	Crown-giant	X	
	A. semilineatus	Grass-bush	X	
	A. shrevei	Trunk-ground	X	
	A. whitemani	Trunk-ground	X	
Jamaica	A. garmani	Crown-giant	X	X
	A. grahami	Trunk-crown	X	X
	A. lineatopus	Trunk-ground	X	X
	A. opalinus	Trunk-crown	X	X
	A. sagrei	Trunk-ground		X
	A. valencienni	Twig	X	X
Puerto Rico	A. cooki	Trunk-ground	X	
	A. cristatellus	Trunk-ground	X	X
	A. cuvieri	Crown-giant	X	X
	A. evermanni	Trunk-crown	X	X
	A. gundlachi	Trunk-ground	X	X

Table 11-1 *(Continued)*

Island	Species	Ecomorph	Size Analysis	Shape Analysis
	A. krugi	Grass-bush	X	X
	A. occultus	Twig	X	X
	A. poncensis	Grass-bush	X	X
	A. pulchellus	Grass-bush	X	X
	A. stratulus	Trunk-crown	X	X

trunk-crown, trunk, crown- giant, grass-bush, and twig (Williams 1972, 1983). Data were available for 38 species from Cuba, Hispaniola, Jamaica, and Puerto Rico. Results of these analyses were published in Butler, Schoener, and Losos (2000).

Shape Dimorphism

To investigate differences in body proportions independent of the effect of size, we measured six variables: SVL, mass (transformed by cube root), forelimb and hindlimb lengths (from the insertion in the body wall to the tip of the longest digit), tail length, and number of subdigital lamellae underlying the third and fourth phalanges of the fourth toe of the hindfoot (measurements as in Losos [1990a]). We considered these variables to be relevant aspects of sexual dimorphism because previous studies on males indicated that interspecific variation in them has important functional and ecological consequences (e.g., Losos 1990a; Glossip and Losos 1997; Irschick and Losos 1999). Data for 30 species were collected in Puerto Rico and Jamaica by M. Butler and in Cuba by J. Losos (Table 11-1). Twenty-three of those species were held in common with the size dimorphism analysis. Mean values for each variable for each sex were calculated and used in subsequent analyses. All mean values of variables were ln-transformed prior to analysis. In this study, we are interested in the relationship between shape variation and habitat, independent of the effects of size; therefore, our methods of creating shape variables should statistically separate size and shape. In another study with somewhat different aims (Butler and Losos, in press), we took a different approach and defined shape geometrically.

Our approach here was to remove the effects of size by analyzing residuals from regressions of measured variables on size. For these purposes, we conducted two sets of analyses utilizing different variables as proxies for size. These residual variables represent size-free measures of morphological variation, which, by convention, we refer to herein as "shape" (we recognize that there are many other variables that we did not measure that one might consider as contributing to overall "shape" differences among species).

The first method used SVL for body size; this is a traditional method in lizard morphometric analyses. Each morphological variable was regressed against SVL using ln-transformed species-sex means as data points. These analyses used SVL values collected for morphometric measurements rather than those compiled by Schoener (1969b, 1970), which were used in the size dimorphism analyses discussed earlier. (However, residuals calculated using the Schoener [1969b, 1970] SVL data are very similar; the correlation between the two sets of SVL measurements is as follows: males, $r = 0.96$; females, $r = 0.98$.)

The second method defined size as a composite of all variables, specifically the first axis of a principal components analysis (PCA), and analyzed residuals of a regression against scores on this (PC I) axis. All variables loaded strongly and positively on PC I (Table 11-2). The rationale for this approach was that all variables in this study increase with body size in interspecific comparisons (Butler 1998). Hence, rather than arbitrarily choosing one variable as size, this approach used a combination of all variables to quantify size. In addition, this approach allowed examination of dimorphism in relative body length, whereas the former approach, by equating size with body length, rendered impossible an examination of dimorphism in relative body length (further discussion of these approaches to removing size effects can be found in Beuttell and Losos [1999]).

To express overall shape variation in a single measure, we took the Euclidean distance between male and female coordinates in "shape space." First, we conducted PCA on residual values calculated as already described for both the SVL and PC I analyses. This analysis used a correlation matrix because of the mixture of different types of variables. Principal components axes were retained in subsequent analyses if they accounted for >5% of the variation or if at least one variable had a loading (absolute value) >0.4. We then calculated dimorphism as the Euclidean distance between sexes of a species in a multidimensional space defined by these PC axes.

More sophisticated analyses of variation in shape dimorphism and of the relationship between dimorphism in shape and dimorphism in habitat use are presented by Butler and Losos (in press).

Table 11-2   PCA using non-size-adjusted data.

Variable	Axis 1
Forelimb length	0.980
Hindlimb length	0.972
Tail length	0.933
SVL	0.981
Mass	0.986
Lamella number	0.890
Variance explained (%)	91.7
Eigenvalue	5.50

Phylogenetic Effects

Data points from closely related species are not phylogenetically independent (Felsenstein 1985). Consequently, statistically significant differences among types of habitat in analyses that ignore phylogeny might be artifactual if closely related species tend to occupy the same type of habitat. In the case of *Anolis,* ecologically similar species on different islands generally are not closely related (Losos et al. 1998). Hence, major phylogenetic confounding of these analyses is unlikely. Nonetheless, our data do include several sets of ecologically similar species from the same island that are closely related, so that some phylogenetic effect may exist. To investigate whether a relationship existed between degree of relatedness and similarity in dimorphism, we used the phylogenetic autocorrelation program of Cheverud, Dow, and Leutenegger (1985). Lack of a significant positive phylogenetic autocorrelation would indicate that closely related species are not particularly similar in dimorphism and thus would indicate that phylogenetic comparative methods are unnecessary (Gittleman and Luh 1992; Losos and Miles 1994; Losos 1999) (One important caveat is that the autocorrelation method may have low statistical power [Martins 1996]; however, the autocorrelation values reported here are not only nonsignificant, but also negative in sign, suggesting that lack of power is not the reason we failed to detect a significant positive autocorrelation.) In addition, Butler, Schoener, and Losos (2000) and Butler and Losos (in press) demonstrated that results of phylogenetic and nonphylogenetic analyses for both size and shape dimorphism are qualitatively indistinguishable. The phylogeny we used in the autocorrelation analyses is from Jackman et al. (1999). Branch lengths were calculated from mitochondrial DNA data assuming a molecular clock (details in Jackman et al. [1999]).

Results

Size Dimorphism

Size dimorphism (ln[(mean male SVL)/(mean female SVL)]) differs among types of habitat (analysis of variance [ANOVA]: $F_{5,32} = 8.40, P <$ 0.001). Trunk-ground and trunk-crown species have relatively high size dimorphism, whereas the other types exhibit considerably less dimorphism, with twig anoles being the least dimorphic (Fig. 11-1). In all species, males are larger than females except in the twig species *A. occultus*. Analysis of covariance (ANCOVA) supplements this finding, showing that the relationship between male and female size (i.e., dimorphism) differs among the ecomorphs even after accounting for size (females as dependent variable: difference in slopes, $F_{5,26} = 3.23, P = 0.021$; males as dependent variable: difference in slopes, $F_{5,26} = 1.71, P = 0.17$; difference in intercepts, $F_{5,31} = 8.42, P < 0.001$; all analyses on ln-transformed values). Indeed, size dimorphism is not related to overall body size, whether a species' size is represented by male size ($r^2 = 0.03, P = 0.33$) or female size ($r^2 = 0.01, P = 0.49$) (size values ln-transformed in these analyses).

Shape Dimorphism

In the analysis of residuals using SVL as an indicator of body size, the first three PC axes accounted for 95.1% of the variation. The first axis loaded strongly for relative fore- and hindlimb length and mass, the second axis for relative tail length and lamella number, and the third axis represented a contrast between relative tail length and lamella number (Table 11-3).

In the analysis using residuals from size defined as PC I, we first had to conduct a PCA on non-size-adjusted data. All morphological variables loaded strongly and positively on PC I, which accounted for 91.7% of the variation (Table 11-2).

Results from PC I–size analyses produced slightly different ordinations of variation than found in the SVL-size analyses. The first axis in this analysis indicated a contrast between relative fore- and hindlimb length versus relative lamella number. The second axis represented a contrast between relative mass and relative SVL versus relative tail length, and the third axis loaded moderately strongly for relative

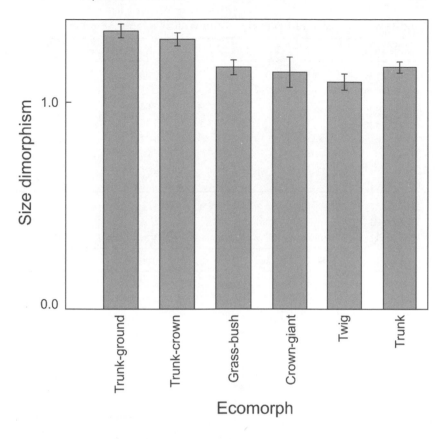

Figure 11-1   Variation in size dimorphism (defined as male SVL/female SVL) among eco-morph classes. Error bars represent ± 1 SE.

tail length and relative SVL versus relative lamella number and fore-limb length (Table 11-4).

However, the manner of size correction had little effect on the shape variables resulting from regressing each morphological variable against size (SVL or PC I). With one exception, residual values calculated using SVL as size were highly correlated with corresponding variables calcu-lated using PC I (Table 11-5). The only exception was mass dimor-phism, for which residuals calculated by the two methods were not strongly correlated. Why this should be is not clear, but it may be related to the extremely high correlation ($r = 0.99$) between SVL and mass, which implies that little variation remains to be explained by mass when residuals are calculated using SVL.

Table 11-3   PCA using residuals calculated with size represented by SVL.

Variable	Axis 1	Axis 2	Axis 3
Relative forelimb length	0.933	−0.083	0.279
Relative hindlimb length	0.954	0.113	−0.153
Relative tail length	0.493	0.683	−0.514
Relative mass	0.845	−0.251	0.331
Relative lamella number	−0.215	0.722	0.657
Variance explained (%)	55.6	21.4	18.1
Eigenvalue	2.78	1.07	0.91

Table 11-4   PCA using residuals calculated with size represented by PC I.

Variable	Axis 1	Axis 2	Axis 3
Relative forelimb length	0.868	0.191	−0.415
Relative hindlimb length	0.866	−0.423	0.010
Relative tail length	−0.165	−0.822	0.519
Relative SVL	−0.218	0.808	0.472
Relative mass	0.393	0.838	0.223
Relative lamella number	−0.829	0.107	−0.546
Variance explained (%)	40.3	37.6	16.9
Eigenvalue	2.42	2.26	1.01

Regardless of whether SVL or PC I was used to represent size, dimorphism in shape varied among types of ecomorphs (SVL: $F_{4,24} = 3.85$, $P = 0.015$; PC I: $F_{4,24} = 4.38$, $P = 0.008$; trunk anoles here and discussed later were excluded from ANOVA and ANCOVA shape analyses because shape data were available for only one species). In both analyses, crown-giant anoles were the most dimorphic, with trunk-ground and trunk-crown anoles also having high levels of dimorphism. By contrast, grass-bush, trunk, and twig anoles were less dimorphic (Fig. 11-2, which includes trunk anoles for illustrative purposes; analyses including them give qualitatively identical results).

The two analyses differ regarding which individual variables contribute to these trends. In the SVL analysis, only dimorphism in relative lamella number differed significantly among the ecomorphs in ANOVAs ($F_{4,24} = 5.64$, $P = 0.002$). Relative lamella number was substantially more female-biased in trunk-ground, trunk-crown, and crown-giant anoles than it was in grass-bush, trunk, and twig anoles. In only three species— *A. alutaceus* (grass-bush), *A. ophiolepis* (grass-bush), and *A. valencienni*

Table 11-5 Bivariate correlations between
residuals calculated using SVL and PC I (data
for 30 species).

Variable	r
Forelimb	0.79
Hindlimb	0.87
Tail	0.92
Lamella number	0.94
Mass	0.33

(twig)—was male residual lamella number greater than female residual lamella number. Nonetheless, the ecomorphs differed in dimorphism in all other variables taken simultaneously: multivariate analysis of variance (MANOVA) in which the dependent variables included all variables except dimorphism in lamella number revealed heterogeneity among ecomorphs (Wilks' $\lambda = 0.213, F_{20,70} = 2.10, P = 0.012$).

In the PC I analysis, only dimorphism in relative mass varied among the ecomorphs ($F_{4,24} = 2.79, P = 0.049$). Males of trunk-ground, trunk-crown, and crown-giant anoles tended to have higher mass residuals than females, whereas in grass-bush, trunk, and twig anoles, the reverse was true. However, as in the SVL size analysis, with MANOVA including all variables except dimorphism in relative mass as dependent variables, variation among the ecomorphs was significant (Wilks' $\lambda = 0.087, F_{25,75} = 2.82, P < 0.001$).

We also investigated whether dimorphism in particular variables was related to the extent of overall shape dimorphism. In the SVL analysis, dimorphism in relative lamella number explained a substantial proportion of the variation in overall dimorphism ($r^2 = 0.87, P < 0.001$). (In this analysis the absolute value of lamella number dimorphism was used because overall dimorphism values are distances, which are always positive; hence, the hypothesis being tested is that the magnitude of dimorphism in lamella number is related to the magnitude of overall dimorphism.) Dimorphism in none of the other variables (tested individually) was significantly related to overall dimorphism ($r^2 < 0.05$ in all cases). By contrast, in the PC I analysis, dimorphism in a number of variables was related (or nearly so) to overall shape dimorphism (relative lamella number: $r^2 = 0.89, P < 0.001$; relative mass: $r^2 = 0.29, P = 0.002$; relative SVL: $r^2 = 0.53, P < 0.002$; relative tail length: $r^2 = 0.12, P = 0.056$).

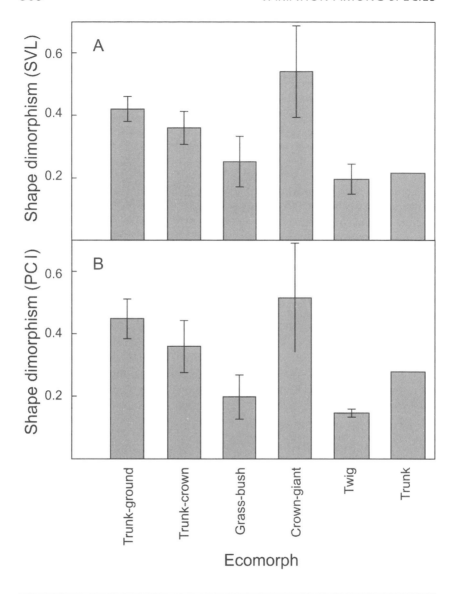

Figure 11-2   Variation in shape dimorphism among ecomorph classes (value for each species is the Euclidean distance in principal components space of males minus females). (*A*) Analysis in which SVL is used as the measure of body size; (*B*) analysis in which scores on the PC I axis are used as the measure of body size.

For both measures of size, degree of size dimorphism was related to degree of shape dimorphism (SVL: $r^2 = 0.27$, $F_{1,21} = 7.75$, $P = 0.011$; PC I: $r^2 = 0.35$, $F_{1,21} = 11.15$, $P = 0.003$) (Fig. 11-3). In both analyses, the relationship between size and shape dimorphism varied among the ecomorphs (ANCOVA, test of homogeneity of slopes; SVL: $F_{4,12} = 10.51$, $P = 0.001$; PC I: $F_{4,12} = 8.77$, $P = 0.001$; results also are significant if the analysis is reversed and shape dimorphism is treated as the independent variable). It is conceivable that this result may have been a phylogenetic artifact of differences resulting from the three closely related Cuban trunk-ground species, which are high in shape dimorphism and low in size dimorphism relative to the other trunk-ground anoles. However, even with trunk-ground anoles removed, differences are still significant or nearly so (ANCOVA, hypothesis of homogeneity of slopes not rejected, difference in intercepts; SVL: $F_{4,11} = 3.86$, $P = 0.034$; PC I: $F_{4,11} = 2.74$, $P = 0.084$).

As well as being related to size dimorphism, shape dimorphism was also related to absolute size, as might be expected given that crown-giants had high shape dimorphism and that the smaller ecomorphs had low shape dimorphism. This was true regardless of how size was defined (using SVL [from morphometric data set] or PC I for either males or females, r^2 values > 0.11 and $P < 0.045$ in all cases). Results of ANCOVA were ambiguous and depended on what variable was used to represent size. Most analyses indicated no difference among ecomorphs in the relationship between shape dimorphism and overall size, but the analysis using female PC I values as a representation of size revealed significant differences among the ecomorphs (difference in intercepts: $F_{4,23} = 3.25$, $P = 0.03$; all other analyses: $P > 0.18$).

Phylogenetic Effects

Figure 11-4 indicates that species similar in size dimorphism are not necessarily closely related. The phylogenetic autocorrelation was non-significant ($r = -0.40$, $z = -0.57$, $P > 0.70$, $n = 23$; calculations based on the log of the dimorphism value). In a similar fashion, no phylogenetic correlation was evident for shape dimorphism (SVL: $r = -0.19$, $z = -0.21$, $P > 0.70$; PC I: $r = -0.15$, $z = -0.15$, $P > 0.70$, $n = 29$).

Discussion
Sexual Dimorphism and Use of Habitat

Sexual dimorphism is an evolutionarily labile trait in Caribbean anoles. Examination of Fig. 11-4, for example, reveals that many closely related

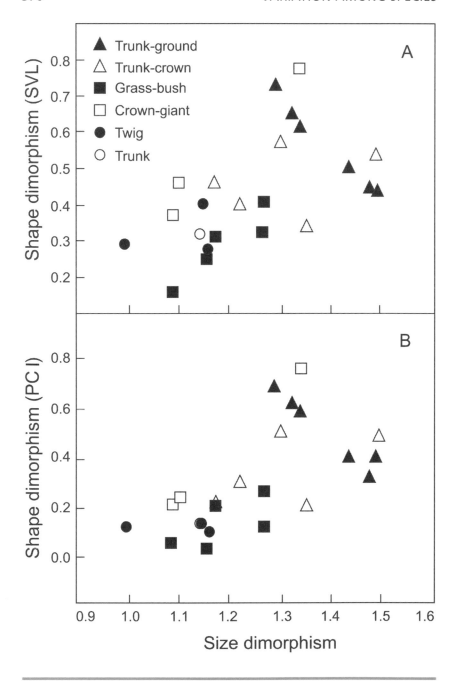

Figure 11-3 Relationship between shape dimorphism and size dimorphism. (*A*) Analysis in which shape dimorphism is calculated using SVL as the measure of body size; (*B*) analysis in which scores on the PC I axis are used as the measure of body size.

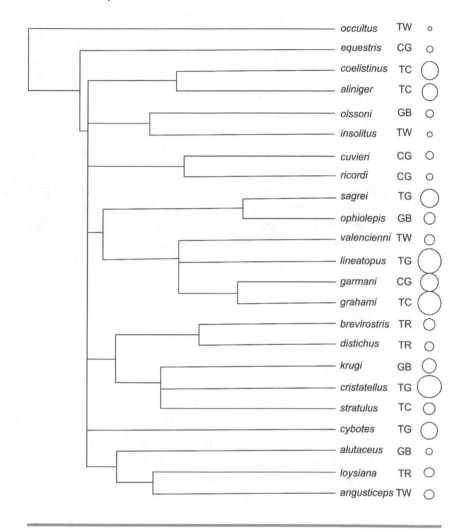

Figure 11-4 Phylogeny of Greater Antillean *Anolis* used in size dimorphism analyses. Size of the circles is proportional to the logarithm of dimorphism of body size. No relationship exists between degree of phylogenetic relationship and similarity in body size dimorphism. CG = crown-giant, GB = grass-bush, TC = trunk-crown, TG = trunk-ground, TR = trunk, TW = twig.

species differ greatly in size dimorphism and that species similar in size dimorphism do not cluster phylogenetically. This kind of impression is confirmed quantitatively by the lack of a phylogenetic autocorrelation for both size and shape dimorphism.

Greater Antillean ecomorphs are characterized by specialization to microhabitat. Interspecific comparisons reveal that both size and shape dimorphism differ among these habitat specialists. Not only do species in different habitats have a predictable difference in size between males and females, but habitat also predicts the magnitude of difference in shape between the sexes. By definition, these differences in shape are independent of differences in size (because size effects have been statistically removed), so this result indicates a remarkable effect of the habitat in structuring morphology of males and females. Moreover, these results indicate that the forces promoting dimorphism vary among types of habitat.

Although there is a correspondence between the degree of size and shape dimorphism among the ecomorph types, it is not perfect (Fig. 11-3). Trunk-ground and trunk-crown anoles tend to be dimorphic for both size and shape, whereas grass-bush, trunk, and twig anoles tend to display relatively little dimorphism in either size or shape. Crown-giant anoles are the exception: highly dimorphic in shape, but not very dimorphic in size. We note, however, that shape dimorphism may be manifest in different ways among shape-dimorphic anoles (e.g., some species may be shape dimorphic in two shape variables, whereas others may be in four), as we discuss later.

Despite the vast body of research on sexual dimorphism, only a few studies have documented relations to habitat use (e.g., Jarman 1974; Cheverud, Dow, and Leutenegger 1985; Geist and Bayer 1988; Ford 1994). Why do anoles differing in habitat use also differ in degree of dimorphism? The answer likely relates to at least one of three possibilities: (1) the opportunity for sexual selection is greater in some habitats than in others; (2) the opportunity for, or profit in, resource partitioning differs among habitats; or (3) the requirements and costs of the sexes' reproductive roles may differ among habitats.

In theory, these possibilities are testable by measuring the strength of sexual selection, the extent of intersexual resource partitioning, and the degree of behavioral and ecological differences directly related to reproductive roles. Given the 40 yr of ecological and behavioral research conducted on anoles (see Losos [1994] and Roughgarden [1995] for recent reviews), one might expect that sufficient data would be on hand to address these issues. However, surprisingly few studies have directly addressed any of them (e.g., Schoener 1968). Indeed, even data on such basic attributes as differences in home range size between the sexes or number of female territories per male territory are available for relatively few species (e.g., Rand 1967a; Schoener and Schoener 1982b; Tokarz 1998; Nicholson and Richards 1999). What is needed are studies of social structure and behavior designed to sample representative species of every ecomorph type. Consequently, general statements are not possible at this time. Nonetheless, we can still examine some specific predictions stemming from the three classes of explanations we have given.

SEXUAL SELECTION

Sexual selection could favor larger size in males if larger males are able to hold better territories. Indeed, many studies document that larger male anoles are dominant to smaller ones and generally win territorial encounters (Rand 1967a; Trivers 1976; Stamps 1977a, 1983b; Andrews 1982, 1985; Tokarz 1985). Shape dimorphism could result if some morphological traits enhance the ability of a male to outcompete another male. For example, males that are more muscular (greater mass per unit size) may be able to overpower less muscular opponents in territorial combat, or males with longer legs may be able to produce push-ups of greater amplitude or chase with greater speed.

Why might habitat use be related to degree of sexual selection and thus to the degree of sexual dimorphism? Variation in habitat may determine whether territoriality is economically feasible. For males to maintain and defend a territory and the females within it, the territory must be sufficiently defensible. An important aspect of defense is visibility; the habitat must be relatively free of visual obstructions so that intruders can be effectively detected. Anoles living on twigs at the periphery of the canopy and anoles in the grass would seem to be in particularly cluttered areas (although grass anoles often perch above the clutter [Garrido 1975; Fleishman 1988c; Garrido and Hedges 1992]). By contrast, males that perch on tree trunks, particularly lower to the ground, may have large and unimpeded vistas to scan for intruding males. Consequently, territoriality may be more feasible in open habitats, resulting in increased opportunity for intrasexual selection and, as a result, sexual dimorphism.

A second aspect of territorial defense is the size of the area defended: small areas are more feasible to defend than large ones. Species of *Anolis* have different foraging methods that imply different foraging areas (Moermond 1979a, b). Thus, species that exploit an abundant or concentrated food source are more likely to have a territorial mating system than are species that actively forage over large distances for rare or dispersed resources (Stamps 1977a). For example, trunk-ground anoles, which are in open habitats, use a sit-and-wait foraging style; move relatively infrequently (Losos 1990a); and, among species in the Bahamas, have smaller and apparently less overlapping home ranges,

which probably indicates a greater degree of territorial defense
(Schoener and Schoener 1982b). By contrast, anoles in other habitats
may have to move greater distances to find prey, which may make territorial defense expensive. The extreme in this case are twig anoles,
which have adopted an active foraging style in which they move slowly
but frequently, covering great distances in search of generally more
cryptic prey (Schoener and Schoener 1982b; Hicks and Trivers 1983;
Irschick and Losos 1996). For species such as this, territorial defense
may be difficult because the range of each animal may be too large to
make territorial defense feasible. As a result, intrasexual selection might
not operate and dimorphism might not occur.

Foraging mode could also be related to sexual dimorphism independent of its effects on sexual selection. Optimal foraging models predict that sit-and-wait predators should be more dimorphic in size than
active foragers (Schoener 1969b).

The data at hand do not allow discrimination between these different ways in which foraging mode may affect sexual dimorphism.
Nonetheless, it is striking that in all but one species of Caribbean anoles,
sexual size dimorphism is male biased. This would be predicted by sexual selection models but is only a prediction of optimal foraging models if the ancestral species were initially at least slightly dimorphic with
males being larger. Unfortunately, the dimorphism of the ancestral
anole species is difficult to infer because basal lineages differ in direction of dimorphism: in many genera closely related to anoles, dimorphism is female biased (Fitch 1981), but in basal lineages within *Anolis,*
dimorphism is male biased (Fitch 1976). Moreover, although almost all
Caribbean anoles have male-biased size dimorphism, many mainland
species are female biased (Fitch 1981). Hence, the dimorphism of the
ancestral anole presently cannot be inferred with confidence.

How well do these predictions hold up? At the extreme, they are very
successful: trunk-ground anoles are sit-and-wait foragers occurring in
open habitats and are, as predicted, highly dimorphic in size, with males
being larger. They are also relatively dimorphic in shape, with
males being relatively heavier (i.e., apparently more muscular), a trait
with obvious advantages in territorial combat. The reverse is true of
twig anoles. However, the case for other ecomorphs is less clear. Crown-
giant and trunk-crown anoles occur in much the same habitat. Both
have relatively large home ranges (Schoener and Schoener 1982b; Losos
et al. 1990; Jenssen, Greenberg, and Hovde 1995; Nicholson and
Richards 1999) through which they "wander" (Trivers 1976; Jenssen,

Greenberg, and Horde 1995), which should favor low dimorphism. Nonetheless, trunk-crown anoles are highly dimorphic in both size and shape, whereas crown-giants are dimorphic in shape, but not size (but see comments on *A. garmani* in Butler, Schoener, and Losos [2000]). Grass-bush anoles are low in both size and shape dimorphism. They move relatively infrequently and appear to be primarily sit-and-wait predators (Losos 1990a; Losos, personal observation; Butler, personal observation), which should favor dimorphism. They occur in cluttered habitats (which should inhibit the evolution of dimorphism) but may be able to partially defend territories in their low-visibility habitats by perching above the clutter. Thus, it is difficult to predict the levels of sexual selection operating in grass-bush anoles.

In summary, some aspects of the data seem consistent with an explanation of sexual selection as a factor promoting sexual dimorphism, but much more research is needed to address these predictions more carefully. In particular, direct measures of the strength of sexual selection would permit tests of the predicted relationship between habitat and sexual selection.

Mate choice, in theory, could also play a role if members of one sex selected for members of the other sex with particular traits. However, evidence for mate choice in anoles is weak; most females have territories that fall within the larger territory of a male and presumably mate primarily with the territory owner (although detailed studies have rarely been conducted [Stamps 1983b; Tokarz 1995b]). However, one anecdotal example of female mate choice is intriguing in that it occurred in the twig anole, *A. valencienni* (Hicks and Trivers 1983), which displays relatively little dimorphism in size or shape.

RESOURCE PARTITIONING

Intersexual resource partitioning resulting from competition could lead to sexual dimorphism if the sexes of a species evolved adaptations for using different resources, such as differences in habitat use or foraging behavior (reviewed in Schoener [1977] and Shine [1989, 1991]). A number of studies have demonstrated that body size and prey size are strongly correlated in anoles (Schoener 1967, 1968; Schoener and Gorman 1968; Roughgarden 1974). Thus, size dimorphism could imply differences in prey-size utilization (Schoener 1967, 1968; Schoener and Gorman 1968). Similarly, differences in shape may allow the sexes to use different habitats or to forage in different ways. Functional studies have demonstrated, for example, that relative limb length is related to maximal sprinting and jumping capabilities (Losos 1990b; Losos and Irschick 1996) and that pad size is related to clinging ability (Irschick and Losos 1996). Moreover, functional abilities depend on the environmental context: long hindlimbs, for example, are beneficial on broad surfaces, but not narrow ones (Losos and Sinervo 1989; Losos and Irschick 1996). Consequently, differences in shape may permit intersexual resource partitioning.

The sexual dimorphism that results from intersexual resource partitioning may be related to habitat in several ways. First, differences in resource availability may determine the degree to which partitioning is profitable and this may vary by habitat (if resources are "superabundant," then one might not expect partitioning to occur at all). Few quantitative data are available to examine this hypothesis.

Second, the resource dimension that is being partitioned may have a greater range in some habitats than in others, making partitioning in the latter more feasible. Again, few quantitative data exist, but in qualitative terms, the hypothesis is supported: trunk-ground and trunk-crown anoles use a wider variety of structural habitats (i.e., a greater range in perch heights and diameters) than the less dimorphic grass and twig anoles (Rand 1964, 1967b; Schoener and Schoener 1971a, b; Irschick and Losos 1999) and thus may have a greater opportunity for intersexual divergence. Not enough data are available for crown-giant anoles to evaluate this prediction.

Third, the presence of other anole species in a habitat might also preclude the ability for sexual dimorphism to occur because increased dimorphism might lead to increased interspecific competition. However, this prediction seems inconsistent with the data, although quantitative data are few (Butler, Schoener, and Losos 2000). Specifically, twig anoles, which use narrow-diameter perches little used by other species, probably interact with the fewest other ecomorph types, yet they have low dimorphism, contrary to expectations. On the other hand, the habitat use of trunk-crown anoles probably overlaps the habitat use of the greatest number of other ecomorph types because they overlap greatly with crown-giant, trunk-ground, and trunk anoles. Despite this overlap, they display high levels of dimorphism.

In summary, there are three possible explanations for differences in the feasiblity of intersexual resource partitioning among different habitats. More detailed and quantitative attempts to distinguish among these hypotheses will require much better data on the availability of resources, particularly relative to the density of the consumers, as well as data on the degree to which the sexes differ in resource use. No precise data are available of the former sort, and less of the latter are available than one would like (Rand 1964, 1967b; Schoener 1967, 1968; Schoener and Gorman 1968; Schoener and Schoener 1971a, b).

REPRODUCTIVE ROLES

Two obvious differences exist in the reproductive roles of anoles. First, females carry eggs, which at least in small species can be quite large relative to body size (Andrews and Rand 1974). Second, males tend to display much more than females to announce their presence, defend their territory, and court females (e.g., Jenssen, Greenberg, and Hovde 1995). To do this, males often choose particularly conspicuous spots from which to display. As a result, males may be more vulnerable to predators than females, as data indicate for *A. sagrei*, a highly territorial species (Schoener and Schoener 1982a).

These reproductive costs may vary among habitats. In females, for example, the burden of carrying an egg may be greater in some habitats than in others, perhaps leading to the evolution of more subdigital lamellae to enhance clinging ability in those habitats. Indeed, some of our analyses indicate that the ecomorphs differ in relative lamella number. These analyses indicate that ecomorphs that are absolutely larger (e.g., crown-giant, trunk-ground) exhibit greater female-biased dimorphism in lamella number, whereas smaller ecomorphs (e.g., twig, grass-bush) are less dimorphic. This result suggests that the burden (either the relative size, or the length of time the female is gravid) imposed by egg bearing is an increasing function of egg size, a hypothesis for which no data presently exist. Other possible relationships exist between dimorphism and reproductive roles, but our data do not reveal any other clear relationships, and no data are currently available to examine whether the costs and functional requirements tied to sex-specific activities differ among habitats.

Aspects of Shape That Are Responsible for the Relationship between Dimorphism and Habitat Use

Even though the ecomorphs differ in the extent of overall shape dimorphism, univariate analyses identify only one trait for which dimorphism varies among the ecomorphs: lamella number in the SVL analysis and mass in the PC I analysis. Nonetheless, the MANOVAs indicate that in both analyses, the ecomorphs can be distinguished by some combination of the other variables taken together, even though

ANOVAs on each of the other variables separately are nonsignificant. In addition, although dimorphism in relative lamella number is the only variable that correlates with extent of overall shape dimorphism in the SVL analyses, several variables correlate with overall dimorphism in the PC I analyses. In summary, the analyses do not give a clear answer about which variables are responsible for overall patterns of shape dimorphism.

Two correlates of shape dimorphism are clear, however. First, species that are dimorphic in shape tend to be dimorphic in size as well. This suggests the possibility that the same selective factor is simultaneously favoring dimorphism in both. Although this is certainly conceivable, the relatively little size dimorphism and great shape dimorphism of crown-giants would then require special explanation. Second, shape dimorphism is related to overall size. That this is the case is perhaps surprising, given that size dimorphism is not related to overall size. Note that the correlation between overall size and size dimorphism is substantially weakened by the largest species, the crown giants, having low size dimorphism; these species, however, have high shape dimorphism, which accounts for the relationship between overall size and shape dimorphism. In addition, the variables that differed among the ecomorphs—dimorphism in lamella number in the SVL analysis and dimorphism in mass in the PC I analysis—also appear related to overall size. These results suggest that, because of the correlation between shape dimorphism and overall size, differences among the ecomorphs in shape might just be a correlated consequence of differences in size. However, support for this supposition is somewhat ambiguous (see the ANCOVA using different measures of size discussed earlier).

A concurrent study (Butler and Losos, in press) examined the relationship among size (defined geometrically), shape, and ecomorph type (and habitat use variables), using some of the data presented here but supplemented with data from additional species. In agreement with the findings here, Butler and Losos (in press) found that crown-giant anoles had the greatest magnitude of shape dimorphism in comparison with the other ecomorphs. In addition, shape dimorphism was largely multivariate and somewhat related to size dimorphism. However, when examining which category of morphological variation was more strongly related to habitat use, Butler and Losos (in press) found that shape generally accounted for greater variation in habitat use than did size (males and females analyzed separately). There were many differences between the studies, including substantial differences in methods of analysis, use of variables (lamella number was excluded in Butler and Losos [in press]), and set of species.

Clearly, a great deal more work is needed to verify patterns of shape dimorphism and then to understand why they occur. In addition, other aspects of shape might profitably be examined. For example, among four species of anoles in the Bahamas (Schoener 1968), jaw-length dimorphism is greatest in the twig species *A. angusticeps,* an intriguing result that runs contrary to the trends reported here.

Before leaving the discussion of shape dimorphism, two potential sources of error must be recognized. First, the shape measurements were collected by two different workers for different sets of species. Although attempts to standardize the methods were made, interindividual differences are always a possibility (Lee 1990). Second, the sample sizes for the Cuban species were often rather low, which may have introduced large error into the data. Whether any important error has been introduced by these procedures can only be evaluated by further data collection.

Sexual Dimorphism and Adaptive Radiation in Caribbean Anoles

Anoles have radiated widely in the Caribbean, independently producing a very similar suite of ecologically and morphologically specialized species on each island of the Greater Antilles (reviewed in Williams [1983] and Jackman et al. [1997]). A large body of research (reviewed in Losos [1994]) has documented how the different ecomorph classes have adapted morphologically and behaviorally to utilize different habitats. For example, long legs, which maximize sprinting and leaping abilities, occur in ecomorphs that sprint and leap often, whereas the shortest legs occur in species that must move with great agility on narrow surfaces (Losos and Sinervo 1989; Losos 1990a, b; Irschick and Losos 1999).

Our results here suggest that habitat has had a much more pervasive influence on the adaptive radiation of anoles than simply affecting locomotor adaptations. Rather, an entire syndrome of interrelated variables seems to evolve in response to habitat use. For example, twig anoles require short legs to avoid falling off the narrow surfaces they use. As a result, they cannot rely on speed to capture prey or escape predators. Instead, they forage actively over large areas in search of hidden prey and rely on crypsis to a much greater extent than other ecomorphs (Hicks and Trivers 1983). These factors, as well as the low visibility of their habitat, preclude the existence of a highly territorial structure and may affect other factors that determine the extent of sexual dimorphism. At the other extreme, the open habitat and long legs of trunk-ground anoles permit conspicuous social behavior and a sit-and-wait foraging style on active prey, factors conducive to the evolution of large dimorphism. Thus, sexual dimorphism is one facet of the remarkable convergence of anole assemblages across the Greater Antilles.

Acknowledgments

We thank the editors for inviting us to participate in the symposium and for their patience with the delivery of this chapter, and Judy Stamps and an anonymous reviewer for useful comments on a previous draft. We gratefully acknowledge the support of the National Science Foundation, the National Geographic Society, and the Society for the Study of Amphibians and Reptiles.

Literature Cited

Abell, A. J. 1997. Estimating paternity with spatial behaviour and DNA fingerprinting in the striped plateau lizard, *Sceloporus virgatus* (Phrynosomatidae). *Behav. Ecol. Sociobiol.* 41:217–226.

———. 1998a. The effect of exogenous testosterone on growth and secondary sexual character development in juveniles of *Sceloporus virgatus*. *Herpetologica* 54:533–543.

———. 1998b. Phenotypic correlates of male survivorship and reproductive success in the striped plateau lizard, *Sceloporus virgatus*. *Herpetol. J.* 8:173–180.

———. 1998c. Reproductive and post-reproductive hormone levels in the lizard *Sceloporus virgatus*. *Acta Zool. Mex. New Ser.* 74:43–57.

———. 1999a. Male-female spacing patterns in the lizard, *Sceloporus virgatus*. *Amphibia–Repilia.* 20:185–194.

———. 1999b. Removal of an evolutionarily-reduced color signal, and restoration of the signal to the ancestral state, influence male-male interactions in the striped plateau lizard, *Sceloporus virgatus*. *Bull. Md. Herpetol. Soc.* 35:115–142.

Abell, A. J., and D. K. Hews. 1999. Physiological stress response in male and female lizards, *Sceloporus virgatus*. *Am. Zool.* 39:28A.

Adams, E. S., and R. L. Caldwell. 1990. Deceptive communication in asymmetric fights of the stomatopod crustacean *Gonodactylus bredini*. *Anim. Behav.* 39:706–716.

Adams, E. S., and M. Mesterton-Gibbons. 1995. The cost of threat displays and the stability of deceptive communication. *J. Theor. Biol.* 175:405–421.

Adkins-Regan, E. 1987. Hormones and sexual differentiation. In *Hormones and reproduction in fishes, amphibians and reptiles,* edited by D. O. Norris and R. E. Jones, 1–29. Plenum, New York.

Adolph, S. C. 1990. Influence of behavioral thermoregulation on microhabitat use by two *Sceloporus* lizards. *Ecology* 71:315–327.

Adolph, S. C., and W. P. Porter. 1993. Temperature, activity, and lizard life histories. *Am. Nat.* 142:273–295.

Albert, D. J., D. M. Petrovic, M. L. Walsh, and R. H. Jonik. 1990. Medial accumbens lesions attenuate testosterone-dependent aggression in male rats. *Physiol. Behav.* 46:625–631.

Albert, D. J., and M. L. Walsh. 1984. Neural systems and the inhibitory modulation of agonistic behavior: a comparison of mammalian species. *Neurosci. Biobehav. Rev.* 8:5–24.

Alberts, A. C. 1994. Dominance hierarchies in male lizards: implications for zoo management programs. *Zoo Biol.* 13:479–490.

Alcock, J. 1987. The effects of experimental manipulation of resources on the behavior of two calopterygid damselflies that exhibit resource-defense polygyny. *Can. J. Zool.* 65:2475– 2482.

———. 1998. *Animal behavior: an evolutionary approach,* 6th ed. Sinauer, Sunderland, Mass.

Alexander, R. D., and G. Borgia. 1979. On the basis of the male-female phenomenon. In *Sexual selection and reproductive competition in insects,* edited by M. S. Blum and N. A Blum, 417–440. Academic, New York.

Alexander, R. D., J. L. Hoogland, R. D. Howard, K. M. Noonan, and P. W. Sherman. 1979. Sexual dimorphism and breeding systems in pinnipeds, ungulates, primates and humans. In *Evolutionary biology and human social behavior: an anthropological perspective*, edited by N. A. Chagnon and W. Irons, 402–435. Duxbury, North Scituate, Mass.

Alonzo, S., and B. Sinervo. 2001. Mate choice games, context-dependent good genes, and genetic cycles in the side-blotched lizard, *Uta stansburiana. Behav. Ecol. Sociobiol.* 49:176–186.

Analytical Software. 1996. *Statistix for Windows.* Analytical Software, Tallahassee, Fla.

Anderson, D. J. 1991. Apparent predator-limited distribution of Galápagos red-footed boobies *Sula sula. Ibis* 133:26–29.

Anderson, R. A., and L. J. Vitt. 1990. Sexual selection versus alternative causes of sexual dimorphism in teiid lizards. *Oecologia* 84:145–157.

Andersson, M. 1982. Female choice selects for extreme tail length in a widowbird. *Nature* 299:818–820.

———. 1986. Evolution of condition-dependent sex ornaments and mating preferences: sexual selection based on viability differences. *Evolution* 40:804–816.

———. 1994. *Sexual selection.* Princeton University Press, Princeton, N.J.

Andersson, M. A., and Y. Iwasa. 1996. Sexual selection. *Trends Ecol. Evol.* 11:53–58.

Andersson, M., and C. G. Wiklund. 1978. Clumping versus spacing out: experiments on nest predation in fieldfares (*Turdus pilaris*). *Anim. Behav.* 26:1207–1212.

Andrews, R. M. 1979. Evolution of life histories: a comparison of *Anolis* lizards from matched island and mainland habitats. *Brevoria* 454:1–51.

———. 1982. Patterns of growth in reptiles. In *Biology of the Reptilia. Vol. 13*, edited by C. Gans and F. H. Pough, 273–320. Academic, London.

———. 1985. Mate choice by females of the lizard, *Anolis carolinensis. J. Herpetol.* 19:284–289.

Andrews, R. M., F. R. Méndez-de la Cruz, M. Villagrán-Santa Cruz, and F. Rodríquez-Romero. 1999. Field and selected body temperatures of the lizards *Sceloporus aeneus* and *Sceloporus bicanthalis. J. Herpetol.* 33:93–100.

Andrews, R. M., and A. S. Rand. 1974. Reproductive effort in anoline lizards. *Ecology* 55:1317–1327.

Aquino-Shuster, A. L., D. W. Duszynski, and H. L. Snell. 1990. Three new coccidia (Apicomplexa) from the Hood Island lizard, *Tropidurus delanonis,* from the Galápagos Archipelago. *J. Parasitol.* 76:313–318.

Arak, A. 1983a. Male-male competition and mate choice in anuran amphibians. In *Mate choice,* edited by P. Bateson, 181–210. Cambridge University Press, Cambridge, U.K.

———. 1983b. Sexual selection by male-male competition in natterjack toad choruses. *Nature* 306:261–262.

———. 1988. Sexual dimorphism in body size: a model and a test. *Evolution* 42:820–825.

Arak, A., and M. Enquist. 1993. Hidden preferences and the evolution of signals. *Philos. Trans. Roy. Soc. London Ser. B* 340:207–213.

Arnold, A. P. 1996. Genetically triggered sexual differentiation of brain and behavior. *Horm. Behav.* 30:495–505.

Arnold, A. P., S. W. Bottjer, E. A. Brenowitz, E. J. Nordeen, and K. W. Nordeen. 1987. Sexual dimorphisms in the neural vocal control system in song birds: ontogeny and phylogeny. *Brain Behav. Evol.* 28:22–31.

Arnold, A. P., and S. M. Breedlove. 1985. Organizational and activational effects of sex steroids on brain and behavior: a reanalysis. *Horm. Behav.* 19:469–498.

Arnold, E. N. 1984. Evolutionary aspects of tail shedding in lizards and their relatives. *J. Nat. Hist.* 18:127–169.

———. 1988. Caudal autotomy as a defense. In *Biology of the Reptilia. Vol. 16. Defense and life history,* edited by C. Gans and R. B. Huey, 236–273. Alan R. Liss, New York.

Arnold, S. J. 1983. Sexual selection: the interface of theory and empiricism. In *Mate choice,* edited by P. Bateson, 67–107. Cambridge University Press, Cambridge, U.K.

Arnold, S. J., and D. Duvall. 1994. Animal mating systems: a synthesis based on selection theory. *Am. Nat.* 143:317–348.

Austad, S. N. 1984. A classification of alternative reproductive behaviors and methods for field-testing of ESS models. *Am. Zool.* 24:308–309.

Ayala, S. C., and R. Hutchings. 1974. Hemogregarines (Protozoa: Sporozoa) as zoogeographical tracers of Galápagos Island lava lizards and marine iguanas. *Herpetologica* 30:128–132.

Bagnara, J. T., and M. E. Hadley. 1973. *Chromatophores and color change.* Prentice-Hall, Englewood Cliffs, N.J.

Baird, T. A., M. A. Acree, and C. L. Sloan. 1996. Age and gender-related differences in the social behavior and mating success of free-living collared lizards, *Crotaphytus collaris. Copeia* 1996:336–347.

Baird, T. A., S. F. Fox, and J. K. McCoy. 1997. Population differences in the roles of size and coloration in intra- and intersexual selection in the collared lizard, *Crotaphytus collaris:* influence of habitat and social organization. *Behav. Ecol.* 8:506–517.

Baird, T. A., and N. R. Liley. 1989. The evolutionary significance of harem polygyny in the sand tilefish, *Malacanthus plumieri:* resource or female defense? *Anim. Behav.* 38:817–829.

Baird, T. A., C. L. Sloan, and D. K. Timanus. 2001. Intra- and interseasonal variation in the socio-spatial behavior of adult male collared lizards, *Crotaphytus collaris* (Reptilia, Crotaphytidae). *Ethology* 106:1–19.

Baird, T. A., and D. K. Timanus. 1998. Social inhibition of territorial behaviour in yearling male collared lizards, *Crotaphytus collaris. Anim. Behav.* 56:989–994.

Bakken, G. S. 1992. Measurement and application of operative and standard operative temperatures in ecology. *Am. Zool.* 32:194–216.

Bakken, G. S., and D. M. Gates. 1975. Heat-transfer analysis of animals: some implications for field ecology, physiology, and evolution. In *Perspectives in biophysical ecology,* edited by D. M. Gates and R. Schmerl, 255–290. Springer-Verlag, New York.

Baldi, R., C. Campagna, S. Pedranza, and B. J. Le Boeuf. 1996. Social effects of space availability on the breeding behaviour of elephant seals in Patagonia. *Anim. Behav.* 51:717–724.

Ballinger, R. E., and T. G. Hipp. 1985. Reproduction in the collared lizard, *Crotaphytus collaris,* in west central Texas. *Copeia* 1985:976–980.

Ballinger, R. E., J. W. Nietfeldt, and J. J. Krupa. 1979. An experimental analysis of the tail in attaining high speed in *Cnemidophorus sexlineatus* (Reptilia: Squamata: Lacertilia). *Herpetologica* 35:114–115.

Balph, M. H., D. F. Balph, and H. C. Romesburg. 1979. Social status signaling in winter flocking birds: an examination of a current hypothesis. *Auk* 96: 78–93.

Bantock, C. R., and J. A. Bayley. 1973. Visual selection for shell size in *Cepaea* (Held.). *J. Anim. Ecol.* 42:247–261.

Barbadillo, L. J., D. Bauwens, F. Barahona, and M. J. Sánchez-Herráiz. 1995. Sexual differences in caudal morphology and its relation to tail autotomy in lacertid lizards. *J. Zool. Lond.* 236:83–93.

Barfield, R. J. 1971. Activation of sexual and aggressive behavior by androgen implanted into the male ring dove brain. *Endocrinology* 89:1470–1476.

Barlow, G. W. 1958. Daily movements of desert pupfish, *Cyprinodon macularius,* in shore pools of the Salton Sea, California. *Ecology* 39:580–587.

———. 1977. Modal action patterns. In *How animals communicate,* edited by T. A. Sebeok, 98–134. Indiana University Press, Bloomington.

———. 1981. Genetics and the development of behavior, with special reference to patterned motor output. In *Behavioral development: the Bielefeld interdisciplinary project,* edited by K. Immelman, G. W. Barlow, L. Petrinovich, and M. Main, 191–251. Cambridge University Press, New York.

Barlow, G. W., W. Rogers, and N. Fraley. 1986. Do Midas cichlids win through prowess or daring? It depends. *Behav. Ecol. Sociobiol.* 19:1–8.

Barrette, C. 1987. Dominance cannot be inherited. *Trends Ecol. Evol.* 2:251.

Bashey, F., and A. E. Dunham. 1997. Elevational variation in the thermal constraints on and microhabitat preferences of the greater earless lizard *Cophosaurus texanus. Copeia* 1997:725–737.

Basolo, A. L. 1990. Female preference predates the evolution of the sword in swordtail fish. *Science* 250:808–810.

———. 1995. Phylogenetic evidence for the role of a pre-existing bias in sexual selection. *Proc. Roy. Soc. Lond. Ser. B* 259:307–311.

Bastock, M. 1956. A gene mutation which changes a behavior pattern. *Evolution* 10:421–439.

Bateman, A. J. 1948. Intra-sexual selection in *Drosophila. Heredity* 2:349–368.

Bateson, P. 1983. *Mate choice.* Cambridge University Press, Cambridge, U.K.

Bauwens, D., A. M. Castilla, and P. F. N. Mouton. 1999. Field body temperatures, activity levels and opportunities for thermoregulation in an extreme microhabitat specialist, the girdled lizard (*Cordylus macropholis*). *J. Zool. Lond.* 249:11–18.

Bauwens, D., P. E. Hertz, and A. M. Castilla. 1996. Thermoregulation in a lacertid lizard: the relative contributions of distinct behavioral mechanisms. *Ecology* 77:1818–1830.

Bauwens, D., and C. Thoen. 1981. Escape tactics and vulnerability to predation associated with reproduction in the lizard *Lacerta vivipara. J. Anim. Ecol.* 50:733–743.

Bayliss, J. R. 1981. The evolution of parental care in fishes, with reference to Darwin's rule of male sexual selection. *Environ. Biol. Fish.* 6:223–251.

Beani, L., and S. Turillazzi. 1999. Stripes display in hover-wasps (Vespidae: Stenogastrinae): a socially costly status badge. *Anim. Behav.* 57:1233–1239.

Becker, J. B., S. M. Breedlove, and D. Crews. 1992. *Behavioral endocrinology.* MIT Press, Cambridge, Mass.

Begon, M., J. L. Harper, and C. R. Townsend. 1990. *Ecology: individuals, populations and communities,* 2nd ed. Blackwell Scientific, Boston.

Beletsky, L. D., and G. H. Orians. 1996. *Red-winged blackbirds: decision-making and reproductive success.* University of Chicago Press, Chicago.

Beletsky, L. D., G. H. Orians, and J. C. Wingfield. 1989. Relationships of steroid hormones and polygyny to territorial status, breeding experience, and reproductive success in male redwinged blackbirds. *Auk* 106:107–117.

———. 1990. Steroid hormones in relation to territoriality, breeding density, and parental behavior in male yellow-headed blackbirds. *Auk* 107:60–68.

Bell, E. L., and A. H. Price. 1996. *Sceloporus occidentalis. Cat. Am. Amphib. Rep.* 631.1–631.17.

Bell, G., and M. J. Lechowicz. 1991. The ecology and genetics of fitness in forest plants. I. Environmental heterogeneity measured by explant trials. *J. Ecol.* 79:663–685.

Belthoff, J. R., A. M. Dufty, Jr., and S. A. Gauthreaux, Jr. 1994. Plumage variation, plasma steroids and social dominance in male house finches. *Condor* 96:614–625.

Berglund, A. 1993. Risky sex: male pipefishes mate at random in the presence of a predator. *Anim. Behav.* 46:169–175.

Berglund, A., A. Bisazza, and A. Pilastro. 1996. Armaments and ornaments: an evolutionary explanation of traits of dual utility. *Biol. J. Linn. Soc.* 58:385–399.

Berry, J. F., and R. Shine. 1980. Sexual size dimorphism and sexual selection in turtles (Order Chelonia). *Oecologia* 44:185–191.

Berry, K. H. 1974. The ecology and social behavior of the chuckwalla (*Sauromalus obesus obesus* Baird). *Univ. Calif. Publ. Zool.* 101:1–60.

Best, T. L., and G. S. Pfaffenberger. 1987. Age and sexual variation in the diet of collared lizards (*Crotaphytus collaris*). *Southwest. Nat.* 32:415–426.

Beuchat, C. A. 1982. Physiological and ecological consequences of viviparity in a lizard. Ph.D. dissertation, Cornell University, Ithaca, N.Y.

———. 1986. Reproductive influences on the thermoregulatory behavior of a livebearing lizard. *Copeia* 1986:971–979.

Beuttell, K., and J. B. Losos. 1999. Ecological morphology of Caribbean anoles. *Herpetol. Monogr.* 13:1–28.

Bischoff, R. L., J. L. Gold, and D. S. Rubenstein. 1985. Tail size and female choice in the guppy (*Poecilia reticulata*). *Behav. Ecol. Sociobiol.* 17:253–256.

Blair, W. F. 1960. *The rusty lizard: a population study.* University of Texas Press, Austin.

Blair, W. F., and A. P. Blair. 1941. Food habits of the collared lizard in northeastern Oklahoma. *Am. Midl. Nat.* 26:230–232.

Boersma, P. D. 1983. An ecological study of the Galápagos marine iguana. In *Patterns of evolution in Galápagos organisms,* edited by R. I. Bowman, M. Berson, and A. E. Leviton, 157–176. Pacific Division, AAAS, San Francisco.

Bontrager, S. K. 1980. Autecology of *Crotaphytus collaris.* M.S. thesis, Oklahoma State University, Stillwater.

Borgia, G. 1979. Sexual selection and the evolution of mating systems. In *Sexual selection and reproductive competition in insects,* edited by M. S. Blum and N. A. Blum, 19–80. Academic, New York.

————. 1985. Bower quality, number of decorations and mating success of male satin bowerbirds (*Ptilonorhynchus violaceus*): an experimental analysis. *Anim. Behav.* 33:266–271.

Bowker, R. W. 1988. A comparative behavioral study and taxonomic analysis of gerrhonotine lizards. Ph.D. dissertation, Arizona State University, Tempe.

Bowman, R. I. 1966. *The Gálapagos: proceedings of the symposia of the Galápagos international scientific project.* University of California Press, Berkeley.

Boyd, S. K., K. D. Wissing, J. E. Heinsz, and G. S. Prins. 1999. Androgen receptors and sexual dimorphisms in the larynx of the bullfrog. *Gen. Comp. Endocrinol.* 113:59–68.

Bradbury, J. W., R. M. Gibson, C. E. McCarthy, and S. L. Vehrencamp. 1989. Dispersion of displaying male sage grouse II. The role of female dispersion. *Behav. Ecol. Sociobiol.* 24:15–24.

Bradbury, J. W., and S. L. Vehrencamp. 1998. *Principles of animal communication.* Sinauer, Sunderland, Mass.

Braña, F. 1993. Shifts in body temperature and escape behaviour of female *Podarcis muralis* during pregnancy. *Oikos* 66:216–222.

Branch, W. R., and M. J. Whiting. 1997. A new *Platysaurus* (Squamata: Cordylidae) from the Northern Cape Province, South Africa. *Afr. J. Herpetol.* 46:124–136.

Brattstrom, B. H. 1974. The evolution of reptilian social behavior. *Am. Zool.* 14:35–49.

————. 1982. A comparison of the social behavior of *Urosaurus auriculatus* and *U. clarionensis* on the Revillagigedo Islands, Mexico. *Herpetol. Rev.* 13:11–12.

Breckenridge, W. J. 1943. The life history of the black-banded skink *Eumeces septentrionalis septentrionalis* (Baird). *Am. Midl. Nat.* 29:607–614.

Brenowitz, E., and A. P. Arnold. 1985. Lack of sexual dimorphism in steroid accumulation in vocal control regions of duetting songbirds. *Brain Res.* 344:172–175.

Breuner, C. W., and M. Orchinik. 1999. Intracellular receptors and membrane corticosteroid receptors in the house sparrow. *Soc. Neurosci. Abstr.* 25:615.

Brillet, C. 1986. Comportement agonistique et structure sociale du lezard nocturne malgache *Paroedura pictus*. *Amphibia-Reptilia* 7:237–258.

Brooks, D. R., and D. A. McLennan. 1991. *Phylogeny, ecology, and behavior: a research program in comparative biology.* University of Chicago Press, Chicago.

Brooks, R., and V. Couldridge. 1999. Multiple sexual ornaments coevolve with multiple mating preferences. *Am. Nat.* 154:37–45.

Brown, G. E., and J. A. Brown. 1996. Kin discrimination in salmonids. *Rev. Fish Biol. Fish.* 6:201–219.

Brown, J. H., and M. V. Lomolino. 1998. *Biogeography,* 2nd ed. Sinauer, Sunderland, Mass.

Brown, J. L. 1964. The evolution of diversity in avian territorial systems. *Wilson Bull.* 76:160–169.

Brown, J. L., and G. H. Orians. 1970. Spacing patterns in mobile animals. *Annu. Rev. Ecol. Syst.* 1:239–269.

Brown, J. T., and N. B. Pavlovic. 1992. Evolution in heterogeneous environments: Effects of migration on habitat specialization. *Evol. Ecol.* 6:360–382.

Brown, K. C., and D. E. Ruby. 1977. Sex-associated variation in the frequencies of tail autotomy in *Sceloporus jarrovi* (Sauria: Iguanidae) at different elevations. *Herpetologica* 33:380–387.

Bruce, L. L., and T. J. Neary. 1995. The limbic system of tetrapods: a comparative analysis of cortical and amygdalar populations. *Brain Behav. Evol.* 46:224–234.

Bryant, D. M., and A. V. Newton. 1994. Metabolic cost of dominance in dippers, *Cinclus cinclus. Anim. Behav.* 48:447–455.

Bull, C. M. 1988. Mate fidelity in an Australian lizard *Trachydosaurus rugosus. Behav. Ecol. Sociobiol.* 23:45–49.

———. 1994. Population dynamics and pair fidelity in sleepy lizards. In *Lizard ecology: historical and experimental perspectives,* edited by L. J. Vitt and E. R. Pianka, 159–174. Princeton University Press, Princeton, N.J.

Bull, C. M., and B. C. Baghurst. 1998. Home range overlap of mothers and their offspring in the sleepy lizard, *Tiliqua rugosa. Behav. Ecol. Sociobiol.* 42: 357–362.

Bull, C. M., M. Doherty, L. R. Schulze, and Y. Pamula. 1994. Recognition of offspring by females of the Australian skink, *Tiliqua rugosa. J. Herpetol.* 28:117–120.

Bull, C. M., C. L. Griffin, E. J. Lanham, and G. R. Johnson. 2000. Recognition of pheremones from group members in a gregarious lizard, *Egernia stokesii. J. Herpetol.* 34:92–99.

Burger, J., and M. Gochfeld. 1981. Discrimination of the threat of direct versus tangential approach to the nest by incubating herring and great black-backed gulls. *J. Comp. Physiol. Psychol.* 95:676–684.

———. 1990. Risk discrimination of direct versus tangential approach by basking black iguanas (*Ctenosaura similis*). *J. Comp. Psychol.* 104:388–394.

———. 1991. Burrow site selection by black iguana (*Ctenosaura similis*) at Palo Verde, Costa Rica. *J. Herpetol.* 25:430–435.

Burger, J., M. Gochfeld, and B. G. Murray, Jr. 1992. Risk discrimination of eye contact and directness of approach in black iguanas (*Ctenosaura similis*). *J. Comp. Psychol.* 106:97–101.

Burns, A. 1970. Temperature of Yarrow's spiny lizard *Sceloporus jarrovi* at high altitudes. *Herpetologica* 26:9–16.

Burns, K. J. 1998. A phylogenetic perspective on the evolution of sexual dichromatism in tanagers (Thraupidae): the role of female versus male plumage. *Evolution* 52:1219–1224.

Burt, W. H. 1943. Territoriality and home range concepts as applied to mammals. *J. Mammal.* 24:346–352.

Butler, M. A. 1998. Evolution of sexual dimorphism and adaptive radiation in *Anolis* lizards. Ph.D. dissertation, Washington University, St. Louis, Mo.

Butler, M. A., and J. B. Losos. Multivariate sexual dimorphism, sexual selection, and adaptation in Greater Antillean *Anolis* lizards. *Ecology,* in press.

Butler, M. A., T. W. Schoener, and J. B. Losos. 2000. The relationship between sexual size dimorphism and habitat use in Greater Antillean *Anolis* lizards. *Evolution* 54:259–272.

Camp, C. L. 1916. The subspecies of *Sceloporus occidentalis. Univ. Calif. Publ. Zool.* 17:63–74.

Candolin, U. 1997. Predation risk affects courtship and attractiveness of competing threespine stickleback males. *Behav. Ecol. Sociobiol.* 41:81–87.

Cardwell, J. R., and N. R. Liley. 1991. Hormonal control of sex and color change in the stoplight parrotfish, *Sparisoma viride. Gen. Comp. Endocrinol.* 81:7–20.

Carey, P. W. 1991. Resource-defense polygyny and male territory quality in the New Zealand fur seal. *Ethology* 88:63–79.

Carlquist, S. 1974. *Island biology.* Columbia University Press, New York.

Caro, T. M., and P. Bateson. 1986. Organization and ontogeny of alternative tactics. *Anim. Behav.* 34:1483–1499.

Carothers, J. H. 1981. Dominance and competition in an herbivorous lizard. *Behav. Ecol. Sociobiol.* 8:261–266.

———. 1984. Sexual selection and sexual dimorphism in some herbivorous lizards. *Am. Nat.* 124:244–254.

Carothers, J. H., S. F. Fox, P. A. Marquet, and F. M. Jaksíc. 1997. Thermal characteristics of ten Andean lizards of the genus *Liolaemus* in central Chile. *Rev. Chilena Hist. Nat.* 70:297–309.

Carothers, J. H., P. A. Marquet, and F. M. Jaksíc. 1998. Thermal ecology of a *Liolaemus* lizard assemblage along an Andean altitudinal gradient in Chile. *Rev. Chilena Hist. Nat.* 71:39–50.

Carpenter, C. C. 1960. Aggressive behavior and social dominance in the six-lined racerunner (*Cnemidophorus sexlineatus*). *Anim. Behav.* 8:61–66.

———. 1961. Patterns of social behaviour in the desert iguana, *Dipsosaurus dorsalis.* *Copeia* 1961:396–405.

———. 1966. Comparative behavior of the Galápagos lava lizards (*Tropidurus*). In *The Galápagos: proceedings of the symposia of the Galápagos international scientific project,* edited by R. I. Bowman, 269–273. University of California Press, Berkeley.

———. 1978a. Comparative display behavior in the genus *Sceloporus* (Iguanidae). *Milwaukee Publ. Mus. Contr. Biol. Geol.* 18:1–71.

———. 1978b. Ritualistic social behaviors in lizards. In *Behavior and neurology of lizards: an interdisciplinary colloquium,* edited by N. Greenberg and P. D. MacLean, 253–267. National Institute of Mental Health, Rockville, Md.

———. 1982. The aggressive displays of iguanine lizards. In *Iguanas of the world: their behavior, ecology, and conservation,* edited by G. M. Burghardt and A. S. Rand, 215–231. Noyes, Park Ridge, N.J.

Carpenter, C. C., and G. W. Ferguson. 1977. Variation and evolution of stereotyped behavior in reptiles. In *Biology of the Reptilia. Vol. 7. Ecology and behaviour A,* edited by C. Gans and D. W. Tinkle, 335–403. Academic, London.

Carpenter, G. C. 1995a. Modeling dominance: the influence of size, coloration, and experience on dominance relations in tree lizards (*Urosaurus ornatus*). *Herpetol. Monogr.* 9:88–101.

———. 1995b. The ontogeny of a variable social badge: throat color development in tree lizards (*Urosaurus ornatus*). *J. Herpetol.* 29:7–13.

Carpenter, L. 1987. Food abundance and territoriality: to defend or not to defend? *Am. Zool.* 27:387–399.

Case, T. J. 1978. A general explanation for insular body size trends in terrestrial vertebrates. *Ecology* 59:1–18.

———. 1982. Ecology and evolution of the insular gigantic chuckwallas, *Sauromalus hispidus* and *Sauromalus varius.* In *Iguanas of the world: their behavior, ecology, and conservation,* edited by G. M. Burghardt and A. S. Rand, 184–212. Noyes, Park Ridge, N.J.

Case, T. J., and D. T. Bolger. 1991a. The role of interspecific competition in the biogeography of island lizards. *Trends Ecol. Evol.* 6:135–139.

———. 1991b. The role of introduced species in shaping the distribution and abundance of island reptiles. *Evol. Ecol.* 5:272–290.

Castillo, M., and M. Halloy. 2000. Un singular despliegue visual en especies de lagartos de *Liolaemus*, Tropiduridae: el pataleo. Poster presentation at XV Reunión de Comunicaciones Herpetológicas de la Asociación Herpetológica Argentina, San Carlos de Bariloche, Río Negro, Argentina, 25–27 October 2000.

Censky, E. J. 1995. Mating strategy and reproductive success in the teiid lizard, *Ameiva plei. Behaviour* 132:529–557.

———. 1997. Female mate choice in the non-territorial lizard *Ameiva plei* (Teiidae). *Behav. Ecol. Sociobiol.* 40:221–225.

Chappell, M. A., M. Zuk, T. H. Kwan, and T. S. Johnsen. 1995. Energy cost of an avian vocal display: crowing in red jungle fowl. *Anim. Behav.* 49:255–257.

Cheverud, J. M., M. M. Dow, and W. Leutenegger. 1985. The quantitative assessment of phylogenetic constraints in comparative analyses: sexual dimorphism in body weight among primates. *Evolution* 39:1335–1351.

Christian, K. A., and C. R. Tracy. 1985. Physical and biotic determinants of space utilization by the Galapagos land iguana (*Conolophus pallidus*). *Oecologia* 66:132–140.

Christian, K., C. R. Tracy, and W. P. Porter. 1983. Seasonal shifts in body temperature and use of microhabitats by Galapagos land iguanas (*Conolophus pallidus*). *Ecology* 64:463–468.

Christman, S. P. 1980. Preliminary observations on the gray-throated form of *Anolis carolinensis* (Reptilia: Iguanidae). *Florida Field Nat.* 8:11–16.

Clutton-Brock, T. H. 1989. Mammalian mating systems. *Proc. Roy. Soc. Lond. Ser. B* 236:339–372.

Clutton-Brock, T. H., and S. D. Albon. 1979. The roaring of red deer and the evolution of honest advertisement. *Behaviour* 69:145–170.

Clutton-Brock, T. H., S. D. Albon, R. M. Gibson, and F. E. Guiness. 1979. The logical stag: adaptive aspects of fighting in red deer (*Cervus elaphus* L.). *Anim. Behav.* 27:211–225.

Clutton-Brock, T. H., S. D. Albon, and F. E. Guinness. 1988. Reproductive success in male and female red deer. In *Reproductive success,* edited by T. H. Clutton-Brock, 325–343. University of Chicago Press, Chicago.

Clutton-Brock, T. H., F. E. Guiness, and S. D. Albon. 1982. *Red deer: behavior of two sexes.* University of Chicago Press, Chicago.

Clutton-Brock, T. H., and P. H. Harvey. 1977. Primate ecology and social organization. *J. Zool. Lond.* 183:1–39.

Collette, B. B. 1961. Correlations between ecology and morphology in anoline lizards from Havana, Cuba and southern Florida. *Bull. Mus. Comp. Zool.* 125:137–162.

Collis, K., and G. Borgia. 1992. Age-related effects of testosterone, plumage, and experience on aggression and social dominance in juvenile male satin bowerbirds (*Ptilonorhynchus violaceus*). *Auk* 109:422–434.

Congdon, J. D., L. J. Vitt, and W. W. King. 1974. Geckos: adaptive significance and energetics of tail autotomy. *Science* 184:1379–1380.

Cooper, W. E., Jr. 1996. Chemosensory recognition of familiar and unfamiliar conspecifics by the scincid lizard *Eumeces laticeps. Ethology* 102:454–464.

———. 1997a. Escape by a refuging prey, the broad-headed skink (*Eumeces laticeps*). *Can. J. Zool.* 75:943–947.

———. 1997b. Factors affecting risk and cost of escape by the broad-headed skink (*Eumeces laticeps*). *Herpetologica* 53:464–474.

———. 1997c. Threat factor affecting antipredatory behavior in the broad-headed skink (*Eumeces laticeps*): repeated approach, change in predator path, and predator's field of view. *Copeia* 1997:613–619.

———. 1998a. Effects of refuge and conspicuousness on escape behavior by the broad-headed skink (*Eumeces laticeps*). *Amphibia-Reptilia* 19:103–108.

———. 1998b. Risk factors and emergence from refuge in the lizard *Eumeces laticeps. Behaviour* 135:1065–1076.

———. 1999a. Escape behavior by prey blocked from entering the nearest refuge. *Can. J. Zool.* 77:671–674.

———. 1999b. Tradeoffs between courtship, fighting, and antipredatory behavior by a lizard, *Eumeces laticeps. Behav. Ecol. Sociobiol.* 47:54–59.

———. 2000. Pursuit deterrence in lizards. *Saudi J. Biol. Sci.* 7:15–29.

———. Multiple roles of tail display by the curly-tailed lizard *Leiocephalus carinatus*: pursuit deterrent and deflective roles of a social signal. *Ethology,* in press.

Cooper, W. E., Jr., and A. C. Alberts. 1991. Tongue-flicking and biting in response to chemical food stimuli by an iguanid lizard (*Dipsosaurus dorsalis*) having sealed vomeronasal ducts: vomerolfaction may mediate these behavioral responses. *J. Chem. Ecol.* 17:135–146.

Cooper, W. E., Jr., and N. Burns. 1987. Social significance of ventrolateral coloration in the fence lizard, *Sceloporus undulatus. Anim. Behav.* 35:526–532.

Cooper, W. E., Jr., and W. R. Garstka. 1987. Lingual responses to chemical fractions of urodaeal glandular pheromone of the skink *Eumeces laticeps. J. Exp. Zool.* 242:249–253.

Cooper, W. E., Jr., W. R. Garstka, and L. J. Vitt. 1986. Female sex pheromone in the lizard *Eumeces laticeps. Herpetologica* 42:361–366.

Cooper, W. E., Jr., and N. Greenberg. 1992. Reptilian coloration and behavior. In *Biology of the Reptilia. Vol. 18. Physiology E. Brain, hormones, and behavior,* edited by C. Gans and D. Crews, 298–422. University of Chicago Press, Chicago.

Cooper, W. E., Jr., M. T. Mendonca, and L. J. Vitt. 1987. Induction of orange head coloration and activation of courtship and aggression by testosterone in the male broad-headed skink (*Eumeces laticeps*). *J. Herpetol.* 21:96–101.

Cooper, W. E., Jr., and L. J. Vitt. 1986a. Lizard pheromones: behavioral responses and adaptive significance in skinks of the genus *Eumeces.* In *Chemical signals in vertebrates 4,* edited by D. Duvall, D. M. Müller-Schwarze, and R. M. Silverstein, 323–340. Plenum, New York.

———. 1986b. Tracking of female conspecific odor trails by male broad-headed skinks (*Eumeces laticeps*). *Ethology* 71:242–248.

———. 1987. Deferred agonistic behavior in a long-lived scincid lizard *Eumeces laticeps*: field and laboratory data on the roles of body size and residence in agonistic strategy. *Oecologia* 72:321–326.

————. 1988. Orange head coloration of the male broad-headed skink (*Eumeces laticeps*), a sexually selected social cue. *Copeia* 1988:1–6.

————. 1989. Sexual dimorphism of head and body size in an iguanid lizard: paradoxical results. *Am. Nat.* 133:729–735.

————. 1993. Female mate choice of large male broad-headed skinks. *Anim. Behav.* 45:683–693.

————. 1997. Maximizing male reproductive success in the broad-headed skink (*Eumeces laticeps*): preliminary evidence for mate guarding, size-assortative pairing, and opportunistic extra-pair mating. *Amphibia-Reptilia* 18:59–73.

Cooper, W. E., Jr., L. J. Vitt, R. Hedges, and R. B. Huey. 1990. Locomotor impairment and defense in gravid lizards (*Eumeces laticeps*): behavioral shift in activity may offset costs of reproduction in an active forager. *Behav. Ecol. Sociobiol.* 27:153–157.

Couch, L., P. A. Stone, D. W. Duszynski, H. L. Snell, and H. M. Snell. 1996. A survey of the coccidian parasites of reptiles from islands of the Galápagos Archipelago: 1990–1994. *J. Parasitol.* 82:432–437.

Cox, C. R. 1981. Agonistic encounters among male elephant seals: frequency, context, and the role of female preference. *Am. Zool.* 21:197–209.

Cox, C. R., and B. J. Le Boeuf. 1977. Female incitation of male competition: a mechanism of sexual selection. *Am. Nat.* 111:317–335.

Crews, D. 1975. Psychobiology of reptilian reproduction. *Science* 189:1059–1065.

————. 1979. The hormonal control of behavior in a lizard. *Sci. Am.* 241:180–187.

————. 1985. Effects of early sex hormone treatment on courtship behavior and sexual attractivity in the red-sided garter snake, *Thamnophis sirtalis parietalis*. *Physiol. Behav.* 35:569–575.

————. 1993. The organizational concept and vertebrates without sex chromosomes. *Brain Behav. Evol.* 42:202–214.

Crews, D., and N. Greenberg. 1981. Function and causation of social signals in lizards. *Am. Zool.* 21:273–294.

Crews, D., J. E. Gustafson, and R. R. Tokarz. 1983. Psychobiology of parthenogenesis. In *Lizard ecology: studies of a model organism*, edited by R. B. Huey, E. R. Pianka, and T. W. Schoener, 205–231. Harvard University Press, Cambridge, Mass.

Crews, D., and R. Silver. 1985. Reproductive physiology and behavior interactions in non-mammalian vertebrates. In *Handbook of behavioral neurobiology, Vol. 7, Reproduction,* edited by N. T. Adler, D. Pfaff, and R. Goy, 1–182. Plenum, New York.

Crews, D., J. Wade, and W. Wilczynski. 1990. Sexually dimorphic areas in the brain of whiptail lizards. *Brain Behav. Evol.* 36:262–270.

Crews, D., and E. E. Williams. 1977. Hormones, reproductive behavior and speciation. *Am. Zool.* 17:271–286.

Crook, J. H. 1964. The evolution of social organisation and visual communication in the weaver birds (Ploceinae). *Behav. Suppl.* 10:1–178.

Crook, J. H., and J. S. Gartlan. 1966. Evolution of primate societies. *Nature* 210:1200–1203.

Crowley, S. R. 1985. Thermal sensitivity of sprint-running in the lizard *Sceloporus undulatus:* support for a conservative view of thermal physiology. *Oecologia* 66:219–225.

Daniels, C. B., S. P. Flaherty, and M. P. Simbotwe. 1986. Tail size and effectiveness of autotomy in a lizard. *J. Herpetol.* 20:93–96.

Darwin, C. 1845 (1909). *The voyage of the Beagle.* The Harvard Classics, P. F. Collier and Son, New York.

———. 1859 (1958). *The origin of species by means of natural selection: or the preservation of favoured races in the struggle for life.* Mentor, New American Library, New York.

———. 1871. *The descent of man, and selection in relation to sex.* Modern Library, New York.

———. 1872. *The expression of the emotions in man and the animals.* John Murray, London.

Davies, N. B. 1991. Mating systems. In *Behavioural ecology: an evolutionary approach,* 3rd ed., edited by J. R. Krebs and N. B. Davies, 263–294. Blackwell Scientific, Oxford, U.K.

———. 1992. *Dunnock behavior and social evolution.* Oxford University Press, Oxford, U.K.

Davies, N. B., and T. R. Halliday. 1978. Deep croaks and fighting assessment in toads *Bufo bufo. Nature* 274:683–685.

Davies, N. B., I. R. Hartley, B. J. Hatchwell, A. Desrochers, J. Skeer, and D. Nebel. 1995. The polygynandrous mating system of the alpine accentor, *Prunella collaris.* I. Ecological causes and reproductive conflicts. *Anim. Behav.* 49:769–788.

Davies, N. B., and A. I. Houston. 1984. Territory economics. In *Behavioural ecology: an evolutionary approach,* 2nd ed., edited by J. R. Krebs and N. B. Davies, 148–169. Sinauer, Sunderland, Mass.

Dawkins, M. S. 1993. Are there general principles of signal design? *Philos. Trans. Roy. Soc. Lond. Ser. B* 340:251–255.

Dawkins, M. S., and T. Guilford. 1991. The corruption of honest signalling. *Anim. Behav.* 41:865–873.

Dawkins, R. 1982. *The extended phenotype: the gene as the unit of selection.* W. H. Freeman, San Francisco.

Dawkins, R., and J. R. Krebs. 1978. Animal signals: information or manipulation? In *Behavioural ecology: an evolutionary approach,* 1st ed., edited by J. R. Krebs and N. B. Davies, 282–309. Blackwell Scientific, Oxford, U.K.

de Boer, B. A. 1980. A causal analysis of the territorial and courtship behaviour of *Chromis cyanea* (Pomacentridae, Pisces). *Behaviour* 73:1–50.

De Groot, P. 1980. Information transfer in a socially roosting weaver bird (*Quelea quelae:* Ploceinae): an experimental study. *Anim. Behav.* 28:1249–1254.

del Pozo, A. H., E. R. Fuentes, E. R. Hajek, and J. D. Molina. 1989. Zonación microclimática por efecto de los manchones de arbustos en el matorral de Chile central. *Rev. Chilena Hist. Nat.* 62:85–94.

Dempsey, G. R., M. J. Reilly, and N. L. Staub. 1996. The distribution of androgen receptor-immunoreactive nuclei in brains of plethodontid salamanders. *Am. Zool.* 36:30A.

DeNardo, D. F., and B. Sinervo. 1994. Effects of corticosterone on activity and home-range size of free-ranging male lizards. *Horm. Behav.* 28:53–65.

de Queiroz, K. 1987. Phylogenetic systematics of the iguanine lizards. *Univ. Calif. Publ. Zool.* 118:1–203.

Deslippe, R. J., and R. T. M'Closkey. 1991. An experimental test of mate defense in an iguanid lizard (*Sceloporus graciosus*). *Ecology* 72:1218–1224.

de Vries, T. 1984. The giant tortoises: a natural history disturbed by man. In *Key environments: Galápagos,* edited by R. Perry, 145–156. Pergamon, Oxford, U.K.

de Vries, T., and J. Black. 1983. Of men, goats & guava—problems caused by introduced species in the Galápagos. *Noticias Galápagos* 38:18–21.

DeWitt, C. B. 1967a. Behavioral thermoregulation in the desert iguana. *Science* 158:809–810.

———. 1967b. Precision of thermoregulation and its relation to environmental factors in the desert iguana, *Dipsosaurus dorsalis. Physiol. Zool.* 40:49–66.

DeWoody, J. A., D. E. Fletcher, S. D. Wilkins, W. S. Nelson, and J. C. Avise. 1998. Molecular genetic dissection of spawning, parentage, and reproductive tactics in a population of redbreast sunfish, *Lepomis auritus. Evolution* 52:1802–1810.

Dewsbury, D. A. 1982. Dominance rank and copulatory behavior and differential reproduction. *Quart. Rev. Biol.* 57:135–159.

———. 1990. Fathers and sons: genetic factors and social dominance in deer mice, *Peromyscus maniculatus. Anim. Behav.* 39:284–289.

Dial, B. E. 1986. Tail display in two species of iguanid lizards: a test of the "predator signal" hypothesis. *Am. Nat.* 127:103–111.

Dial, B. E., and L. C. Fitzpatrick. 1981. The energetic costs of tail autotomy to reproduction in the lizard *Coleonyx brevis* (Sauria: Gekkonidae). *Oecologia* 51: 310–317.

———. 1984. Predator escape success in tailed versus tailless *Scincella lateralis* (Sauria: Scincidae). *Anim. Behav.* 32:301–302.

Diamond, J. M. 1975. Assembly of species communities. In *Ecology and evolution of communities,* edited by M. L. Cody and M. Diamond, 342–444. Belknap, Cambridge, Mass.

Díaz, J. A. 1997. Ecological correlates of the thermal quality of an ectotherm's habitat: a comparison between two temperate lizard populations. *Funct. Ecol.* 11:79–89.

Díaz-Uriarte, R. 1999. Anti-predator behaviour changes following an aggressive encounter in the lizard *Tropidurus hispidus. Proc. Roy. Soc. Lond., Ser. B* 266:2457–2464.

di Castri, F., and E. R. Hajek. 1976. *Bioclimatología de Chile.* Ediciones de la Universidad de Chile, Santiago, Chile.

Digby, P. G. N., and J. C. Gower. 1986. *Ordination and classification.* Les Presses de l'Université de Montreal, Montreal, Quebec.

Distel, H., and J. Veazey. 1982. The behavioral inventory of the green iguana. In *Iguanas of the world: their behavior, ecology, and conservation,* edited by G. M. Burghardt and A. S. Rand, 252–270. Noyes, Park Ridge, N.J.

Dixon, K. A. 1993. Microgeographic variation in sexual selection in the mountain spiny lizard. Ph.D. dissertation, University of Chicago, Chicago.

Donoso-Barros, R. 1966. *Reptiles de Chile.* Ediciones de la Universidad de Chile, Santiago, Chile.

Doughty, P., and B. Sinervo. 1994. The effects of habitat, time of hatching, and body size on dispersal in *Uta stansburiana. J. Herpetol.* 28:485–490.

Duellman, W. E., and A. Schwartz. 1958. Amphibians and reptiles of southern Florida. *Bull. Florida State Mus. Biol. Sci.* 3:181–324.

Dugan, B. 1982. The mating behavior of the green iguana, *Iguana iguana*. In *Iguanas of the world: their behavior, ecology, and conservation,* edited by G. M. Burghardt and A. S. Rand, 320–341. Noyes, Park Ridge, N.J.

Dugan, B., and T. V. Wiewandt. 1982. Socio-ecological determinants of mating strategies in iguanine lizards. In *Iguanas of the world: their behavior, ecology, and conservation,* edited by G. M. Burghardt and A. S. Rand, 303–319. Noyes, Park Ridge, N.J.

Dunham, A. E. 1978. Food availability as a proximate factor influencing individual growth rates in the iguanid lizard *Sceloporus merriami. Ecology* 59:770–778.

Dunham, A. E., D. W. Tinkle, and J. W. Gibbons. 1978. Body size in island lizards: a cautionary tale. *Ecology* 59:1230–1238.

Duvall, D. 1982. Western fence lizard (*Sceloporus occidentalis*) chemical signals. III. An experimental ethogram of conspecific body licking. *J. Exp. Zool.* 221:23–26.

Eadie, J. M., and J. M. Fryxell. 1992. Density dependence, frequency dependence, and alternative nesting strategies in goldeneyes. *Am. Nat.* 140:621–641.

Eason, P. K., and J. A. Stamps. 1992. The effect of visibility on territory size and shape. *Behav. Ecol. Sociobiol.* 3:166–172.

Eberhard, W. G. 1979. The function of horns in *Podischnus agenor* (Dynastinae) and other beetles. In *Sexual selection and reproductive competition in insects,* edited by M. S. Blum and N. A. Blum, 231–258. Academic, New York.

Eckhardt, R. C. 1972. Introduced plants and animals in the Galápagos Islands. *Bioscience* 22:585–590.

Edwards, S. V., and S. Naeem. 1993. The phylogenetic component of cooperative breeding in perching birds. *Am. Nat.* 141:754–789.

Ellis, L. 1995. Dominance and reproductive success among nonhuman animals: a cross-species comparison. *Ethol. Sociobiol.* 16:257–333.

Emerson, S. B. 1994. Testing pattern predictions of sexual selection: a frog example. *Am. Nat.* 143:848–869.

———. 1996. Phylogenies and physiological processes—the evolution of sexual dimorphism in southeast Asian frogs. *Syst. Biol.* 45:278–289.

Emerson, S. B., C. N. Rowsemitt, and D. L. Hess. 1993. Androgen levels in a Bornean voiceless frog, *Rana blythi. Can. J. Zool.* 71:196–203.

Emlen, S. T. 1980. Ecological determinism and sociobiology. In *Sociobiology: beyond nature/nurture? Reports, definitions and debate, AAAS selected symposium 35,* edited by G. W. Barlow and J. Silverberg, 125–150.Westview, Boulder, Colo.

Emlen, S. T., and L. W. Oring. 1977. Ecology, sexual selection, and the evolution of mating systems. *Science* 197:215–223.

Endler, J. A. 1978. A predator's view of animal color patterns. *Evol. Biol.* 11:319–364.

———. 1980. Natural selection on color patterns in *Poecilia reticulata. Evolution* 34:76–91.

———. 1983. Natural and sexual selection on color patterns in poeciliid fishes. *Environ. Biol. Fish.* 9:173–190.

———. 1986. *Natural selection in the wild.* Princeton University Press, Princeton, N.J.

———. 1992. Signals, signal conditions, and the direction of evolution. *Am. Nat.* 139:S125–S153.

———. 1993. Some general comments on the evolution and design of animal communication systems. *Philos. Trans. Roy. Soc. Lond., Ser. B* 340:215–225.

Endler, J. A., and A. E. Houde. 1995. Geographic variation in female preferences for male traits in *Poecilia reticulata. Evolution* 49:456–468.

Enquist, M. 1985. Communication during aggressive interactions with particular reference to variation in choice of behaviour. *Anim. Behav.* 33:1152–1161.

Enquist, M., and O. Leimar. 1983. Evolution of fighting behaviour: decision rules and assessment of relative strength. *J. Theor. Biol.* 102:387–410.

Enquist, M., O. Leimar, T. Ljungberg, Y. Mallner, and N. Segerdahl. 1990. A test of the sequential assessment game: fighting in the cichlid fish *Nannacara anomala. Anim. Behav.* 40:1–14.

Erbelding, D. C., J. H. Schroeder, M. Schartl, I. Nanda, M. Schmid, and J. T. Epplen. 1994. Male polymorphism in *Limia perugiae* (Pisces: Poeciliidae). *Behav. Genet.* 24:95–101.

Evans, L. T. 1951. Field study of the social behavior of the black lizard, *Ctenosaura pectinata. Am. Mus. Novit.* 1493:1–26.

———. 1959. A motion picture study of maternal behavior of the lizard, *Eumeces obsoletus* Baird and Girard. *Copeia* 1959:103–110.

Evans, M. R. 1991. The size of adornments of male scarlet-tufted malachite sunbirds varies with environmental conditions, as predicted by handicap theories. *Anim. Behav.* 42:797–803.

Evans, M. R., A. R. Goldsmith, and S. R. A. Norris. 2000. The effects of testosterone on antibody production and plumage coloration in male house sparrows (*Passer domesticus*). *Behav. Ecol. Sociobiol.* 47:156–163.

Evans, M. R., and K. Norris. 1996. The importance of carotenoids in signaling during aggressive interactions between male firemouth cichlids (*Cichlasoma meeki*). *Behav. Ecol.* 7:1–6.

Fairbairn, D. J. 1997. Allometry for sexual size dimorphism: Pattern and process in the coevolution of body size in males and females. *Annu. Rev. Ecol. Syst.* 28:659–687.

Fairbanks, R. G. 1989. A 17,000-year glacio-eustatic sea level record: influence of glacial melting rates on the Younger Dryas event and deep-ocean circulation. *Nature* 342:637–642.

Farr, J. A. 1975. The role of predation in the evolution of social behavior of natural populations of the guppy, *Poecilia reticulata* (Pisces: Poecilidae). *Evolution* 29:151–158.

Felsenstein, J. 1976. Theoretical population genetics of variable selection and migration. *Annu. Rev. Genet.* 10:253–280.

———. 1985. Phylogenies and the comparative method. *Am. Nat.* 125:1–15.

Ferguson, G. W., and S. F. Fox. 1984. Annual variation of survival advantage of large juvenile side-blotched lizards, *Uta stansburiana:* its causes and evolutionary significance. *Evolution* 38:342–349.

Ferguson, G. W., J. L. Hughes, and K. L. Brown. 1983. Food availability and territorial establishment of juvenile *Sceloporus undulatus.* In *Lizard ecology: studies of a model organism,* edited by R. B. Huey, E. R. Pianka, and T. W. Schoener, 134–148. Harvard University Press, Cambridge, Mass.

Fernandez, A. S., C. Pieau, J. Reperan, E. Boncinelli, and M. Wassef. 1998. Expression of the *Emx-1* and *Dlx-1* homeobox genes defines three molecularly distinct domains in the telencephalon of mouse, chick, turtle and frog embryos: implications for the evolution of telencephalic subdivisions in amniotes. *Development* 125:2099–2111.

Fisher, R. A. 1930. *The genetical theory of natural selection.* Clarendon, Oxford, U.K.

Fitch, H. S. 1954. Life history and ecology of the five-lined skink, *Eumeces fasciatus. Univ. Kans. Publ. Mus. Nat. Hist.* 8:1–156.

———. 1955. Habits and adaptations of the great plains skink (*Eumeces obsoletus*). *Ecol. Monogr.* 25:59–83.

———. 1956. An ecological study of the collared lizard (*Crotaphytus collaris*). *Univ. Kans. Publ. Mus. Nat. Hist.* 8:213–274.

———. 1976. Sexual size differences in the mainland anoles. *Occas. Pap. Mus. Nat. Hist. Univ. Kans.* 50:1–21.

———. 1981. Sexual size differences in reptiles. *Misc. Publ. Mus. Nat. Hist. Univ. Kans.* 70:1–72.

Fitch, H. S., and J. Hackforth-Jones. 1983. *Ctenosaura similis.* In *Costa Rican natural history,* edited by D. H. Janzen, 394–396. University of Chicago Press, Chicago.

Fitch, H. S., and R. W. Henderson. 1978. Ecology and exploitation of *Ctenosaura similis. Univ. Kans. Sci. Bull.* 51:483–500.

Fitch, H. S., and P. L. von Achen. 1977. Spatial relationships and seasonality in the skinks *Eumeces fasciatus* and *Scincella laterale* in northeastern Kansas. *Herpetologica* 33:303–313.

FitzGibbon, C. D., and J. H. Fanshawe. 1988. Stotting in Thomson's gazelles: an honest signal of condition. *Behav. Ecol. Sociobiol.* 23:69–74.

Fleishman, L. J. 1986. Motion detection in the presence and absence of background motion in an *Anolis* lizard. *J. Comp. Physiol. A, Sensory, Neural Behav. Physiol.* 159:711–720.

———. 1988a. Sensory and environmental influences on display form in *Anolis auratus,* a grass anole from Panama. *Behav. Ecol. Sociobiol.* 22:309–316.

———. 1988b. Sensory influences on physical design of a visual display. *Anim. Behav.* 36:1420–1424.

———. 1988c. The social behavior of *Anolis auratus,* a grass anole from Panama. *J. Herpetol.* 22:13–23.

———. 1992. The influence of the sensory system and the environment on motion patterns in the visual displays of anoline lizards and other vertebrates. *Am. Nat.* 139:S36–S61.

Flemming, A. F., and J. H. van Wyk. 1992. The female reproductive cycle of the lizard *Cordylus polyzonus* (Sauria: Cordylidae) in the Southwestern Cape Province, South Africa. *J. Herpetol.* 26:121–127.

Folstad, I., and A. J. Karter. 1992. Parasites, bright males, and the immunocompetence handicap. *Am. Nat.* 139:603–622.

Ford, S. M. 1994. Evolution of sexual dimorphism in body weight in platyrrhines. *Am. J. Primatol.* 34:221–244.

Formanowicz, D. R., Jr., E. D. Brodie, Jr., and P. J. Bradley. 1990. Behavioural compensation for tail loss in the ground skink, *Scincella lateralis. Anim. Behav.* 40:782–784.

Formanowicz, D. R., Jr., E. D. Brodie, Jr., and J. A. Campbell. 1990. Intraspecific aggression in *Abronia vasconcelosii* (Sauria, Anguidae), a tropical, arboreal lizard. *Biotropica* 22:391–396.

Foster, S. A., and J. A. Endler. 1999. *Geographic variation in behavior: perspectives on evolutionary mechanisms.* Oxford University Press, New York.

Fox, D. L. 1976. *Animal biochromes and structural colors.* University of California Press, Berkeley.

Fox, S. F. 1975. Natural selection on morphological phenotypes of the lizard *Uta stansburiana. Evolution* 29:95–107.

———. 1978. Natural selection on behavioral phenotypes of the lizard *Uta stansburiana. Ecology* 59:834–847.

———. 1983. Fitness, home range quality and aggression in *Uta stansburiana.* In *Lizard ecology: studies of a model organism,* edited by R. B. Huey, E. R. Pianka, and T. W. Schoener, 134–148. Harvard University Press, Cambridge, Mass.

Fox, S. F., and T. A. Baird. 1992. The dear enemy phenomenon in collared lizards, *Crotaphytus collaris,* with a cautionary note on experimental methodology. *Anim. Behav.* 44:780–782.

Fox, S. F., J. M. Conder, and A. E. Smith. 1998. Sexual dimorphism in the ease of tail autotomy: *Uta stansburiana* with and without previous tail loss. *Copeia* 1998:376–382.

Fox, S. F., N. A. Heger, and L. S. DeLay. 1990. Social cost of tail loss in *Uta stansburiana:* lizard tails as status-signalling badges. *Anim. Behav.* 39:549–554.

Fox, S. F., S. S. Perea-Fox, and R. Castro Franco. 1994. Development of the tail autotomy adaptation at high and low elevations in Mexico under disparate levels of predation. *Southwest. Nat.* 39:311–322.

Fox, S. F., E. Rose, and R. Myers. 1981. Dominance and the acquisition of superior home ranges in the lizard *Uta stansburiana. Ecology* 62:888–893.

Fox, S. F., and M. A. Rostker. 1982. Social cost of tail loss in *Uta stansburiana. Science* 218:692–693.

Francis, R. C. 1988. On the relationship between aggression and social dominance. *Ethology* 78:223–237.

Franklin, W. L. 1982. Biology, ecology, and relationship to man of the South American camelids. In *Mammalian biology in South America,* edited by M. A. Mares and H. H. Genoways, 457–489. Special Publication Series, Vol. 6, Pymatunig Laboratory of Ecology, University of Pittsburgh.

———. 1983. Contrasting socioecologies of South America's wild camelids: the vicuña and the guanaco. In *Advances in the study of mammalian behavior,* edited by J. F. Eisenberg and D. G. Kleiman, 573–629. American Society of Mammalogists Special Publication 7, Shippensburg, Pa.

Franklin, W. L., W. E. Johnson, R. J. Sarno, and J. A. Iriarty. 1999. Ecology of the Patagonian puma *Felis concolor patagonica* in southern Chile. *Biol. Conserv.* 90:33–40.

Freeman, L. M., B. A. Padgett, G. S. Prins, and S. M. Breedlove. 1995. Distribution of androgen receptor immunoreactivity in the spinal cord of wild-type, androgen-insensitive and gonadectomized male rats. *J. Neurobiol.* 27:51–59.

Frost, D. R. 1992. Phylogenetic analysis and taxonomy of the *Tropidurus* group of lizards (Iguania: Tropiduridae). *Am. Mus. Novit.* 3033:1–68.

Frost, D. R., and R. Etheridge. 1989. A phylogenetic analysis and taxonomy of iguanian lizards. *Misc. Publ. Mus. Nat. Hist. Univ. Kans.* 81:1–65.

Frutos, N., L. C. Belver, and L. J. Avila. 2000. Dominio vital ("home range") de *Liolaemus koslowskyi* Etheridge, 1993 (Squamata: Iguania: Tropiduridae) en el norte de

La Rioja, Argentina. Poster presentation at XV Reunión de Comunicaciones Herpetológicas de la Asociación Herpetológica Argentina, San Carlos de Bariloche, Río Negro, Argentina, 25–27 October 2000.

Fuentes, E. R., and F. M. Jaksíc. 1979. Activity temperatures of eight *Liolaemus* (Iguanidae) species in central Chile. *Copeia* 1979:546–548.

Furlow, B., R. T. Kimball, and M. C. Marshall. 1998. Are rooster crows honest signals of fighting ability? *Auk* 115:763–766.

Gardner, M. G., C. M. Bull, S. J. B. Cooper, and G. A. Duffield. 2001. Genetic evidence for a family structure in stable social aggregations of the Australian lizard *Egernia stokesii*. *Mol. Ecol.* 10:175–183.

Garland, T., Jr. 1984. Physiological correlates of locomotor performance in a lizard: an allometric approach. *Am. J. Physiol.* 247:R806–R815.

Garrido, O. H. 1975. Distribución y variación del complejo *Anolis cyanopleurus* (Lacertilia: Iguanidae) en Cuba. *Poeyana* 143:1–58.

Garrido, O. H., and S. B. Hedges. 1992. Three new grass anoles from Cuba (Squamata: Iguanidae). *Carib. J. Sci.* 28:21–29.

Garson, P. J., and M. L. Hunter. 1979. Effects of temperature and time of year on singing behavior of wrens (*Troglodytes troglodytes*) and great tits (*Parus major*). *Ibis* 121:481–487.

Geist, D. 1996. On the emergence and submergence of the Galápagos Islands. *Noticias Galápagos* 56:5–9.

Geist, V. 1978. On weapons, combat, and ecology. In *Aggression, dominance, and individual spacing,* edited by L. Kramer, P. Pliner, and T. Alloway. 1–30. Plenum, NewYork.

Geist, V., and M. Bayer. 1988. Sexual dimorphism in the Cervidae and its relation to habitat. *J. Zool. Lond.* 214:45–53.

George, F. W., J. F. Nobel, and J. D. Wilson. 1981. Female feathering in Sebright cocks is due to conversion of testosterone to estradiol in the skin. *Science* 213:557–559.

Getty, T. 1998. Handicap signalling: when fecundity and viability do not add up. *Anim. Behav.* 56:127–130.

Gibbons, J. W., and J. E. Lovich. 1990. Sexual dimorphism in turtles with emphasis on the slider turtle (*Trachemys scripta*). *Herpetol. Monogr.* 4:1–29.

Gil, M., V. Perez-Mellado, and F. Guerrero. 1990. Habitat structure and home ranges of *Podarcis hispanica* (Steindachner, 1870). *Misc. Zool.* 12:273–282.

Gillespie, J. 1974. Polymorphism in patchy environments. *Am. Nat.* 108:145–151.

———. 1975. The role of migration in the genetic structure of populations in temporally and spatially varying environments. I. Conditions for polymorphism. *Am. Nat.* 109:127–135.

Gittleman, J. L., and H. K. Luh. 1992. On comparing comparative methods. *Annu. Rev. Ecol. Syst.* 23:383–404.

Gittleman, J. L., and B. Van Valkenburgh. 1997. Sexual dimorphism in the canines and skulls of carnivores: effects of size, phylogeny, and behavioral ecology. *J. Zool. Lond.* 242:97–117.

Glickman, S. E., L. G. Frank, S. Pavgi, and P. Licht. 1992. Hormonal correlates of 'masculinization' in female spotted hyaenas (*Crocuta crocuta*): I. Infancy through sexual maturity. *J. Reprod. Fertil.* 95:451–462.

Gliwicz, J. 1980. Island populations of rodents: their organization and functioning. *Biol. Rev.* 55:109–138.

Glossip, D., and J. B. Losos. 1997. Ecological correlates of number of subdigital lamellae in anoles. *Herpetologica* 53:192–199.

Godfray, H. C. J. 1995. Signalling of need between parents and young: Parent-offspring conflict and sibling rivalry. *Am. Nat.* 146:1–24.

Godwin, J., and D. Crews. 1997. Sex differences in the nervous system of reptiles. *Cell. Mol. Neurobiol.* 17:649–669.

Goin, O. B. 1957. An observation of mating in the broad-headed skink, *Eumeces laticeps. Herpetologica* 13:155–156.

Gonzalez, G., G. Sorci, and F. de Lope. 1999. Seasonal variation in the relationship between cellular immune response and badge size in male house sparrows (*Passer domesticus*). *Behav. Ecol. Sociobiol.* 46:117–122.

Gonzalez, G., G. Sorci, A. P. Møller, P. Ninni, C. Haussy, and F. de Lope. 1999. Immunocompetence and condition-dependent sexual advertisement in male house sparrows (*Passer domesticus*). *J. Anim. Ecol.* 68:1225–1234.

Goodwin, N. B., S. Balshine-Earn, and J. D. Reynolds. 1998. Evolutionary transitions in parental care in cichlid fish. *Proc. Roy. Soc. Lond., Ser. B* 265:2265–2272.

Grafen, A. 1984. Natural selection, kin selection and group selection. In *Behavioural ecology: An evolutionary approach,* 2nd ed., edited by J. R. Krebs and N. B. Davies, 62–89. Sinauer, Sunderland, Mass.

———. 1990. Biological signals as handicaps. *J. Theor. Biol.* 144:517–546.

Grafen, A., and R. A. Johnstone. 1993. Why we need ESS signalling theory. *Philos. Trans. Roy. Soc. Lond., Ser. B* 340:245–250.

Grant, B. R., and P. R. Grant. 1989. *Evolutionary dynamics of a natural population: the large cactus finch of the Galápagos.* University of Chicago Press, Chicago.

Grant, B. W. 1990. Trade-offs in activity time and physiological performance for thermoregulating desert lizards, *Sceloporus merriami. Ecology* 71:2323–2333.

Grant, B. W., and A. E. Dunham. 1988. Thermally imposed time constraints on the activity of the desert lizard *Sceloporus merriami. Ecology* 69:167–176.

———. 1990. Elevational covariation in environmental constraints and life histories of the desert lizard *Sceloporus merriami. Ecology* 71:1765–1776.

Grant, P. R. 1975. Four Galápagos islands. *Geogr. J.* 141:76–87.

———. 1986. *Ecology and evolution of Darwin's finches.* Princeton University Press, Princeton, N.J.

———. 1998. *Evolution on islands.* Oxford University Press, Oxford, U.K.

Grant, P. R., and P. T. Boag. 1980. Rainfall on the Galápagos and the demography of Darwin's finches. *Auk* 97:227–244.

Graves, B. M., and D. Duvall. 1983. Occurrence and function of prairie rattlesnake mouth gaping in a non-feeding context. *J. Exp. Zool.* 227:471–474.

———. 1985. Mouth gaping and head shaking by prairie rattlesnakes are associated with vomeronasal organ olfaction. *Copeia* 1985:496–497.

Greeff, J. M., and M. J. Whiting. 1999. Dispersal of Namaqua fig seeds by the lizard *Platysaurus broadleyi* (Sauria: Cordylidae). *J. Herpetol.* 33:328–330.

———. 2000. Foraging-mode plasticity in the lizard *Platysaurus broadleyi. Herpetologica* 56:402–407.

Greenberg, B. 1943. Social behavior of the western banded gecko *Coleonyx variegatus* Baird. *Physiol. Zool.* 16:110–121.

Greenberg, N., M. Scott, and D. Crews. 1985. The role of the amygdala in the reproductive and aggressive behavior of the lizard *Anolis carolinensis. Physiol. Behav.* 32:147–151.

Greene, H. W. 1988. Antipredator mechanisms in reptiles. In *Biology of the Reptilia. Vol. 16. Defense and life history,* edited by C. Gans and R. B. Huey, 1–152. Alan R. Liss, New York.

Greenfield, M. D. 1997. Sexual selection and the evolution of advertisement signals. In *Perspectives in ethology. Vol. 12,* edited by D. H. Owings, M. D. Beecher, and N. S. Thompson, 145–177. Plenum, New York.

Greenfield, M. D., and R. L. Minckley. 1993. Acoustic dueling in tarbush grasshoppers: settlement of territorial contests via alternation of reliable signals. *Ethology* 95:309–326.

Greer, A. E. 1970. A subfamilial classification of scincid lizards. *Bull. Mus. Comp. Zool.* 139:151–183.

Griffith, H. 1991. Heterochrony and evolution of sexual dimorphism in the *fasciatus* group of the scincid genus *Eumeces. J. Herpetol.* 25:24–30.

Gross, M. R. 1982. Sneakers, satellite and parentals: polymorphic mating strategies in North American sunfishes. *Z. Tierpsychol.* 60:1–26.

———. 1983. Sunfish, salmon, and the evolution of alternative reproductive strategies and tactics in fishes. In *Fish reproduction: strategies and tactics,* edited by R. J. Wootten and G. Potts, 55–75. Academic, New York.

———. 1991. Evolution of alternative reproductive strategies: frequency-dependent sexual selection in male bluegill sunfish. *Philos. Trans. Roy. Soc. Lond., Ser. B* 332:59–66.

———. 1996. Alternative reproductive strategies and tactics: diversity within the sexes. *Trends Ecol. Evol.* 11:92–98.

Gross, M. R., and E. L. Charnov. 1980. Alternative male life history strategies in bluegill sunfish. *Proc. Natl. Acad. Sci. USA* 77:6937–6940.

Guilford, T., and M. S. Dawkins. 1991. Receiver psychology and the evolution of animal signals. *Anim. Behav.* 42:1–14.

———. 1995. What are conventional signals? *Anim. Behav.* 49:1689–1695.

Guillette, L. J., Jr. 1983. Notes concerning reproduction of the montane skink, *Eumeces copei. J. Herpetol.* 17:144–148.

Guyer, C. 1988a. Food supplementation in a tropical mainland anole, *Norops humilis:* demographic effects. *Ecology* 69:350–361.

———. 1988b. Food supplementation in a tropical mainland anole, *Norops humilis:* effects on individuals. *Ecology* 69:362–369.

———. 1994. Mate limitation in male *Norops humilis.* In *Lizard ecology: historical and experimental perspectives,* edited by L. J. Vitt and E. R. Pianka, 145–158. Princeton University Press, Princeton, N.J.

Hall, R. J. 1971. Ecology of a population of the Great Plains skink (*Eumeces obsoletus*). *Univ. Kans. Sci. Bull.* 49:357–388.

Halliday, T. R. 1983. The study of mate choice. In *Mate choice,* edited by P. Bateson, 3–32. Cambridge University Press, Cambridge, U.K.

Halloy, M. 1996. Behavioral patterns in *Liolaemus quilmes* (Tropiduridae), a South American lizard. *Bull. Md. Herpetol. Soc.* 32:43–57.

Halloy, M., and S. Halloy. 1997. An indirect form of parental care in a high altitude viviparous lizard, *Liolaemus huacahuasicus* (Tropiduridae). *Bull. Md. Herpetol. Soc.* 33:139–155.

Halperin, J. R. P., T. Giri, J. Elliott, and D. W. Dunham. 1998. Consequences of hyperaggressiveness in Siamese fighting fish: cheaters seldom prospered. *Anim. Behav.* 55:87–96.

Hamilton, W. D. 1964. The genetical evolution of social behaviour. I, II. *J. Theor. Biol.* 7:1–52.

Hamilton, W. D., and M. Zuk. 1982. Heritable true fitness and bright birds: a role for parasites? *Science* 218:384–387.

Hanley, K. A., M. L. Elliott, and J. A. Stamps. 1999. Chemical recognition of familiar vs. unfamiliar conspecifics by juvenile lizards, *Ctenosaura similis. Ethology* 105:641–650.

Harris, V. A. 1964. *The life of the rainbow lizard.* Hutchinson and Company, London.

Harvey, P. H., and M. D. Pagel. 1991. *The comparative method in evolutionary biology.* Oxford University Press, Oxford, U.K.

Hasegawa, M. 1984. Biennial reproduction in the lizard *Eumeces okadae* on Miyake-jima, Japan. *Herpetologica* 40:194–199.

———. 1985. Effect of brooding on egg mortality in the lizard *Eumeces okadae* on Miyake-jima. *Copeia* 1985:497–500.

———. 1990a. Demography of an island population of the lizard, *Eumeces okadae,* on Miyake-jima, Izu Islands. *Res. Pop. Ecol.* 32:119–133.

———. 1990b. The thrush, *Turdus celaenops,* as an avian predator of juvenile *Eumeces okadae* on Miyake-jima, Izu Islands. *Jap. J. Herpetol.* 33:65–69.

———. 1991. Life history of the lizard *Eumeces okadae* on the Izu Islands, Japan. Ph.D. dissertation, Tokyo Metropolitan University, Tokyo, Japan.

———. 1994a. Demography, social structure and sexual dimorphism of the lizard *Eumeces okadae.* In *Animal societies, individuals, interactions and organization,* edited by P. J. Jarman and A. Rossiter, 248–263. Kyoto University Press, Kyoto, Japan.

———. 1994b. Insular radiation in life history of the lizard *Eumeces okadae* in the Izu Islands, Japan. *Copeia* 1994:732–747.

———. 1997. Density effects on life history traits of an island lizard population. *Ecol. Res.* 12:111–118.

———. 1999. Impacts of introduced weasel on the insular food web. In *Diversity of reptiles, amphibians and other terrestrial animals on tropical islands: origin, current status and conservation,* edited by H. Ohta, 129–154. Elsevier, New York.

Hasegawa, M. and H. Moriguchi. 1989. Geographic variation in food habits, body size and life history traits of the snakes on the Izu Islands. In *Current herpetology in East Asia,* edited by M. Matui, T. Hikida, and R. C. Goris, 414–432. Herpetological Society of Japan, Kyoto.

Hasson, O. 1991. Sexual displays as amplifiers: practical examples with an emphasis on feather decorations. *Behav. Ecol.* 2:189–197.

Hasson, O., R. Hibbard, and G. Ceballos. 1989. The pursuit deterrent function of tail-wagging in the zebra-tailed lizard (*Callisaurus draconoides*). *Can. J. Zool.* 67:1203–1209.

Hayashi, F., and M. Hasegawa. 1984a. Infestation level, attachment site and distribution pattern of the lizard tick, *Ixodes asanumai* (Acarina:Ixodidae) in Aoga-shima, Izu Islands. *Appl. Ent. Zool.* 19:299–305.

———. 1984b. Selective parasitism of the tick, *Ixodes asanumai* (Acarina:Ixodidae) and its influence on the host lizard *Eumeces okadae* in Miyake-jima, Izu Islands. *Appl. Ent. Zool.* 19:181–191.

Hayes, T. B. 1997. Hormonal mechanisms as potential constraints on evolution: examples from the Anura. *Am. Zool.* 27:482–490.

Heath, J. E. 1965. Temperature regulation and diurnal activity in horned lizards. *Univ. Calif. Publ. Zool.* 64:97–129.

Hebrard, J. J., and T. Madsen. 1984. Dry season intersexual habitat partitioning by flap-necked chameleons (*Chamaeleo dilepis*) in Kenya. *Biotropica* 16:69–72.

Hedrick, A. V., and L. M. Dill. 1993. Mate choice by female crickets is influenced by predation risk. *Anim. Behav.* 46:193–196.

Hedrick, A. V., and E. J. Temeles. 1989. The evolution of sexual dimorphism in animals: hypotheses and tests. *Trends Ecol. Evol.* 4:136–138.

Hedrick, P. W. 1986. Genetic polymorphism in heterogeneous environments, a decade later. *Annu. Rev. Ecol. Syst.* 17:535–566.

Hedrick, P. W., M. W. Ginevan, and E. P. Ewing. 1976. Genetic polymorphism in heterogeneous environments. *Annu. Rev. Ecol. Syst.* 7:1–32.

Herrel, A., L. Spithoven, R. Van Damme, and F. de Vree. 1999. Sexual dimorphism of head size in *Gallotia galloti:* testing the niche divergence hypothesis by functional analyses. *Funct. Ecol.* 13:289–297.

Hertz, P. E. 1992. Temperature regulation in Puerto Rican *Anolis* lizards: a field test using null hypotheses. *Ecology* 73:1405–1417.

Hertz, P. E., R. B. Huey, and R. D. Stevenson. 1993. Evaluating temperature regulation by field-active ectotherms: the fallacy of the inappropriate question. *Am. Nat.* 142:796–818.

Hews, D. K. 1990. Examining hypotheses generated by field measures of sexual selection on male lizards, *Uta palmeri. Evolution* 44:1956–1966.

———. 1993. Food resources affect female distribution and male mating opportunities in the iguanian lizard *Uta palmeri. Anim. Behav.* 46:279–291.

Hews, D. K., and M. F. Benard. Negative association between visual display and chemosensory behaviors in two phrynosomatid lizards. *Ethology,* in press.

Hews, D. K., R. Knapp, and M. C. Moore. 1994. Early exposure to androgens affects adult expression of alternative male types in tree lizards. *Horm. Behav.* 28:96–115.

Hews, D. K., M. M. Moga, and G. S. Prins. 1999. Distribution of androgen receptor immunoreactivity in the brain of the eastern fence lizard, *Sceloporus undulatus. Am. Zool.* 39:28A.

Hews, D. K., and M. C. Moore. 1995. Influence of androgens on differentiation of secondary sex characters in tree lizards, *Urosaurus ornatus. Gen. Comp. Endocrinol.* 97:86–102.

————. 1996. A critical period for the organization of alternative male phenotypes of tree lizards by exogenous testosterone? *Physiol. Behav.* 62:425–429.

Hicks, R. A., and R. L. Trivers. 1983. The social behavior of *Anolis valencienni*. In *Advances in herpetology and evolutionary biology: essays in honor of Ernest E. Williams,* edited by A. G. J. Rhodin and K. Miyata, 570–595. Museum of Comparative Zoology, Cambridge, Mass.

Hikida, T. 1978. Postembryonic development of the skull of the Japanese skink, *Eumeces latiscutatus* (Scincidae). *Jap. J. Herpetol.* 7:56–72.

————. 1981. Reproduction of the Japanese skink (*Eumeces latiscutatus*) in Kyoto. *Zool. Mag.* 40:85–92.

————. 1993. Phylogenetic relationships of the skinks of the genus *Eumeces* (Scincidae:Reptilia) from East Asia. *Jap. J. Herpetol.* 12:10–15.

Hill, G. E., and W. R. Brawner, III. 1998. Melanin–based plumage coloration in the house finch is unaffected by coccidial infection. *Proc. Roy. Soc. Lond., Ser. B* 265:1105–1109.

Hill, G. E., and R. Montgomerie. 1994. Plumage colour signals nutritional condition in the house finch. *Proc. Roy. Soc. Lond., Ser. B* 258:47–52.

Hillard, S. 1996. The importance of the thermal environment to juvenile desert tortoises. M.S. thesis, Colorado State University, Fort Collins.

Hinde, R. A. 1970. *Animal behaviour: a synthesis of ethology and comparative psychology,* 2nd ed. McGraw-Hill, New York.

Hixon, M. A. 1987. Territory area as a determinant of mating systems. *Am. Zool.* 27:229–243.

Hoeck, H. N. 1984. Introduced fauna. In *Key environments: Galápagos,* edited by R. Perry, 233–245. Pergamon, Oxford, U.K.

Hoffman, S. G. 1983. Sex-related foraging behavior in sequentially hermaphroditic hogfishes (*Bodianus* spp.). *Ecology* 64:798–808.

————. 1985. Effects of size and sex on the social organization of reef-associated hogfishes, *Bodianus* spp. *Environ. Biol. Fish.* 14:185–197.

Hoglund, J. 1989. Size and plumage dimorphism in lek-breeding birds: a comparative analysis. *Am. Nat.* 134:72–87.

Hogstad, O. 1987. It is expensive to be dominant. *Auk* 104:333–336.

Holberton, R. L., K. P. Able, and J. C. Wingfield. 1989. Status signalling in dark-eyed juncos, *Junco hyemalis:* plumage manipulations and hormonal correlates of dominance. *Anim. Behav.* 37:681–689.

Horn, A. G., M. L. Leonard, and D. M. Weary. 1995. Oxygen consumption during crowing by roosters: talk is cheap. *Anim. Behav.* 50:1171–1175.

Houde, A. E. 1997. *Sex, color, and mate choice in guppies.* Princeton University Press, Princeton, N.J.

Houde, A. E., and J. A. Endler. 1990. Correlated evolution of female mating preferences and male color patterns in the guppy *Poecilia reticulata. Science* 248:1405–1408.

Howard, R. D. 1978a. The evolution of mating strategies in bullfrogs, *Rana catesbeiana. Evolution* 32:850–871.

————. 1978b. The influence of male-defended oviposition sites on early embryo mortality in bullfrogs. *Ecology* 59:789–798.

————. 1979. Estimating reproductive success in natural populations. *Am. Nat.*
114:221–231.

————. 1983. Sexual selection and variation in reproductive success in a long-lived
organism. *Am. Nat.* 122:301–325.

————. 1988. Sexual selection on male body size and mating behaviour in American
toads, *Bufo americanus. Anim. Behav.* 36:1796–1808.

Hranitz, J. M., and T. A. Baird. 2000. Effective population size and genetic structure of
a population of collared lizards, *Crotaphytus collaris,* in central Oklahoma. *Copeia*
2000:786–791.

Huey, R. B. 1982. Temperature, physiology, and the ecology of reptiles. In *Biology
of the Reptilia. Vol. 13*, edited by C. Gans and F. H. Pough, 25–91. Academic,
London.

Huey, R. B., E. R. Pianka, and T. W. Schoener. 1983. Introduction. In *Lizard ecology:
studies of a model organism*, edited by R. B. Huey, E. R. Pianka, and T. W.
Schoener, 1–6. Harvard University Press, Cambridge, Mass.

Huhta, E., and R. V. Alatalo. 1993. Plumage colour and male-male interactions in the
pied flycatcher. *Anim. Behav.* 45:511–518.

Hunsaker, D. 1962. Ethological isolating mechanisms in the *Sceloporus torquatus*
group of lizards. *Evolution.* 16:62–74.

Hurd, P. L. 1997a. Cooperative signalling between opponents in fish fights. *Anim.
Behav.* 54:1309–1315.

————. 1997b. Is signalling of fighting ability costlier for weaker individuals? *J. Theor.
Biol.* 184:83–88.

Husak, J. F., and J. K. McCoy. 2000. Diet composition of the collared lizard (*Crotaphytus collaris*) in West-Central Texas. *Texas J. Sci.* 52:93–100.

Huxley, J. 1963. Lorenzian ethology. *Z. Tierpsychol.* 20:402–409.

Ibargüengoytía, N., and V. E. Cussac. 1996. Reproductive biology of the viviparous
lizard, *Liolaemus pictus* (Tropiduridae): biennial female reproductive cycle? *Herpetol. J.* 6:137–143.

————. 1998. Reproduction of the viviparous lizard *Liolaemus elongatus* in the high-
lands of southern South America: plastic cycles in response to climate? *Herpetol. J.*
8:99–105.

Iguchi, K., and T. Hino. 1996. Effect of competitor abundance on feeding territoriality
in a grazing fish, the ayu *Plecoglossus altivelis. Ecol. Res.* 11:165–173.

Ims, R. A. 1987. Responses in spatial organisation and behaviour to manipulations of
the food resource in the vole *Clethrionomys rufocanus. J. Anim. Ecol.* 56:585–596.

Irschick, D. J., and J. B. Losos. 1996. Morphology, ecology, and behavior of the twig
anole, *Anolis angusticeps.* In *Contributions to West Indian herpetology: a tribute to
Albert Schwartz*, edited by R. Powell and R. W. Henderson, 291–301. Society for
the Study of Amphibians and Reptiles, Ithaca, N.Y.

————. 1999. Do lizards avoid habitats in which performance is submaximal? The relationship between sprinting capabilities and structural habitat use in Caribbean
anoles. *Am. Nat.* 154:293–305.

Irwin, R. E. 1994. The evolution of plumage dichromatism in the New World blackbirds: social selection on female brightness? Am. Nat. 144:890–907.

Iverson, J. B. 1979. Behavior and ecology of the rock iguana *Cyclura carinata. Bull. Fla.
State Mus. Biol. Sci.* 24:175–358.

————. 1982. Adaptations to herbivory in iguanine lizards. In *Iguanas of the world: their behavior, ecology, and conservation,* edited by G. M. Burghardt and A. S. Rand, 60–76. Noyes, Park Ridge, N.J.

Iwasa, Y., A. Pomiankowski, and S. Nee. 1991. The evolution of costly mate preferences II. The "handicap" principle. *Evolution* 45:1431–1442.

Jackman, T. R., A. Larson, K. de Queiroz, and J. B. Losos. 1999. Phylogenetic relationships and tempo of early diversification in *Anolis* lizards. *Syst. Biol.* 48:254–285.

Jackman, T. R., J. B. Losos, A. Larson, and K. de Queiroz. 1997. Phylogenetic studies of convergent adaptive radiation in Caribbean *Anolis* lizards. In *Molecular evolution and adaptive radiation,* edited by T. J. Givnish and K. J. Sytsma, 535–557. Cambridge University Press, Cambridge, U.K.

Jakobsson, S., O. Brick, and C. Kullberg. 1995. Escalated fighting behavior incurs increased predation risk. *Anim. Behav.* 49:235–239.

Jaksíc, F. M., and H. W. Greene. 1984. Empirical evidence of non-correlation between tail loss frequency and predation intensity on lizards. *Oikos* 42:407–411.

Jaksíc, F. M., H. W. Greene, K. Schwenk, and R. L. Seib. 1982. Predation upon reptiles in Mediterranean habitats of Chile, Spain, and California: a comparative analysis. *Oecologia* 53:152–159.

Jarman, P. J. 1974. The social organization of antelope in relation to their ecology. *Behaviour* 48:215–267.

————. 1988. On being thick skinned: dermal shields in large mammalian herbivores. *Biol. J. Linn. Soc.* 36:169–191.

Järvi, T., and M. Bakken. 1984. The function of the variation in the breast stripe of the great tit (*Parus major*). *Anim. Behav.* 32:590–596.

Järvi, T., Ø. Walsø, and M. Bakken. 1987. Status signalling by *Parus major:* an experiment in deception. *Ethology* 76:334–342.

Jenssen, T. A. 1970. The ethoecology of *Anolis nebulosus* (Sauria, Iguanidae). *J. Herpetol.* 4:1- 38.

Jenssen, T. A., N. Greenberg, and K. A. Hovde. 1995. Behavioral profile of free-ranging male lizards, *Anolis carolinensis,* across breeding and post-breeding seasons. *Herpetol. Monogr.* 9:41–62.

Jenssen, T. A., and S. C. Nunez. 1998. Spatial and breeding relationships of the lizard, *Anolis carolinensis:* evidence of intrasexual selection. *Behaviour* 135:981–1003.

John-Alder, H. B., S. McMann, S. Katz, A. Gross, and D. S. Barton. 1996. Social modulation of exercise endurance in a lizard (*Sceloporus undulatus*). *Physiol. Zool.* 69:547–567.

Johnstone, R. A. 1995a. Honest advertisement of multiple qualities using multiple signals. *J. Theor. Biol.* 177:87–94.

————. 1995b. Sexual selection, honest advertisement and the handicap principle: reviewing the evidence. *Biol. Rev.* 70:1–65.

————. 1996. Multiple displays in animal communication: 'backup signals' and 'multiple messages.' *Philos. Trans. Roy. Soc. Lond., Ser. B* 351:329–338.

————. 1997. The evolution of animal signals. In *Behavioural ecology: an evolutionary approach,* 4th ed., edited by J. R. Krebs and N. B. Davies, 155–178. Blackwell Scientific, Oxford, U.K.

Johnstone, R. A., and A. Grafen. 1993. Dishonesty and the handicap principle. *Anim. Behav.* 46:759–764.

Johnstone, R. A., and K. Norris. 1993. Badges of status and the cost of aggression. *Behav. Ecol. Sociobiol.* 32:127–134.

Jones, A. G., and J. C. Avise. 1997. Microsatellite analysis of maternity and the mating system in the Gulf pipefish *Syngnathus scovelli*, a species with male pregnancy sex-role reversal. *Mol. Ecol.* 6:203-213.

Jones, A. G., S. Ostlund-Nilsson, and J. C. Avise. 1998. A microsatellite assessment of sneaked fertilizations and egg thievery in the fifteenspine stickleback. *Evolution* 52:848–858.

Jones, G. P. 1981. Spawning-site choice by female *Pseudolabrus celidotus* (Pisces: Labridae) and its influence on the mating system. *Behav. Ecol. Sociobiol.* 8:129–142.

Jones, I. L. 1990. Plumage variability functions for status signalling in least auklets. *Anim. Behav.* 39:967–975.

Jordan, M. A. 1999. Phenotypic plasticity in the reproduction of Galápagos lava lizards (*Microlophus delanonis*). Ph.D. dissertation, University of New Mexico, Albuquerque.

Kaiser, B. W., and H. R. Mushinsky. 1994. Tail loss and dominance in captive adult male *Anolis sagrei*. *J. Herpetol.* 28:342–346.

Kato, J., Ota, H., and T. Hikida. 1994. Biochemical systematics of the *latiscutatus* species-group of the genus *Eumeces* (Scincidae: Reptilia) from East Asian islands. *Biochem. Syst. Ecol.* 22:419–425.

Kavanau, J. L. 1990. Conservative behavioural evolution, the neural substrate. *Anim. Behav.* 39:758–767.

Keating, E. G., L. A. Korman, and J. A. Horel. 1970. The behavioral effects of stimulating and ablating the reptilian amygdala. *Physiol. Behav.* 5:55–59.

Kelley, D. B. 1988. Sexually dimorphic behaviors. *Annu. Rev. Neurosci.* 11:225–251.

Kelley, D., D. Sasson, N. Segil, and M. Scudder. 1989. Development and hormone regulation of androgen receptor levels in the sexually dimorphic larynx of *Xenopus laevis*. *Dev. Biol.* 131:111–118.

Kendrick, A. M., and B. A. Schlinger. 1996. Independent differentiation of sexual and social traits. *Horm. Behav.* 30:600–610.

Kennedy, C. F. J., A. Endler, S. L. Poynton, and H. McMinn. 1987. Parasite load predicts mate choice in guppies. *Behav. Ecol. Sociobiol.* 21:291–295.

Ketterson, E. D., and V. Nolan, Jr. 1994. Hormones and life histories: an integrative approach. *Am. Nat.* 140:S33–S62.

Ketterson, E. D., V. Nolan, Jr., L. Wolf, C. Ziegenfus, A. M. Dufty, Jr., G. F. Ball, and T. S. Johnsen. 1991. Testosterone and avian life histories: the effect of experimentally elevated testosterone on corticosterone and body mass in dark–eyed juncos. *Horm. Behav.* 25:489–503.

Keys, G. C., and S. I. Rothstein. 1991. Benefits and cost of dominance and subordinance in white-crowned sparrows and the paradox of status signalling. *Anim. Behav.* 42:899–912.

Kim, Y.-G. 1995. Status signalling games in animal contests. *J. Theor. Biol.* 176:221–231.

Kimball, F. E., and M. J. Erpino. 1971. Hormonal control of pigmentary sexual dimorphism in *Sceloporus occidentalis*. *Gen. Comp. Endocrinol.* 16:375–384.

Kimball, R. T., and J. D. Ligon. 1999. Evolution of avian plumage dichromatism from a proximate perspective. *Am. Nat.* 154:182–193.

Klukowski, M., and C. E. Nelson. 1998. The challenge hypothesis and seasonal changes in aggression and steroids in male northern fence lizards (*Sceloporus undulatus hyacinthinus*). *Horm. Behav.* 33:197–204.

Klump, G. M., and H. C. Gerhardt. 1987. Use of non-arbitrary acoustic criteria in mate choice by female tree frogs. *Nature* 326:286–288.

Kodric-Brown, A. 1985. Female preference and sexual selection for male coloration in the guppy (*Poecilia reticulata*). *Behav. Ecol. Sociobiol.* 17:199–205.

———. 1989. Dietary carotenoids and male mating success in the guppy: an environmental component to female choice. *Behav. Ecol. Sociobiol.* 25:393–401.

———. 1995. Does past reproductive history predict competitive interactions and male mating success in pupfish? *Anim. Behav.* 50:1433–1440.

Kodric-Brown, A., and J. H. Brown. 1984. Truth in advertising: the kinds of traits favoured by sexual selection. *Am. Nat.* 124:309–323.

Konecny, M. J. 1987. Food habits and energetics of feral house cats in the Galápagos Islands. *Oikos* 50:24–32.

Kortlandt, A. 1940. Eine Übersicht der angeborenen Verhaltungsweisen des Mittel-Europäischen Kormorans (*Phalacrocorax carbo sinensis* Shaw & Nodd.), ihre Funktion, ontogenetische Entwicklung und phylogenetische Herkunft. *Arch. Neerland. Zool.* 4:401–442.

Krakauer, D. C., and M. Pagel. 1995. Spatial structure and the evolution of honest cost-free signalling. *Proc. Roy. Soc. Lond., Ser. B* 260:365–372.

Kramer, P. 1984. Man and other introduced organisms. In *Evolution in the Galápagos Islands*, edited by R. J. Berry, 253–258. Academic, London.

Krebs, J. R. 1982. Territorial defense in the great tit (*Parus major*): do residents always win? *Behav. Ecol. Sociobiol.* 11:185–194.

Krebs, J. R., and N. B. Davies. 1993. *An introduction to behavioural ecology*, 3rd ed. Blackwell Scientific, Oxford, U.K.

Krebs, J. R., and R. Dawkins. 1984. Animal signals: mind reading and manipulation. In *Behavioural ecology: an evolutionary approach*, 2nd ed., edited by J. R. Krebs and N. B. Davies, 380–402. Blackwell Scientific, Oxford, U.K.

Krebs, J. R., M. H. MacRoberts, and J. M. Cullen. 1972. Flocking and feeding in the great tit *Parus major*: an experimental study. *Ibis* 114:507–530.

Krekorian, C. O. 1976. Home range size and overlap and their relationship to food abundance in the desert iguana, *Dipsosaurus dorsalis*. *Herpetologica* 32:405–412.

———. 1984. Life history of the desert iguana, *Dipsosaurus dorsalis*. *Herpetologica* 40:415–424.

Kruuk, H. 1964. Predators and anti-predator behaviour of the black-headed gull (*Larus ridibundus* L.). *Behav. Suppl.* 11:1–29.

———. 1979. Ecology and control of feral dogs in Galápagos. Technical report to Charles Darwin Research Station, Galápagos, Ecuador.

Külling, D., and M. Milinski. 1992. Size-dependent predation risk and partner quality in predator inspection of sticklebacks. *Anim. Behav.* 44:949–955.

Labra, A., and H. M. Niemeyer. 1999. Interspecific chemical recognition in the lizard *Liolaemus tenuis*. *J. Chem. Ecol.* 25:1799–1811.

Lack, D. 1947. *Darwin's finches*. Cambridge University Press, Cambridge, U.K.

———. 1968. *Ecological adaptations for breeding in birds*. Methuen, London.

Lanctot, R. B., P. J. Weatherhead, B. Kempenaers, and K. T. Scribner. 1998. Male traits, mating tactics and reproductive success in the buff-breasted sandpiper, *Tryngites subruficollis. Anim. Behav.* 56:419–432.

Langbein, J., and S. J. Thirgood. 1989. Variation in mating system of fallow deer (*Dama dama*) in relation to ecology. *Ethology* 83:195–214.

Lank, D. B., C. M. Smith, O. Hancotte, T. Burke, and F. Cooke. 1995. Genetic polymorphism for alternative mating behaviour in lekking male ruff *Philomachus pugnax. Nature* 378:59– 62.

Lappin, A. K., and E. J. Swinney. 1999. Sexual dimorphism as it relates to natural history of leopard lizards (Crotaphytidae: *Gambelia*). *Copeia* 1999:649–660.

Laurie, W. A., and D. Brown. 1990. Population biology of marine iguanas (*Amblyrhynchus cristatus*). II. Changes in annual survival rates and the effects of size, sex, age and fecundity in a population crash. *J. Anim. Ecol.* 59:529–544.

Leal, M. 1999. Honest signalling during prey-predator interactions in the lizard *Anolis cristatellus. Anim. Behav.* 58:521-526.

Leal, M., and J. A. Rodríguez-Robles. 1997. Signalling displays during predator-prey interactions in a Puerto Rican anole, *Anolis cristatellus. Anim. Behav.* 54:1147–1154.

Le Boeuf, B. J. 1974. Male-male competition and reproductive success in elephant seals. *Am. Zool.* 14:163–176.

Le Boeuf, B. J., and J. Reiter. 1988. Lifetime reproductive success in northern elephant seals. In *Reproductive success,* edited by T. H. Clutton Brock, 344–362. University of Chicago Press, Chicago.

Lee, J. C. 1990. Sources of extraneous variation in the study of meristic characters: the effect of size and of inter-observer variability. *Syst. Zool.* 39:31–39.

Lemos-Espinal, J. A., and R. E. Ballinger. 1995. Comparative thermal ecology of the high-altitude lizard *Sceloporus grammicus* on the eastern slope of the Iztaccihuatl Volcano, Puebla, Mexico. *Can. J. Zool.* 73:2184–2191.

Lemos-Espinal, J. A., R. E. Ballinger, S. S. Sarabia, and G. R. Smith. 1997. Aggregation behavior of the lizard *Sceloporus mucronatus mucronatus* in Sierra del Ajusco, México. *Herpetol. Rev.* 28:126–127.

Lemos-Espinal, J. A., G. R. Smith, and R. E. Ballinger. 1997. Neonate-female associations in *Xenosaurus newmanorum:* A case of parental care in a lizard? *Herpetol. Rev.* 28:22–23.

Lena, J. P., and M. De Fraipont. 1998. Kin recognition in the common lizard. *Behav. Ecol. Sociobiol.* 42:341–347.

Leonard, M. L., and A. G. Horn. 1995. Crowing in relation to status in roosters. *Anim. Behav.* 49:1283–1290.

Levene, H. 1953. Genetic equilibrium when more than one niche is available. *Am. Nat.* 87:311–313.

Levins, R. 1968. Evolution in changing environments. Princeton University Press, Princeton, N.J.

Lewis, A. R., G. Tirado, and J. Sepulveda. 2000. Body size and paternity in a teiid lizard (*Ameiva exsul*). *J. Herpetol.* 34:110–120.

Leyton, V., and J. Valencia. 1992. Follicular population dynamics: its relation to clutch and litter size in Chilean *Liolaemus* lizards. In *Reproductive biology of South American vertebrates,* edited by W. C. Hamlett, 123–134. Springer-Verlag, New York.

Licht, P. 1974. Response of *Anolis* lizards to food supplementation in nature. *Copeia* 1974:215–221.

Licht, P., and G. C. Gorman. 1970. Reproductive and fat cycles in Caribbean *Anolis* lizards. *Univ. Calif. Publ. Zool.* 95:1–52.

Licht, P., H. Papkoff, S. W. Farmer, C. H. Muller, H. W. Tsui, and D. Crews. 1977. Evolution of gonadotropin structure and function. *Recent Prog. Horm. Res.* 33:169–248.

Lieb, C. S. 1985. Systematics and distribution of skinks allied to *Eumeces tetragrammus* (Sauria: Scincidae). *Los Angeles Co. Mus. Contr. Sci.* 357: 1–19.

Lifson, N., and R. McClintock. 1966. Theory of use of the turnover rates of body water for measuring energy and material balance. *J. Theor. Biol.* 12:46–74.

Lima, S. L., and L. M. Dill. 1990. Behavioral decisions made under the risk of predation: a review and prospectus. *Can. J. Zool.* 68:619–640.

Lorenz, K. Z. 1941. Vergleichende Bewegungsstudien an Anatinen. *J. Ornithol.* 3:194–293.

———. 1966. *On aggression.* Harcourt Brace Jovanovich, New York.

Losos, J. B. 1985a. An experimental demonstration of the species-recognition role of *Anolis* dewlap color. *Copeia* 1985:905–910.

———. 1985b. Male aggressive behavior in a pair of sympatric sibling species. *Breviora* 484:1–30.

———. 1990a. Ecomorphology, performance capability, and scaling of West Indian *Anolis* lizards: an evolutionary analysis. *Ecol. Monogr.* 60:369–388.

———. 1990b. The evolution of form and function: morphology and locomotor performance in West Indian *Anolis* lizards. *Evolution* 44:1189–1203.

———. 1990c. A phylogenetic analysis of character displacement in Caribbean *Anolis* lizards. *Evolution* 44:558–569.

———. 1994. Integrative approaches to evolutionary ecology: *Anolis* lizards as model systems. *Annu. Rev. Ecol. Syst.* 25:467–493.

———. 1999. Uncertainty in the reconstruction of ancestral character states and limitations on the use of phylogenetic comparative methods. *Anim. Behav.* 58:1319–1324.

Losos, J. B., M. R. Gannon, W. J. Pfeiffer, and R. B. Waide. 1990. Notes on the ecology and behavior of the lagarto verde, *Anolis cuvieri*, in Puerto Rico. *Carib. J. Sci.* 26:65–66.

Losos, J. B., and D. J. Irschick. 1996. The effect of perch diameter on escape behaviour of *Anolis* lizards: laboratory-based predictions and field tests. *Anim. Behav.* 51:593–602.

Losos, J. B., T. R. Jackman, A. Larson, K. de Queiroz, and L. Rodriguez-Schettino. 1998. Contingency and determinism in replicated adaptive radiations of island lizards. *Science* 279:2115–2118.

Losos, J. B., and D. B. Miles. 1994. Adaptation, constraint, and the comparative method: phylogenetic issues and methods. In *Ecological morphology: integrative organismal biology,* edited by P. C. Wainwright and S. Reilly, 60–98. University of Chicago Press, Chicago.

Losos, J. B., and B. Sinervo. 1989. The effect of morphology and perch diameter on sprint performance of *Anolis* lizards. *J. Exp. Biol.* 145:23–30.

Lott, D. F. 1991. *Intraspecific variation in the social systems of wild vertebrates.* Cambridge University Press, Cambridge, U.K.

Luyten, P. H., and N. R. Liley. 1985. Geographic variation in the sexual behavior of the guppy, *Poecilia reticulata* (Peters). *Behaviour* 95:164–179.

MacArthur, R. H., and E. O. Wilson. 1967. *The theory of island biogeography.* Princeton University Press, Princeton, N.J.

Madsen T., R. Shine, J. Loman, and T. Hakansson. 1992. Why do females copulate so frequently? *Nature* 355:440–441.

Magnhagen, C. 1991. Predation risk as a cost of reproduction. *Trends Ecol. Evol.* 6:183–186.

Mahrt, L. A. 1998a. Response to intruders and the dear enemy phenomenon in female lizards, *Urosaurus ornatus,* in relation to age and reproductive condition. *J. Herpetol.* 32:162–168.

———. 1998b. Territorial establishment and maintenance by female tree lizards, *Urosaurus ornatus. J. Herpetol.* 32:176–182.

Manzur, M. I., and E. R. Fuentes. 1979. Polygyny and agonistic behavior in the tree-dwelling lizard *Liolaemus tenuis* (Iguanidae). *Behav. Ecol. Sociobiol.* 6: 23–28.

Marler, P. 1999. Nature, nurture and the instinct to learn. In *Proceedings of the 22nd International Ornithological Congress, Durban,* edited by N. J. Adams and R. H. Slotow, 2379–2393. Birdlife South Africa, Johannesburg.

Marler, C. A., and M. C. Moore. 1988. Evolutionary cost of aggression revealed by testosterone manipulations in free-living male lizards. *Behav. Ecol. Sociobiol.* 23:21–26.

———. 1989. Time and energy costs of aggression in testosterone-implanted free-living male spiny lizards (*Sceloporus jarrovi*). *Physiol. Zool.* 62:1334–1350.

———. 1991. Supplementary feeding compensates for testosterone-induced costs of aggression in male mountain spiny lizards, *Sceloporus jarrovi. Anim. Behav.* 42:209–219.

Marler, C. A., G. Walsberg, M. L. White, and M. C. Moore. 1995. Increased energy expenditure due to increased territorial defense in male lizards after phenotypic manipulation. *Behav. Ecol. Sociobiol.* 37:225–231.

Marquet, P. A., J. C. Ortíz, F. Bozinovic, and F. M. Jaksíc. 1989. Ecological aspects of thermoregulation at high altitudes: the case of Andean *Liolaemus* lizards in northern Chile. *Oecologia* 81:16–20.

Martín, J., and A. Forsman. 1999. Social costs and development of nuptial coloration in male *Psammodromus algirus* lizards: an experiment. *Behav. Ecol.* 10:396–400.

Martín, J., and A. Salvador. 1993a. Tail loss reduces mating success in the Iberian rock lizard, *Lacerta monticola. Behav. Ecol. Sociobiol.* 32:185–189.

———. 1993b. Thermoregulatory behavior of rock lizards in response to tail loss. *Behaviour* 124:123–136.

———. 1995. Effects of tail loss on activity patterns of rock lizards, *Lacerta monticola. Copeia* 1995:984–988.

———. 1997. Effects of tail loss on the time-budgets, movements, and spacing patterns of Iberian rock lizards, *Lacerta monticola. Herpetologica* 53:117–125.

Martín, J., and P. López. 1999a. An experimental test of the costs of antipredatory refuge use in the wall lizard, *Podarcis muralis. Oikos* 84:499–505.

———. 1999b. Nuptial coloration and mate guarding affect escape decisions of male lizards *Psammodromus algirus*. *Ethology* 105:439–447.

———. 1999c. When to come out of refuge: risk-sensitive and state-dependent decisions in an alpine lizard. *Behav. Ecol.* 10:487–492.

Martín, J., and R. A. Avery. 1998. Effects of tail loss on the movement patterns of the lizard, *Psammodromus algirus*. *Funct. Ecol.* 12:794–802.

Martins, E. P. 1993. Contextual use of the push-up display by the sagebrush lizard, *Sceloporus graciosus*. *Anim. Behav.* 45:25–36.

———. 1994. Phylogenetic perspectives on the evolution of lizard territoriality. In *Lizard ecology: historical and experimental perspectives,* edited by L. J. Vitt and E. R. Pianka, 117–144. Princeton University Press, Princeton, N.J.

———. 1996. Phylogenies, spatial autoregression, and the comparative method: a computer simulation study. *Evolution* 50:1750–1765.

———. 2000. Adaptation and the comparative method. *Trends Ecol. Evol.* 15:296–299.

Martins, E. P., and T. F. Hansen. 1996. The statistical analysis of interspecific data: a review and evaluation of phylogenetic comparative methods. In *Phylogenies and the comparative method in animal behavior,* edited by E. P. Martins, 22–27. Oxford University Press, New York.

———. 1997. Phylogenies and the comparative method: a general approach to incorporating phylogenetic information into the analysis of interspecific data. *Am. Nat.* 149:646–667.

Mason, R. T., and D. Crews. 1985. Female mimicry in garter snakes. *Nature* 316: 59–60.

Mateos, C., and J. Carranza. 1997. The role of bright plumage in male-male interactions in the ring-necked pheasant. *Anim. Behav.* 54:1205–1214.

Mautz, W. J., and K. A. Nagy. 1987. Ontogenetic changes in diet, field metabolic rate, and water flux in the herbivorous lizard *Dipsosaurus dorsalis*. *Physiol. Zool.* 60:640–658.

Maynard Smith, J. 1974. The theory of games and the evolution of animal conflict. *J. Theor. Biol.* 47:209–221.

———. 1977. Parental investment: a prospective analysis. *Anim. Behav.* 25:1–9.

———. 1978. Optimization theory in evolution. *Annu. Rev. Ecol. Syst.* 9:31–56.

———. 1982. *Evolution and the theory of games.* Cambridge University Press, Cambridge, U.K.

———. 1989. *Did Darwin get it right? Essays on games, sex, and evolution.* Chapman & Hall, New York.

———. 1991. Honest signalling: the Philip Sydney game. *Anim. Behav.* 42:1034–1035.

———. 1994. Must reliable signals always be costly? *Anim. Behav.* 47:1115–1120.

Maynard Smith, J., and D. G. Harper. 1988. The evolution of aggression: can selection generate variability? *Philos. Trans. Roy. Soc. Lond., Ser. B* 319:557–570.

———. 1995. Animal signals: models and terminology. *J. Theor. Biol.* 177:305–311.

Maynard Smith, J., and R. Hoekstra. 1980. Polymorphism in a varied environment: how robust are the models? *Genet. Res.* 35:45–57.

Maynard Smith, J., and G. A. Parker. 1976. The logic of asymmetric contests. *Anim. Behav.* 24:159–175.

Maynard Smith, J., and G. R. Price. 1973. The logic of animal conflict. *Nature* 246:15–18.

McCoy, J. K. 1995. Mechanisms of selection for the evolution of sexual dimorphism in the collared lizard (*Crotaphytus collaris*). Ph.D. dissertation, Oklahoma State University, Stillwater.

McCoy, J. K., S. F. Fox, and T. A. Baird. 1994. Geographic variation in sexual dimorphism in the collared lizard, *Crotaphytus collaris* (Sauria: Crotaphytidae). *Southwest. Nat.* 39:328–335.

McCoy, J. K., H. J. Harmon, T. A. Baird, and S. F. Fox. 1997. Geographic variation in sexual dichromatism in the collared lizard, *Crotaphytus collaris* (Sauria: Crotaphytidae). *Copeia* 1997:565–571.

McEwen, B. S., V. N. Luine, and C. T. Fischette. 1988. Developmental actions of hormones: from receptors to function. In *From message to mind: directions in developmental neurobiology,* edited by S. S. Easter, Jr., K. F. Barald, and B. M. Carlson, 272–287. Sinauer, Sunderland, Mass.

McGregor, A., and J. Herbert. 1992. Differential effects of excitotoxic basolateral and corticomedial lesions of the amygdala on the behavioral and endocrine responses to either sexual or aggression-promoting stimuli in the male rat. *Brain Res.* 574:9–20.

McGregor, P. K. 1993. Signalling in territorial systems: a context for individual identification, ranging and eavesdropping. *Philos. Trans. Roy. Soc. Lond., Ser. B* 340:237–244.

McGuire, J. A. 1996. Phylogenetic systematics of crotaphytid lizards (Reptilia: Iguania: Crotaphytidae). *Bull. Carnegie Mus. Nat. Hist.* 32:1–143.

McKinney, R. B., and K. R. Marion. 1985. Plasma androgens and their association with the reproductive cycle of the male fence lizard *Sceloporus undulatus. Comp. Biochem. Physiol.* 82A:515–519.

McLaughlin, J. F., and J. Roughgarden. 1989. Avian predation on *Anolis* lizards in the northeastern Caribbean: an inter-island contrast. *Ecology* 70:617–628.

M'Closkey, R. T., R. J. Deslippe, and C. P. Szpak. 1990. Tree lizard distribution and mating system: the influence of habitat and food resources. *Can. J. Zool.* 68:2083–2089.

McPhaul, M. J., M. Marcelli, S. Zoppi, J. E. Griffin, and J. D. Wilson. 1993. Genetic basis of endocrine disease: the spectrum of mutations in the androgen receptor gene that causes androgen resistance. *J. Clin. Endocrinol. Metab.* 76: 17–23.

McVey, M. E. 1988. The opportunity for sexual selection in a territorial dragonfly, *Erythemis simplicollis.* In *Reproductive success,* edited by T. H. Clutton Brock, 44–58. University of Chicago Press, Chicago.

Medel, R. G., J. E. Jiménez, S. F. Fox, and F. M. Jaksíc. 1988. Experimental evidence that high population frequencies of lizard tail autotomy indicate inefficient predation. *Oikos* 53:321–324.

Medel, R. G., P. A. Marquet, S. F. Fox, and F. M. Jaksíc. 1990. Depredación sobre lagartijas en Chile central: importancia relativa de atributos ecológicos y morfológicos. *Rev. Chilena Hist. Nat.* 63:261–266.

Metz, K. J., and P. J. Weatherhead. 1992. Seeing red: uncovering coverable badges in red-winged blackbirds. *Anim. Behav.* 43:223–229.

Miles, D. B. 1994. Covariation between morphology and locomotory performance in sceloporine lizards. In *Lizard ecology: historical and experimental perspectives,* edited by L. J. Vitt and E. R. Pianka, 207–235. Princeton University Press, Princeton, N.J.

Miles, D. B., and A. E. Dunham. 1993. Historical perspectives in ecology and evolutionary biology: the use of phylogenetic comparative analysis. *Annu. Rev. Ecol. Syst.* 24:587–619.

———. 1996. The paradox of the phylogeny: character displacement of analyses of body size in island *Anolis. Evolution* 50:594–603.

Millar, J. S., and R. M. Zammuto. 1983. Life histories of mammals: an analysis of life tables. *Ecology* 64:631–635.

Milstead, W. W. 1961. Competitive relations in lizard populations. In *Vertebrate speciation: a University of Texas symposium,* edited by W. F. Blair, 460–489. University of Texas Press, Austin.

Minnich, J. E., and V. H. Shoemaker. 1970. Diet, behavior and water turnover in the desert iguana, *Dipsosaurus dorsalis. Am. Midl. Nat.* 84:496–509.

Mitchell-Olds, T. 1992. Does environmental variation maintain genetic variation? A question of scale. *Trends Ecol. Evol.* 7:397–398.

Moermond, T. C. 1979a. Habitat constraints on the behavior, morphology, and community structure of *Anolis* lizards. *Ecology* 60:152–164.

———. 1979b. The influence of habitat structure on *Anolis* foraging behavior. *Behaviour* 70:147–167.

Møller, A. P. 1987a. Social control of deception among status signalling house sparrows *Passer domesticus. Behav. Ecol. Sociobiol.* 20:307–311.

———. 1987b. Variation in badge size in male house sparrows *Passer domesticus:* evidence for status signalling. *Anim. Behav.* 35:1637–1644.

———. 1988a. Badge size in the house sparrow *Passer domesticus:* effects of intra- and intersexual selection. *Behav. Ecol. Sociobiol.* 22:373–378.

———. 1988b. Female mate choice selects for male sexual tail ornaments in the monogamous swallow. *Nature* 332:640–642.

Møller, A. P., and F. de Lope. 1994. Differential costs of a secondary sexual character: an experimental test of the handicap principle. *Evolution* 48:1676–1683.

Møller, A. P., R. Dufva, and J. Erritzøe. 1998. Host immune function and sexual selection in birds. *J. Evol. Biol.* 11:703–719.

Møller, A. P., and J. Erritzøe. 1992. Acquisition of breeding coloration depends on badge size in male house sparrows *Passer domesticus. Behav. Ecol. Sociobiol.* 31:271–277.

Møller, A. P., R. T. Kimball, and J. Erritzøe. 1996. Sexual ornamentation, condition, and immune defence in the house sparrow *Passer domesticus. Behav. Ecol. Sociobiol.* 39:317–322.

Møller, A. P., and A. Pomiankowski. 1993. Why have birds got multiple sexual ornaments? *Behav. Ecol. Sociobiol.* 32:167–176.

Moodie, G. E. E. 1972. Predation, natural selection and adaptation in an unusual threespine stickleback. *Heredity* 28:155–167.

Moore, A. J. 1990. The evolution of sexual dimorphism by sexual selection: the separate effects of intrasexual selection and intersexual selection. *Evolution* 44:315–331.

Moore, C. L. 1995. Maternal contributions to mammalian reproductive development and the divergence of males and females. *Adv. Study Behav.* 24:47–118.

Moore, M. C. 1987. Castration affects territorial and sexual behaviour of free-living male lizards, *Sceloporus jarrovi*. *Anim. Behav.* 35:1193–1199.

———. 1991. Application of the organization-activation theory to alternative male reproductive strategies: a review. *Horm. Behav.* 25:154–179.

Moore, M. C., D. K. Hews, and R. Knapp. 1998. Hormonal control and evolution of alternative male phenotypes: generalizations of models for sexual differentiation. *Am. Zool.* 38:133–151.

Moore, M. C., and J. Lindzey. 1992. The physiological basis of sexual behavior in male reptiles. In *Biology of the Reptilia. Vol. 18. Physiology E. Brain, hormones, and behavior*, edited by C. Gans and D. Crews, 70-113. University of Chicago Press, Chicago.

Moore, M. C., and C. A. Marler. 1987. Effects of testosterone on non-breeding season territorial aggression in free-living male lizards, *Sceloporus jarrovi*. *Gen. Comp. Endocrinol.* 65:225–232.

Moore, M. C., and C. W. Thompson. 1990. Field endocrinology of reptiles: hormonal control of alternative male reproductive tactics. In *Progress in comparative endocrinology*, edited by A. Epple, C. G. Scanes, and M. H. Stetson, 685–690. Wiley-Liss, New York.

Mora, J. M. 1989. Eco-behavioral aspects of two communally nesting iguanines and the structure of their shared nesting burrows. *Herpetologica* 45: 293–298.

Morris, M. R., M. Mussel, and M. J. Ryan. 1995. Vertical bars on male *Xiphophorus multilineatus:* a signal that deters rival males and attracts females. *Behav. Ecol.* 6:274–279.

Morris, M. R., and M. J. Ryan. 1996. Sexual differences in signal-receiver coevolution. *Anim. Behav.* 52:1017–1024.

Morrison, R. L. 1995. A transmission electron microscopic (TEM) method for determining structural colors reflected by lizard iridophores. *Pigment Cell. Res.* 8:23–36.

Morrison, R. L., and S. K. Frost-Mason. 1991. Ultrastructural analysis of iridophore organellogenesis in a lizard, *Sceloporus graciosus*. *J. Morphol.* 209:229–239.

Morrison, R. L., W. C. Sherbrooke, and S. K. Frost-Mason. 1996. Temperature-sensitive, physiologically active iridophores in the lizard *Urosaurus ornatus:* an ultrastructural analysis of color change. *Copeia* 1996:804–812.

Mosely, K. T. 1963. Behavior patterns of the collared lizard, *Crotaphytus collaris collaris*. M.S. thesis, University of Oklahoma, Norman.

Mount, R. H. 1963. The natural history of the red-tailed skink, *Eumeces egregius* Baird. *Am. Midl. Nat.* 70:356–385.

Mousseau, T. A., and C. W. Fox. 1998. The adaptive significance of maternal effects. *Trends Ecol. Evol.* 13:403–407.

Mouton, P. le F. N., A. F. Flemming, and E. M. Kanga. 1999. Grouping behaviour, tail-biting behaviour and sexual dimorphism in the armadillo lizard (*Cordylus cataphractus*) from South Africa. *J. Zool. Lond.* 249:1–10.

Mouton, P. le F. N., and J. H. van Wyk. 1993. Sexual dimorphism in cordylid lizards: a case study of the Drakensberg crag lizard, *Pseudocordylus melanotus*. *Can. J. Zool.* 71:1715–1723.

Murphy, R. W., W. E. Cooper, Jr., and W. S. Richardson. 1983. Phylogenetic relationships and genetic variability of the North American five-lined skinks, genus *Eumeces* (Sauria: Scincidae). *Herpetologica* 39:200–211.

Nagy, K. A. 1980. CO_2 production in animals: analysis of potential errors in the doubly labeled water method. *Am. J. Physiol.* 238:R466–R473.

———. 1983. *The doubly labeled water ($^3HH^{18}O$) method: a guide to its use.* University of California Los Angeles Publication 12—1417:1–45.

———. 1989. Doubly labeled water studies of vertebrate physiological ecology. In *Stable isotopes in ecological research,* edited by P. W. Rundel, J. R. Ehleringer, and K. A. Nagy, 270–287. Springer-Verlag, New York.

Nefdt, R. J. C., and S. J. Thirgood. 1997. Lekking, resource defense, and harassment in two subspecies of lechwe antelope. *Behav. Ecol.* 8:1–9.

Nelson, J. B. 1978. *The Sulidae: gannets and boobies.* Aberdeen University Studies, Vol. 154. Oxford University Press, Oxford, U.K.

Nelson, R. J. 2000. *An introduction to behavioral endocrinology,* 2nd ed. Sinauer, Sunderland, Mass.

Nicholson, K. E., and P. M. Richards. 1999. Observations of a population of the Cuban knight anole, *Anolis equestris.* In *Anolis newsletter V,* edited by J. B. Losos and M. Leal, 95–98. Washington University, St. Louis, Mo.

Niewiarowski, P. H., J. D. Congdon, A. E. Dunham, L. J. Vitt, and D. W. Tinkle. 1997. Tales of lizard tails: effects of tail autotomy on subsequent survival and growth of free-ranging hatchling *Uta stansburiana. Can. J. Zool.* 75:542–548.

Nilsson, S. G. 1977. Density compensation and competition among birds breeding on small islands in a South Sweden lake. *Oikos* 28:170–176.

Noble, G. K. 1939. The role of dominance in the social life of birds. *Auk* 56:263–273.

Noble, G. K., and E. R. Mason. 1933. Experiments on the brooding habits of the lizards *Eumeces* and *Ophisaurus. Am. Mus. Novit.* 619:1–29.

Nol, E., K. Cheng, and C. Nichols. 1996. Heritability and phenotypic correlations of behaviour and dominance rank of Japanese quail. *Anim. Behav.* 52: 813–820.

Nolan, V., Jr. 1978. Ecology and behavior of the prairie warbler *Dendroica discolor. Ornithol. Monogr.* 26:1–595.

Norris, K. S. 1953. The ecology of the desert iguana *Dipsosaurus dorsalis. Ecology* 34:265–287.

Nunez, S. C., T. A. Jenssen, and K. Ersland. 1997. Female activity profile of a polygynous lizard (*Anolis carolinensis*): evidence of intrasexual selection. *Behaviour* 135:981–1003.

O'Bryant, E. L., and J. Wade. 1999. Sexual dimorphisms in a neuromuscular system regulating courtship in the green anole lizard: effects of season and androgen treatment. *J. Neurobiol.* 40:202–213.

Olson, V. A., and I. P. F. Owens. 1998. Costly sexual signals: are carotenoids rare, risky or required? *Trends Ecol. Evol.* 13:510–514.

Olsson, M. 1992. Contest success in relation to size and residency in male sand lizards, *Lacerta agilis. Anim. Behav.* 44:386–388.

———. 1993a. Contest success and mate guarding in male sand lizards, *Lacerta agilis. Anim. Behav.* 46:408–409.

————. 1993b. Nuptial coloration and predation risk in model sand lizards, *Lacerta agilis. Anim. Behav.* 46:410–412.

————. 1994a. Nuptial coloration in the sand lizard, *Lacerta agilis:* an intra-sexually selected cue to fighting ability. *Anim. Behav.* 48:607–613.

————. 1994b. Why are sand lizard males (*Lacerta agilis*) not equally green? *Behav. Ecol. Sociobiol.* 35:169–173.

————. 1995. Forced copulation and costly female resistance behavior in the Lake Eyre dragon, *Ctenophorus maculosus. Herpetologica* 51:19–24.

Olsson, M., A. Gullberg, and H. Tegelstrom. 1994. Sperm competition in the sand lizard. *Anim. Behav.* 48:193–200.

Olsson, M., and T. Madsen. 1995. Female choice on male quantitative traits in lizards—why is it so rare? *Behav. Ecol. Sociobiol.* 36:179–184.

————. 1998. Sexual selection and sperm competition in reptiles. In *Sperm competition and sexual selection,* edited by T. R. Birkhead and A. P. Møller, 503–564. Academic, San Diego.

Olsson, M., and R. Shine. 1998. Chemosensory mate recognition may facilitate prolonged mate guarding by male snow skinks, *Niveoscincus microlepidotus. Behav. Ecol. Sociobiol.* 43:359–363.

Olsson, M., and B. Silverin. 1997. Effects of growth rate on variation in breeding coloration in male sand lizards (*Lacerta agilis:* Sauria). *Copeia* 1997:456–460.

Omland, K. E. 1997. Examining two standard assumptions of ancestral reconstructions: repeated loss of dichromatism in dabbling ducks (Anatini). *Evolution* 51:1636–1646.

Orchinik, M., T. F. Murray, and F. L. Moore. 1991. A corticosteroid receptor in neuronal membranes. *Science* 252:1848–1851.

Orians, G. H. 1969. On the evolution of mating systems in birds and mammals. *Am. Nat.* 103:589–602.

————. 1980. *Some adaptations of marsh-nesting blackbirds.* Princeton University Press, Princeton, N.J.

Orians, G. H., and J. F. Wittenberger. 1991. Spatial and temporal scales in habitat selection. *Am. Nat.* 137:29–49.

Otte, D., and K. Stayman. 1979. Beetle horns: some patterns in functional morphology. In *Sexual selection and reproductive competition in insects,* edited by M. S. Blum and N. A. Blum, 259–292. Academic, New York.

Owens, I. P. F., and I. R. Hartley. 1991. "Trojan sparrows": evolutionary consequences of dishonest invasion for the badges-of-status model. *Am. Nat.* 138:1187–1205.

Owens, I. P. F., and R. V. Short. 1995. Hormonal basis of sexual dimorphism in birds: implications for new theories of sexual selection. *Trends Ecol. Evol.* 10:44–47.

Packer, C. 1983. Sexual dimorphism: the horns of African antelopes. *Science* 221:1191–1193.

Palmer, M. W. 1993. Putting things in even better order: the advantages of canonical correspondence analysis. *Ecology* 74:2215–2230.

Panov, E. N., and L. Y. Zykova. 1993. Social organization and demography of Caucasian agama, *Stellio caucasius* (Squamata, Agamidae). *Zool. Zh.* 72:74–93.

Parker, G. A. 1974. Assessment strategy and the evolution of fighting behaviour. *J. Theor. Biol.* 47:223–243.

Parker, W. S., and E. R. Pianka. 1973. Notes on the ecology of the iguanid lizard *Sceloporus magister*. *Herpetologica* 29:143–152.

Patterson, I. J. 1965. Timing and spacing of broods in the black-headed gull *Larus ridibundus*. *Ibis* 107:433–459.

Pearson, O. P. 1954. Habits of the lizards *Liolaemus multiformis multiformis* at high altitudes in southern Peru. *Copeia* 1954:111–116.

Pearson, O. P., and D. F. Bradford. 1976. Thermoregulation of lizards and toads at high altitudes in Peru. *Copeia* 1976:155–170.

Perrill, S. A. 1973. Social communication in *Eumeces inexpectatus* (Scincidae). Ph.D. dissertation, North Carolina State University, Raleigh.

Perry, G. 1996. The evolution of sexual dimorphism in the lizard *Anolis polylepis* (Iguania): evidence from intraspecific variation in foraging behavior and diet. *Can. J. Zool.* 74:1238–1245.

Perry, R. 1984. The islands and their history. In *Key environments: Galápagos,* edited by R. Perry, 1–14. Pergamon, Oxford, U.K.

Peterson, A. T. 1996. Geographic variation in sexual dichromatism in birds. *Bull. Br. Ornithol. Club* 116:156–172.

Petersen, C. W., and R. R. Warner. 1998. Sperm competition in fishes. In *Sperm competition and sexual selection,* edited by T. R. Birkhead and A. P. Møller, 435–458. Academic, San Diego.

Peterson, E. 1980. Behavioral studies of telencephalic function in reptiles. In *Comparative neurology of the telencephalon,* edited by S. E. O. Ebbesson, 343–388. Plenum, New York.

Phoenix, C. H., R. W. Goy, A. A. Gerall, and W. C. Young. 1959. Organizing action of prenatally administered testosterone propionate on the tissues mediating mating behavior in the female guinea pig. *Endocrinology* 65:369–382.

Pianka, E. R. 1966. Convexity, desert lizards, and spatial heterogeneity. *Ecology* 47:1055–1059.

———. 1967. On lizard species diversity: North American flatland deserts. *Ecology* 48:333–351.

———. 1970. Comparative autecology of the lizard *Cnemidophorus tigris* in different parts of its geographic range. *Ecology* 51:703–720.

———. 1973. The structure of lizard communities. *Annu. Rev. Ecol. Syst.* 4:53–73.

Pianka, E. R., and H. D. Pianka. 1976. Comparative ecology of twelve species of nocturnal lizards (Gekkonidae) in the Western Australian desert. *Copeia* 1976:125–142.

Pleszczynska, W. K. 1978. Microgeographic prediction of polygyny in the lark bunting. *Science* 201:935–937.

Poiani, A., A. R. Goldsmith, and M. R. Evans. 2000. Ectoparasites of house sparrows (*Passer domesticus*): an experimental test of the immunocompetence handicap hypothesis and a new model. *Behav. Ecol. Sociobiol.* 47:230–242.

Pomiankowski, A. 1987. Sexual selection: the handicap principle does work—sometimes. *Proc. Roy. Soc. Lond., Ser. B* 231:123–145.

Pomiankowski, A., and A. P. Møller. 1995. A resolution of the lek paradox. *Proc. Roy. Soc. Lond., Ser. B* 260:21–29.

Porter, D. M. 1976. Geography and dispersal of Galápagos Islands vascular plants. *Nature* 264:745–746.

Porter, W. P., J. W. Mitchell, W. A. Beckman, and C. B. DeWitt. 1973. Behavioral implications of mechanistic ecology: thermal and behavioral modeling of desert ectotherms and their microenvironment. *Oecologia* 13:1–54.

Post, E., R. Langvatn, M. C. Forchhammer, and N. C. Stenseth. 1999. Environmental
variation shapes sexual dimorphism in red deer. *Proc. Natl. Acad. Sci. USA*
96:4467–4471.

Pough, F. H. 1973. Lizard energetics and diet. *Ecology* 54:837–844.

Pouyaud, L., E. Desmarais, A. Chenuil, J. F. Agnese, and F. Bonhomme. 1999. Kin co-
hesiveness and possible inbreeding in the mouthbrooding tilapia *Sarotherodon
melanotheron* (Pisces Cichlidae). *Mol. Ecol.* 8:803–812.

Powell, G. L., and A. P. Russell. 1992. Locomotor correlates of ecomorph designation
in *Anolis:* an examination of three sympatric species from Jamaica. *Can. J. Zool.*
70:725–739.

Pratt, N. C., J. A. Phillips, A. C. Alberts, and K. S. Bolda. 1994. Functional versus physiological puberty: an analysis of sexual bimaturism in the green iguana, *Iguana
iguana. Anim. Behav.* 47:1101–1114.

Price, T. D., and G. L. Birch. 1996. Repeated evolution of sexual color dimorphism in
birds. *Auk* 113:842–848.

Propper, C. R., R. E. Jones, and K. H. López. 1992. Distribution of arginine vasotocin
in the brain of the lizard *Anolis carolinensis. Cell. Tissue Res.* 267:391–398.

Punzo, F. 1982. Tail autotomy and running speed in the lizards *Cophosaurus texanus*
and *Uma notata. J. Herpetol.* 16:329–331.

Quinn, V. S., and D. K. Hews. 2000. Signals and behavioural responses are not coupled
in males: aggression affected by replacement of an evolutionarily lost color signal.
Proc. Roy. Soc. Lond., Ser. B 267:755–758.

Qvarnström, A. 1997. Experimentally increased badge size increases male competition
and reduces male parental care in the collared flycatcher. *Proc. Roy. Soc. Lond., Ser.
B* 264:1225–1231.

Qvarnström, A., and E. Forsgren. 1998. Should females prefer dominant males? *Trends
Ecol. Evol.* 13:498–501.

Ramírez-Bautista, A., J. Barba-Torres, and L. J. Vitt. 1998. Reproductive cycle and
brood size of *Eumeces lynxe* from Pinal de Amoles, Querétaro, México. *J. Herpetol.*
32:18–24.

Ramírez Pinilla, M. P. 1991. Estudio histológico de los tractos reproductivos y activi-
dad cíclica anual reproductiva de machos y hembras de dos especies del género
Liolaemus (Reptilia: Sauria: Iguanidae). Ph.D. dissertation, Universidad Nacional
de Tucumán, Facultad de Ciencias Naturales e Instituto Miguel Lillo, Tucumán,
Argentina.

Rand, A. S. 1954. Variation and predator pressure in an island and a mainland population of lizards. *Copeia* 1954:260–262.

———. 1964. Ecological distribution in anoline lizards of Puerto Rico. *Ecology*
45:745–752.

———. 1967a. The adaptive significance of territoriality in iguanid lizards. In *Lizard
ecology: a symposium,* edited by W. W. Milstead, 106–156. University of Missouri
Press, Columbia.

———. 1967b. The ecological distribution of anoline lizards around Kingston, Jamaica. *Breviora* 272:1–18.

———. 1967c. Ecology and social organization in the iquanid lizard, *Anolis lineatopus*. *Proc. U.S. Natl. Mus.* 122:1–79.

Rand, M. S. 1990. Polymorphic sexual coloration in the lizard *Sceloporus undulatus erythrocheilus*. *Am. Midl. Nat.* 125:352–359.

———. 1992. Hormonal control of polymorphic and sexually dimorphic coloration in the lizard *Sceloporus undulatus erythrocheilus*. *Gen. Comp. Endocrinol.* 88:461–468.

Read, A. F., and D. M. Weary. 1990. Sexual selection and the evolution of bird song: a test of the Hamilton-Zuk hypothesis. *Behav. Ecol. Sociobiol.* 26:47–56.

Read, A. F., and P. H. Harvey. 1989. Reassessment of comparative evidence for Hamilton and Zuk theory on the evolution of secondary sexual characters. *Nature* 339:618–620.

Reeder, T. W., and J. J. Wiens. 1996. Evolution of the lizard family Phrynosomatidae as inferred from diverse types of data. *Herpetol. Monogr.* 10:43–84.

Reid, M. L. 1987. Costliness and reliability in the singing vigour of Ipswich sparrows. *Anim. Behav.* 35:1735–1744.

Reyer, H.-U., W. Fischer, P. Steck, T. Nabulon, and P. Kessler. 1998. Sex-specific nest defense in house sparrows (*Passer domesticus*) varies with badge size of males. *Behav. Ecol. Sociobiol.* 42:93–99.

Reynolds, J. D., and P. H. Harvey. 1994. Sexual selection and the evolution of sex differences. In *The differences between the sexes,* edited by R. V. Short and E. Balaban, 53–70. Cambridge University Press, Cambridge, U.K.

Reznick, D. N., and H. Bryga. 1987. Life-history evolution in guppies (*Poecilia reticulata*): 1. Phenotypic and genetic changes in an introduction experiment. *Evolution* 41:1370–1385.

Reznick, D. N., and J. A. Endler. 1982. The impact of predation on life history evolution in Trinidadian guppies (*Poecilia reticulata*). *Evolution* 36:160–177.

Rhen, T., and D. Crews. 2000. Organization and activation of sexual and agonistic behavior in the leopard gecko, *Eublepharis macularis. Neuroendocrinology* 71:252–261.

Rice, W. R. 1990. A consensus combined P-value test and the family-wide significance of component tests. *Biometrics* 46:303–308.

Robertson, D. R. 1972. Social control of sex-reversal in a coral reef fish. *Science* 177:1007–1009.

Robles, C., and M. Halloy. 2000. Areas de acción en machos y hembras de lagartos de *Liolaemus quilmes,* Tropiduridae. Poster presentation at XV Reunión de Comunicaciones Herpetológicas de la Asociación Herpetológica Argentina, San Carlos de Bariloche, Río Negro, Argentina, 25–27 October 2000.

Rocha, C. F. D. 1996. Sexual dimorphism in the sand lizard *Liolaemus lutzae* of southeastern Brazil. In *Herpetología neotropical,* edited by J. E. Péfaur, 131–141. Actas II Congreso Latinoamericano de Herpetología, Vol. II, Universidad de los Andes, Mérida, Venezuela.

———. 1999. Home range of the tropidurid lizard *Liolaemus lutzae:* sexual and body size differences. *Rev. Brasil. Biol.* 59:125–130.

Rodda, G. H. 1992. The mating behavior of *Iguana iguana. Smithsonian Contr. Zool.* 534:1–40.

Rohwer, S. 1975. The social significance of avian winter plumage variability. *Evolution* 29:593–610.

———. 1977. Status signalling in Harris sparrows: Some experiments in deception. *Behaviour* 61:107–129.

———. 1982. The evolution of reliable and unreliable badges of fighting ability. *Am. Zool.* 22:531–546.

———. 1985. Dyed birds achieve higher social status than control in Harris sparrows. *Anim. Behav.* 33:1325–1331.

Rohwer, S., and P. W. Ewald. 1981. The cost of dominance and advantage of subordination in a badge signalling system. *Evolution* 35:441–454.

Rohwer, S., and F. C. Rohwer. 1978. Status signalling in Harris sparrows: experimental deceptions achieved. *Anim. Behav.* 26:1012–1022.

Roper, T. 1986. Badges of status in avian societies. *N. Scientist* 109(1494):38–40.

Rose, B. 1981. Factors affecting activity in *Sceloporus virgatus. Ecology* 62:706–716.

———. 1982. Lizard home ranges: methodology and functions. *J. Herpetol.* 16:253–269.

Røskaft, E., T. Järvi, M. Bakken, C. Bech, and R. E. Reinertsen. 1986. The relationship between social status and resting metabolic rate in great tits (*Parus major*) and pied flycatchers (*Ficedula hypoleuca*). *Anim. Behav.* 34:838–842.

Rostker, M. A. 1983. An experimental study of collared lizards: effects of habitat and male quality on fitness. Ph.D. dissertation, Oklahoma State University, Stillwater.

Roughgarden, J. 1974. Niche width: biogeographic patterns among *Anolis* lizard populations. *Am. Nat.* 108:429–442.

———. 1995. *Anolis* lizards of the Caribbean: ecology, evolution, and plate tectonics. Oxford University Press, Oxford, U.K.

Rozzi, R., J. D. Molina, and P. Miranda. 1989. Microclima y períodos de floración en laderas de exposición ecuatorial y polar en los Andes de Chile central. *Rev. Chilena Hist. Nat.* 62:75–84.

Ruby, D. E. 1978. Seasonal changes in territorial behavior of the iguanid lizard, *Sceloporus jarrovi. Copeia* 1978:430–438.

———. 1981. Phenotypic correlates of male reproductive success in the lizard, *Sceloporus jarrovi.* In *Natural selection and social behavior,* edited by R. D. Alexander and D. W. Tinkle, 96–107. Chiron, Roche Harbor, Wash.

———. 1984. Male breeding success and differential access to females in *Anolis carolinensis. Herpetologica* 40:272–280.

Ruby, D. E., and D. I. Baird. 1994. Intraspecific variation in behavior: comparisons between populations at different altitudes of the lizard *Sceloporus jarrovi. J. Herpetol.* 28:70–78.

Ruby, D. E., and A. E. Dunham. 1987. Variation in home range size along an elevational gradient in the iguanid lizard *Sceloporus merriami. Oecologia* 71:473–480.

Ruth, S. B. 1977. A comparison of the demography and female reproduction in sympatric western fence lizards (*Sceloporus occidentalis*) and sage–brush lizards (*Sceloporus graciosus*) on Mount Diablo, California. Ph.D. dissertation, University of California, Berkeley.

Ryan, M. J. 1982. Variation in iguanine social structure: mating systems in chuckwallas. In *Iguanas of the world: their behavior, ecology, and conservation,* edited by G. M. Burghardt and A. S. Rand, 380–390. Noyes, Park Ridge, N.J.

———. 1985. *The túngara frog: a study in sexual selection and communication.* University of Chicago Press, Chicago.

———. 1994. Mechanisms underlying sexual selection. In *Behavioral mechanisms in evolutionary ecology,* edited by L. A. Real, 190–215. University of Chicago Press, Chicago.

———. 1997. Sexual selection and mate choice. In *Behavioural ecology: an evolutionary approach,* 4th ed., edited by J. R. Krebs and N. B. Davies, 179–202. Blackwell Scientific, Oxford, U.K.

Ryan, M. J., K. Autumn, and D. B. Wake. 1998. Integrative biology and sexual selection. *Integr. Biol.* 1:68–72.

Ryan, M. J., J. H. Fox, W. Wilczynski, and A. S. Rand. 1990. Sexual selection for sensory exploitation in the frog *Physalaemus pustulosus. Nature* 343:66–67.

Ryan, M. J., and A. S. Rand. 1990. The sensory bias of sexual selection for complex calls in the túngara frog, *Physalaemus pustulosus* (sexual selection for sensory exploitation). *Evolution* 44:305–314.

———. 1993. Sexual selection and signal evolution: the ghost of biases past. *Philos. Trans. Roy. Soc. Lond., Ser. B* 340:187–195.

———. 1995. Female responses to ancestral advertisement calls in túngara frogs. *Science* 269:390–392.

Ryan, M. J., M. D. Tuttle, and A. S. Rand. 1982. Bat predation and sexual advertisement in a neotropical anuran. *Am. Nat.* 119:136–139.

Sadler, L. M., and M. A. Elgar. 1994. Cannibalism among amphibian larvae: a case of good taste. *Trends Ecol. Evol.* 9:5–6.

Salvador, A., J. Martín, and P. López. 1995. Tail loss reduces home range size and access to females in male lizards, *Psammodromus algirus. Behav. Ecol.* 6:382–387.

Salvador, A., J. Martín, P. López, and J. P. Veiga. 1996. Long-term effect of tail loss on home-range size and access to females in male lizards (*Psammodromus algirus*). *Copeia* 1996:208–209.

Salvador, A., J. P. Veiga, J. Martín, and P. López. 1997. Testosterone supplementation in subordinate, small male lizards: consequences for aggressiveness, color development, and parasite load. *Behav. Ecol.* 8:135–139.

Sato, T. 1994. Active accumulation of spawning substrate: a determinant of extreme polygyny in a shell-brooding cichlid fish. *Anim. Behav.* 48:669–678.

Schall, J. J., C. R. Bromwich, Y. L. Werner, and J. Midlege. 1989. Clubbed regenerated tails in *Agama agama* and their possible use in social interactions. *J. Herpetol.* 23:303–305.

Schall, J. J., and M. D. Dearing. 1987. Malarial parasitism and male competition for mates in the western fence lizard, *Sceloporus occidentalis. Oecologia* 73:389–392.

Schartl, M., D. C. Erbelding, S. Holter, I. Nanda, M. Schmid, J. H. Schroeder, and J. T. Epplen. 1993. Reproductive failure of dominant males in the poeciliid fish *Limia perugiae* determined by DNA fingerprinting. *Proc. Natl. Acad. Sci. USA* 90:7064–7068.

Schäuble, C. S., and G. C. Grigg. 1998. Thermal ecology of the Australian agamid *Pogona barbata. Oecologia* 114:461–470.

Schlichting, C. D., and M. Pigliucci. 1998. *Phenotypic evolution: a reaction norm perspective.* Sinauer, Sunderland, Mass.

Schluter, D. 1984. Body size, prey size and herbivory in the Galápagos lava lizard, *Tropidurus*. *Oikos* 43:291–300.

Schoener, T. W. 1967. The ecological significance of sexual dimorphism in size in the lizard *Anolis conspersus*. *Science* 155:474–477.

———. 1968. The *Anolis* lizards of Bimini: resource partitioning in a complex fauna. *Ecology* 49:704–726.

———. 1969a. Models of optimal size for solitary predators. *Am. Nat.* 103:277–313.

———. 1969b. Size patterns in West Indian *Anolis* lizards. I. Size and species diversity. *Syst. Zool.* 18:386–401.

———. 1970. Size patterns in West Indian *Anolis* lizards. II. Correlations with the size of particular sympatric species—displacement and convergence. *Am. Nat.* 104:155–174.

———. 1977. Competition and the niche. In *Biology of the Reptilia. Vol. 7. Ecology and behaviour A,* edited by C. Gans and D. W. Tinkle, 35–136. Academic, London.

———. 1979. Inferring the properties of predation and other injury-producing agents from injury frequencies. *Ecology* 60:1110–1115.

Schoener, T. W., and G. C. Gorman. 1968. Some niche differences in three Lesser-Antillean lizards of the genus *Anolis*. *Ecology* 49:819–830.

Schoener, T. W., and A. Schoener. 1971a. Structural habitats of West Indian *Anolis* lizards. I. Jamaican lowlands. *Breviora* 368:1–53.

———. 1971b. Structural habitats of West Indian *Anolis* lizards. II. Puerto Rican uplands. *Breviora* 375:1–39.

———. 1978. Inverse relation of survival of lizards with island size and avifaunal richness. *Nature* 274:685–687.

———. 1980. Densities, sex ratios and population structure in four species of Bahamian *Anolis* lizards. *J. Anim. Ecol.* 49:19–53.

———. 1982a. The ecological correlates of survival in some Bahamian *Anolis* lizards. *Oikos* 39:1–16.

———. 1982b. Intraspecific variation in home-range size in some *Anolis* lizards. *Ecology* 63:809–823.

Schulte, J. A., J. R. Macey, R. E. Espinoza, and A. Larson. 2000. Phylogenetic relationships in the iguanid lizard genus *Liolaemus:* multiple origins of viviparous reproduction and evidence for recurring Andean vicariance and dispersal. *Biol. J. Linn. Soc.* 69:75–102.

Schwarzkopf, L., and R. Shine. 1992. Costs of reproduction in lizards: escape tactics and susceptibility to predation. *Behav. Ecol. Sociobiol.* 31:17–25.

Searcy, W. A. 1979. Female choice of mates: a general model for birds and its application to red-winged blackbirds (*Agelaius phoeniceus*). *Am. Nat.* 114:77–100.

Searcy, W. A., and Yasukawa, K. 1995. *Polygyny and sexual selection in red-winged blackbirds.* Princeton University Press, Princeton, N.J.

Selander, R. K. 1966. Sexual dimorphism and differential niche utilization in birds. *Condor* 68:113–151.

———. 1972. Sexual selection and sexual dimorphism in birds. In *Sexual selection and the descent of man, 1871–1971,* edited by B. Campbell, 180–230. Aldine-Atherton, Chicago.

Semler, D. E. 1971. Some aspects of adaptation in a polymorphism for breeding colours in the threespine stickleback (*Gasterosteus aculeatus*). *J. Zool. Lond.* 165:291–302.

Semple, S., and K. McComb. 1996. Behavioural deception. *Trends Ecol. Evol.* 11:434–437.

Senar, J. C. 1999. Plumage colouration as a signal of social status. In *Proceedings of the 22nd international ornithological congress, Durban,* edited by N. J. Adams and R. H. Slotow, 1669–1686. BirdLife South Africa, Johannesburg.

Senar, J. C., and M. Camerino. 1998. Status signalling and the ability to recognize dominants: an experiment with siskins (*Carduelis spinus*). *Proc. Roy. Soc. Lond., Ser. B* 265:1515–1520.

Senar, J. C., M. Camerino, J. L. Copete, and N. B. Metcalfe. 1993. Variation in black bib of the Eurasian siskin (*Carduelis spinus*) and its role as a reliable badge of dominance. *Auk* 110:924–927.

Senar, J. C., J. L. Copete, and A. J. Martin. 1998. Behavioural and morphological correlates of variation in the extent of postjuvenile moult in the siskin *Carduelis spinus.* *Ibis* 140:661–669.

Senar, J. C., V. Polo, F. Uribe, and M. Camerino. 2000. Status signalling, metabolic rate and body mass in the siskin: the cost of being a subordinate. *Anim. Behav.* 59:103–110.

Shapiro, D. Y. 1991. Intraspecific variability in social systems of coral reef fishes. In *The ecology of fishes on coral reefs,* edited by P. F. Sale, 331–355. Academic, San Diego.

Sherbrooke, W. C., and S. K. Frost. 1989. Integumental chromatophores of a color-change, thermoregulating lizard, *Phrynosoma modestum* (Iguanidae: Reptilia). *Am. Mus. Novit.* 2943:1–14.

Shields, W. M. 1977. The social significance of avian winter plumage variability: a comment. *Evolution* 31:905–907.

Shine, R. 1978. Sexual size dimorphism and male combat in snakes. *Oecologia* 33:269–278.

———. 1979. Sexual selection and sexual dimorphism in Amphibia. *Copeia* 1979:297–306.

———. 1980. "Costs" of reproduction in reptiles. *Oecologia* 46:92–100.

———. 1985. The evolution of viviparity in reptiles: an ecological analysis. In *Biology of the Reptilia. Vol. 15,* edited by C. Gans and F. Billet, 605–694. John Wiley, New York.

———. 1989. Ecological causes for the evolution of sexual dimorphism: a review of the evidence. *Quart. Rev. Biol.* 64:419–461.

———. 1991. Intersexual dietary divergence and the evolution of sexual dimorphism in snakes. *Am. Nat.* 138:103–122.

Shine, R., S. Keogh, P. Doughty, and H. Giragossyan. 1998. Costs of reproduction and the evolution of sexual dimorphism in a 'flying lizard' *Draco melanopogon* (Agamidae). *J. Zool. Lond.* 246:203–213.

Shine, R., and L. Schwarzkopf. 1992. The evolution of reproductive effort in lizards and snakes. *Evolution* 46:62–75.

Shuster, S. M., and M. J. Wade. 1992. Equal mating success among male reproductive strategies in a marine isopod. *Nature* 350:606–661.

Sigmund, W. R. 1983. Female preferences for *Anolis carolinensis* males as a function of dewlap color and background coloration. *J. Herpetol.* 17:137–143.

Sih, A. 1992. Prey uncertainty and the balancing of antipredator and feeding needs. *Am. Nat.* 139:1052–1069.

Simerly, R. B., C. Chang, M. Muramatsu, and L. W. Swanson. 1990. Distribution of androgen and estrogen receptor mRNA-containing cells in the rat brain: an *in situ* hybridization study. *J. Comp. Neurol.* 294:76–95.

Simkin, T. 1984. Geology of Galápagos Islands. In *Key environments: Galápagos,* edited by R. Perry, 15–41. Pergamon, Oxford, U.K.

Simon, C. A. 1975. The influence of food abundance on territory size in the iguanid lizard *Sceloporus jarrovi. Ecology* 56:993–998.

———. 1983. A review of lizard chemoreception. In *Lizard ecology: studies of a model organism,* edited by R. B. Huey, E. R. Pianka, and T. W. Schoener, 119–133. Harvard University Press, Cambridge, Mass.

Simon, N. G., and R. E. Whalen. 1986. Hormonal regulation of aggression: evidence for a relationship among genotype, receptor binding, and behavioral sensitivity to androgen and estrogen. *Aggress. Behav.* 12:255–267.

Sinervo, B. 1990. The evolution of thermal physiology and growth rate between populations of the western fence lizard (*Sceloporus occidentalis*). *Oecologia* 83:228–237.

———. 1996. *MacTurf, manual and software* (distributed by the author, http://www.biology.ucsc.edu/~barrylab).

———. 1999. Mechanistic analysis of natural selection and a refinement of Lack's and Williams's principles. *Am. Nat.* 154:S26–S42.

———. 2000. Adaptation, natural selection, and optimal life history allocation in the face of genetically-based trade-offs. In *Adaptive genetic variation in the wild,* edited by T. Mousseau, B. Sinervo, and J. A. Endler, 41–64. Oxford University Press, Oxford, U.K.

Sinervo, B., and S. C. Adolph. 1989. Thermal sensitivity of hatchling growth in *Sceloporus* lizards: environmental, behavioral and genetic aspects. *Oecologia* 78:411–417.

———. 1994. Growth plasticity and thermal opportunity in *Sceloporus* lizards. *Ecology* 75:776–790.

Sinervo, B., and D. F. DeNardo. 1996. Costs of reproduction in the wild: path analysis of natural selection and experimental tests of causation. *Evolution* 50:1299–1313.

Sinervo, B., and P. Doughty. 1996. Interactive effects of offspring size and timing of reproduction on offspring reproduction: experimental, maternal, and quantitative genetic aspects. *Evolution* 50:1314–1327.

Sinervo, B., P. Doughty, R. B. Huey, and K. Zamudio. 1992. Allometric engineering: a causal analysis of natural selection on offspring size. *Science* 258:1927–1930.

Sinervo, B., and P. Licht. 1991a. The physiological and hormonal control of clutch size, egg size and egg shape in *Uta stansburiana:* constraints on the evolution of lizard life histories. *J. Exp. Zool.* 257:252–264.

———. 1991b. Proximate constraints on the evolution of egg size, egg number, and total clutch mass in lizards. *Science* 252:1300–1302.

Sinervo, B., and C. M. Lively. 1996. The rock–paper–scissors game and the evolution of alternative male strategies. *Nature* 380:240–243.

Sinervo, B., and J. B. Losos. 1991. Walking the tight rope: arboreal sprint performance among *Sceloporus occidentalis* lizard populations. *Ecology* 72:1225–1233.

Sinervo, B., D. B. Miles, D. DeNardo, T. Frankino, and M. Klukowski. 2000. Testosterone, endurance, and Darwinian fitness: natural and sexual selection on the physiological bases of alternative male behaviors in side-blotched lizards. *Horm. Behav.* 38:222–223.

Sinervo, B., and E. Svensson. 1998. Mechanistic and selective causes of life history trade-offs and plasticity. *Oikos* 83:432–442.

Sinervo, B., E. Svensson, and T. Comendant. 2000. Density cycles and an offspring quantity and quality game driven by natural selection. *Nature* 406:985–988.

Sinervo, B., and K. R. Zamudio. 2001. The evolution of alternative reproductive strategies: fitness differential, heritability, and genetic correlation beween the sexes. *J. Hered.* 92:198–205.

Sites, J. W., J. W. Archie, C. J. Cole, and O. F. Villela. 1992. A review of phylogenetic hypotheses for lizards in the genus *Sceloporus* (Phrynosomatidae): implications for ecological and evolutionary studies. *Bull. Am. Mus. Nat. Hist.* 213:1–110.

Slagsvold, T., S. Dale, and A. Kruszewicz. 1995. Predation favours cryptic coloration in breeding male pied flycatchers. *Anim. Behav.* 50:1109–1121.

Slatkin, M. 1984. Ecological causes of sexual dimorphism. *Evolution* 38:622–630.

Sloan, C. L. 1997. Intrasexual aggression, competition, and social dominance in female collared lizards, *Crotaphytus collaris*. M.S. thesis, University of Central Oklahoma, Edmond.

Sloan, C. L., and T. A. Baird. 1999. Is heightened post-ovipositional aggression in female collared lizards (*Crotaphytus collaris*) nest defense? *Herpetologica* 55:516–522.

Slotow, R., J. Alcock, and S. I. Rothstein. 1993. Social status signalling in white-crowned sparrows: an experimental test of the social control hypothesis. *Anim. Behav.* 46:977–989.

Smith, D. G. 1997. Ecological factors influencing the antipredator behaviors of the ground skink, *Scincella lateralis*. *Behav. Ecol.* 8:622–629.

Smith, L. C., and H. B. John-Alder. 1999. Seasonal specificity of hormonal, behavioral, and coloration responses to within- and between-sex encounters in male lizards (*Sceloporus undulatus*). *Horm. Behav.* 36:39–52.

Snell, H. L., R. D. Jennings, H. M. Snell, and S. Harcourt. 1988. Intrapopulation variation in predator-avoidance performance of Galápagos lava lizards: the interaction of sexual and natural selection. *Evol. Ecol.* 2:353–369.

Snell, H. M., P. A. Stone, and H. L. Snell. 1996. A summary of the geographical characteristics of the Galápagos Islands. *J. Biogeogr.* 23:619–624.

Solberg, E. J., and T. H. Ringsby. 1997. Does male badge size signal status in small island populations of house sparrows, *Passer domesticus? Ethology* 103:177–186.

Somma, L. A. 1985. Notes on maternal behavior and post-brooding agression [sic] in the prairie skink *Eumeces septentrionalis*. *Nebr. Herpetol. Newslett.* 6:9–12.

———. 1987. Reproduction of the prairie skink, *Eumeces septentrionalis*, in Nebraska. *Great Basin Nat.* 47:373–374.

Sorci, G., J. Colbert, and S. Belichon. 1996. Phenotypic plasticity of growth and survival in the common lizard *Lacerta vivipara*. *J. Anim. Ecol.* 65:781–790.

Stamps, J. A. 1973. Displays and social organization in female *Anolis aeneus*. *Copeia* 1973:264–272.

———. 1977a. The relationship between resource competition, risk and aggression in a tropical territorial lizard. *Ecology* 58:349–358.

————. 1977b. Social behavior and spacing patterns in lizards. In *Biology of the Reptilia. Vol. 7. Ecology and behaviour A,* edited by C. Gans and D. W. Tinkle, 265–334. Academic, London.

————. 1978. A field study of the ontogeny of social behavior in the lizard *Anolis aeneus. Behaviour* 66:1–31.

————. 1983a. The relationship between ontogenetic habitat shifts, competition, and predator avoidance in a juvenile lizard (*Anolis aeneus*). *Behav. Ecol. Sociobiol.* 12:19–33.

————. 1983b. Sexual selection, sexual dimorphism, and territoriality. In *Lizard ecology: studies of a model organism,* edited by R. B. Huey, E. R. Pianka, and T. W. Schoener, 169–204. Harvard University Press, Cambridge, Mass.

————. 1988. Conspecific attraction and aggregation in territorial species. *Am. Nat.* 131:329–347.

————. 1990. The effects of contender pressure on territory size and overlap in seasonally territorial species. *Am. Nat.* 135:614–632.

————. 1992. Simultaneous versus sequential settlement in territorial species. *Am. Nat.* 139:1070–1088.

————. 1994. Territorial behavior: testing the assumptions. In *Advances in the study of behavior. Vol. 23,* edited by P. J. B. Slater, J. S. Rosenblatt, C. T. Snowden, and M. Milinski, 173–232. Academic, San Diego.

Stamps, J. A., and G. W. Barlow. 1973. Variation and stereotypy in the displays of *Anolis aeneus* (Sauria: Iguanidae). *Behaviour* 47:67–94.

Stamps, J. A., and M. Buechner. 1985. The territorial defense hypothesis and the ecology of insular vertebrates. *Quart. Rev. Biol.* 60:155–181.

Stamps, J. A., and V. V. Krishnan. 1994. Territory acquisition in lizards. II. Establishing social and spatial relationships. *Anim. Behav.* 47:1387–1400.

————. 1995. Territory acquisition in lizards. III. Competing for space. *Anim. Behav.* 49:679–693.

————. 1998. Territory acquisition in lizards. IV. Obtaining high status and exclusive home ranges. *Anim. Behav.* 55:461–472.

Stamps, J., J. B. Losos, and R. M. Andrews. 1997. A comparative study of population density and sexual size dimorphism in lizards. *Am. Nat.* 149:64–90.

Stamps, J. A., and S. K. Tanaka. 1981a. The influence of food and water on growth rates in a tropical lizard (*Anolis aeneus*). *Ecology* 62:33–40.

————. 1981b. The relationship between food and social behavior in juvenile lizards (*Anolis aeneus*). *Copeia* 1981:422–434.

Steadman, D. W. 1986. Holocene vertebrate fossils from Isla Floreana, Galápagos. *Smithsonian Contrib. Zool.* 413:1–103.

Stearns, S. C. 1983. The influence of size and phylogeny on patterns of covariation among life-history traits in mammals. *Oikos* 41:173–187.

Stebbins, R. C. 1985. *A field guide to western reptiles and amphibians,* 2nd ed. Houghton Mifflin, Boston.

Stebbins, R. C., J. M. Lowenstein, and N. W. Cohen. 1967. A field study of the lava lizard (*Tropidurus albemarlensis*) in the Galápagos Islands. *Ecology* 48:839–851.

Stepien, C. A., and T. D. Kocher. 1997. Molecules and morphology in studies of fish evolution. In *Molecular systematics of fishes,* edited by T. D. Kocher and C. A. Stepien, 1–11. Academic, San Diego.

Stone, P. A. 1995. Sexual selection in Galápagos lava lizards (*Tropidurus*). Ph.D. dissertation, University of New Mexico, Albuquerque.

Stone, P. A., H. L. Snell, and H. M. Snell. 1994. Behavioral diversity as biological diversity: Introduced cats and lava lizard wariness. *Conserv. Biol.* 8: 569–573.

Stoner, G., and F. Breden. 1988. Phenotypic differentiation in female preference related to geographic variation in male predation risk in the Trinidad guppy (*Poecilia reticulata*). *Behav. Ecol. Sociobiol.* 22:285–291.

Stratton, D. 1994. Genotype-by-environment interactions for fitness of *Erigeron annuus* show fine-scale selective heterogeneity. *Evolution* 48:1607–1618.

Stratton, D., and C. C. Bennington. 1996. Measuring spatial variation in fitness of *Arabidopsis thaliana* using randomly sown seeds. *J. Evol. Biol.* 9:215–228.

———. 1998. Fine-grained spatial and temporal variation in selection does not maintain genetic variation in *Erigeron annuus*. *Evolution* 52:678–691.

Studd, M. V., and R. J. Robertson. 1985a. Evidence for reliable badges of status in territorial yellow warblers (*Dendroica petechia*). *Anim. Behav.* 33:1102–1113.

———. 1985b. Sexual selection and variation in reproductive strategy in male yellow warblers (*Dendroica petechia*). *Behav. Ecol. Sociobiol.* 17:101–109.

Sugerman, R. A., and L. S. Demski. 1978. Agonistic behavior elicited by electrical stimulation of the brain in western collard lizards, *Crotaphytus collaris*. *Brain Behav. Evol.* 15:446–469.

Sullivan, B. K. 1983. Sexual selection in woodhouse's toad (*Bufo woodhousei*). II. Female choice. *Anim. Behav.* 31:1011–1017.

Swenson, R. O. 1999. The ecology, behavior, and conservation of the tidewater goby, *Eucyclogobius newberryi*. *Environ. Biol. Fish.* 55:99–114.

Számadó, S. 2000. Cheating as a mixed strategy in a simple model of aggressive communication. *Anim. Behav.* 59:221–230.

Tarr, R. S. 1977. Role of the amygdala in the intraspecies aggressive behavior of the iguanid lizard *Sceloporus occidentalis*. *Physiol. Behav.* 18:1153–1158.

Taylor, E. H. 1935. A taxonomic study of the cosmopolitan scincoid lizards of the genus *Eumeces* with an account of the distribution and relationships of its species. *Univ. Kans. Sci. Bull.* 23:1–643.

Taylor, J. A. 1984. Ecology of the lizard, *Ctenotus taeniolatus:* interaction of life history, energy storage and tail autotomy. Ph.D. dissertation, University of New England, Armidale, Australia.

Taylor, J. D., and M. E. Hadley. 1970. Chromatophores and color change in the lizard, *Anolis carolinensis*. *Z. Zellforsh. Mikrosk. Anat.* 140:282–294.

Temeles, E. J. 1990. Northern harriers on feeding territories respond more aggressively to neighbors than to floaters. *Behav. Ecol. Sociobiol.* 26:57–63.

ter Braak, C. J. F. 1986. Canonical correspondence analysis: a new eigenvector technique for multivariate direct gradient analysis. *Ecology* 67:1167–1179.

ter Braak, C. J. F., and P. Šmilauer. 1998. *CANOCO reference manual and user's guide to Canoco for Windows: software for canonical community ordination, Version 4.* Microcomputer Power, Ithaca, N.Y.

Thomaz, D., E. Beall, and T. Burke. 1997. Alternative reproductive tactics in Atlantic salmon: Factors affecting mature parr success. *Proc. Roy. Soc. Lond., Ser. B* 264:219–226.

Thompson, C. W., and M. C. Moore. 1991a. Syntopic occurrence of multiple dewlap color morphs in male tree lizards. *Copeia* 1991:492–503.

———. 1991b. Throat color reliably signals status in male tree lizards, *Urosaurus ornatus. Anim. Behav.* 42:745–753.

Thompson, N. 1999. Specific hypotheses on the geographic mosaic of coevolution. *Am. Nat.* 153:S1–S14.

Thornhill, R., and J. Alcock. 1983. *The evolution of insect mating systems.* Harvard University Press, Cambridge, Mass.

Thorpe, R. S., and R. P. Brown. 1989. Microgeographic variation in the colour pattern of the lizard *Gallotia galloti* within the island of Tenerife: distribution, pattern and hypothesis testing. *Biol. J. Linn. Soc.* 38:303–322.

Throckmorton, L. 1975. The phylogeny, ecology and geography of *Drosophila*. In *Handbook of genetics,* Vol. 3, edited by R. C. King, 421–469. Plenum, New York.

Timanus, D. K. 1999. Phenotypic variation among adult male collared lizards, *Crotaphytus collaris:* influence of morphology and space use on mating success. M.S. thesis, University of Central Oklahoma, Edmond.

Tinbergen, N. 1956. The functions of territory. *Ibis* 98:14–27.

———. 1959. Comparative studies of the behaviour of gulls (Laridae): a progress report. *Behaviour* 15:1–70.

———. 1964. The evolution of signaling devices. In *Social behavior and organization among vertebrates,* edited by W. Etkin, 206–230. University of Chicago Press, Chicago.

Tinbergen, N., G. J. Broekhuysen, F. Feekes, J. C. W. Houghton, H. Kruuk, and E. Szulc. 1962. Egg shell removal by the black-headed gull, *Larus ridibundus* L.: a behaviour component of camouflage. *Behaviour* 19:74–117.

Tinbergen, N., M. Impekoven, and D. Franck. 1967. An experiment on spacing-out as a defense against predation. *Behaviour* 28:307–321.

Tinkle, D. W. 1967a. Home range, density, dynamics, and structure of a Texas population of the lizard *Uta stansburiana.* In *Lizard ecology: a symposium,* edited by W. W. Milstead, 5–29. University of Missouri Press, Columbia.

———. 1967b. The life and demography of the side-blotched lizard, *Uta stansburiana. Misc. Publ. Mus. Zool. Univ. Mich.* 132:1–182.

Tinkle, D. W., and R. E. Ballinger. 1972. *Sceloporus undulatus:* a study of the intraspecific comparative demography of a lizard. *Ecology* 53:570–584.

Tinkle, D. W., D. McGregor, and S. Dana. 1962. Home range ecology of *Uta stansburiana stejnegeri. Ecology* 43:223–229.

Tokarz, R. R. 1985. Body size as a factor determining dominance in staged agonistic encounters between male brown anoles (*Anolis sagrei*). *Anim. Behav.* 33:746–753.

———. 1995a. Importance of androgens in male territorial acquisition in the lizard *Anolis sagrei:* an experimental test. *Anim. Behav.* 49:661–669.

———. 1995b. Mate choice in lizards: a review. *Herpetol. Monogr.* 9:17–40.

———. 1998. Mating pattern in the lizard *Anolis sagrei:* implications for mate choice and sperm competition. *Herpetologica* 54:388–394.

Tokarz, R. R., and D. Crews. 1981. Effects of prostaglandins on sexual receptivity in the lizard, *Anolis carolinensis. Endocrinology* 109:451–457.

Torr, G. A., and R. Shine. 1993. Experimental analysis of thermally dependent behavior patterns in the scincid lizard *Lampropholis guichenoti. Copeia* 1993:850–854.

Tousignant, A., and D. Crews. 1994. Effects of exogenous estradiol applied at different embryonic stages on sex determination, growth, and mortality in the leopard gecko (*Eublepharis macularius*). *J. Exp. Zool.* 268:17–21.

———. 1995. Incubation temperature and gonadal sex affect growth and physiology in the leopard gecko (*Eublepharis macularius*), a lizard with temperature–dependent sex determination. *J. Morphol.* 224:159–170.

Tracy, C. R., and K. A. Christian. 1986. Ecological relations among space, time, and thermal niche axes. *Ecology* 67:609–615.

Trauth, S. E. 1979. Testicular cycle and timing of reproduction in the collared lizard (*Crotaphytus collaris*) in Arkansas. *Herpetologica* 35:184–192.

Trillmich, K. G. K. 1983. The mating system of the marine iguana (*Amblyrhynchus cristatus*). *Z. Tierpsychol.* 63:141–172.

Trivers, R. L. 1972. Parental investment and sexual selection. In *Sexual selection and the descent of man, 1871–1971,* edited by B. Cambell,136–179. Aldine-Atherton, Chicago.

———. 1976. Sexual selection and resource-accruing abilities in *Anolis garmani. Evolution* 30:253–269.

———. 1985. *Social evolution.* Benjamin/Cummings Publishing, Menlo Park, Calif.

Tuttle, M. D., and M. J. Ryan. 1981. Bat predation and the evolution of vocalisations in the neotropics. *Science* 214:677–678.

Van Damme, R., D. Bauwens, A. M. Castilla, and R. F. Verheyen. 1989. Altitudinal variation of the thermal biology and running performance in the lizard *Podarcis tiliguerta. Oecologia* 80:516–524.

Van Denburgh, J., and J. R. Slevin. 1913. Expedition of the California Academy of Sciences to the Galápagos Islands, 1905–1906. IX. The Galapagoan lizards of the genus *Tropidurus;* with notes on the iguanas of the genera *Conolophus* and *Amblyrhynchus. Proc. Calif. Acad. Sci.,* Fourth Series 2:133–202.

Vehrencamp, S. L., and J. W. Bradbury. 1984. Mating systems and ecology. In *Behavioural ecology: an evolutionary approach,* 2nd ed., edited by J. R. Krebs and N. B. Davies, 251–278. Sinauer, Sunderland, Mass.

Vehrencamp, S. L., J. B. Bradbury, and R. M. Gibson. 1989. The energetic cost of displaying in male sage grouse. *Anim. Behav.* 38:885–896.

Veiga, J. P. 1993. Badge size, phenotypic quality, and reproductive success in the house sparrow: a study of honest advertisement. *Evolution* 47:1161–1170.

———. 1995. Honest signaling and the survival cost of badges in the house sparrow. *Evolution* 49:570–572.

Veiga, J. P., and M. Puerta. 1996. Nutritional constraints determine the expression of a sexual trait in the house sparrow, *Passer domesticus. Proc. Roy. Soc. Lond., Ser. B* 263:229–234.

Viljugrein, H. 1997. The cost of dishonesty. *Proc. Roy. Soc. Lond., Ser. B* 264:815–821.

Vincent, A. C. J. 1992. Prospects for sex role reversal in teleost fishes. *Neth. J. Zool.* 2:392–399.

Vinegar, M. B. 1975. Comparative aggression in *Sceloporus virgatus, S. undulatus consobrinus,* and *S. u. tristichus* (Sauria: Iguanidae). *Anim. Behav.* 23:279–286.

Vinnedge, B., and P. Verrell. 1998. Variance in male mating success and female choice for persuasive courtship displays. *Anim. Behav.* 56:443–448.

Vitt, L. J. 1983. Reproduction and sexual dimorphism in the tropical teiid lizard *Cnemidophorus ocellifer. Copeia* 1983:359–366.

Vitt, L. J., J. D. Congdon, A. C. Hulse, and J. F. Platz. 1974. Territorial aggressive encounters and tail breaks in the lizard *Sceloporus magister. Copeia* 1974: 990–993.

Vitt, L. J., and W. E. Cooper, Jr. 1985a. The evolution of sexual dimorphism in the skink *Eumeces laticeps:* an example of sexual selection. *Can. J. Zool.* 63:995–1002.

———. 1985b. The relationship between reproduction and lipid cycling in the skink *Eumeces laticeps* with comments on brooding ecology. *Herpetologica* 41:419–432.

———. 1986. Skink reproduction and sexual dimorphism: *Eumeces fasciatus* in the southeastern United States, with notes on *Eumeces inexpectatus. J. Herpetol.* 20:65–76.

———. 1989. Maternal care in skinks (*Eumeces*). *J. Herpetol.* 23:29–34.

Vitt, L. J., P. A. Zani, and R. D. Durtsche. 1995. Ecology of the lizard *Norops oxylophus* (Polychrotidae) in lowland forest of southeastern Nicaragua. *Can. J. Zool.* 73:1918–1927.

vom Saal, F. S. 1979. Prenatal exposure to androgens influences morphology and aggressive behavior of male and female mice. *Horm. Behav.* 12:1–11.

von Neumann, J., and O. Morgenstern. 1953. *Theory of games and economic behavior.* Princeton University Press, Princeton, N.J.

Wade, J. 1997. Androgen metabolism in the brain of the green anole lizard (*Anolis carolinensis*). *Gen. Comp. Endocrinol.* 106:127–137.

———. 1999. Sexual dimorphisms in avian and reptilian courtship: two systems that do not play by mammalian rules. *Brain Behav. Evol.* 54:15–27.

Wade, J., and D. Crews. 1991a. The effects of intracranial implantation of estrogen on receptivity in sexually and asexually reproducing female whiptail lizards, *Cnemidophorus inornatus* and *Cnemidophorus uniparens. Horm. Behav.* 25:342–353.

———. 1991b. The relationship between reproductive state and "sexually" dimorphic brain areas in sexually reproducing and parthenogenetic whiptail lizards. *J. Comp. Neurol.* 309:507–514.

———. 1992. Sexual dimorphism in the soma size of neurons in the brain of whiptail lizards. *Brain Res.* 594:311–314.

Wade, J., J. M. Huang, and D. Crews. 1993. Hormonal control of sex differences in the brain, behavior and accessory sex structures of whiptail lizards (*Cnemidophorus* species). *J. Neuroendocrinol.* 5:81–93.

Wade, M. J. 1979. Sexual selection and variance in reproductive success. *Am. Nat.* 114:742– 747.

———. 1987. Measuring sexual selection. In *Sexual selection: testing the alternatives,* edited by J. W. Bradbury and M. B. Andersson, 197–207. Wiley, Chichester, U.K.

———. 1995. The ecology of sexual selection: mean crowding of females and resource defense polygyny. *Evol. Ecol.* 9:118–124.

Wade, M. J., and S. J. Arnold. 1980. The intensity of sexual selection in relation to male sexual behavior, female choice and sperm precedence. *Anim. Behav.* 28:446–461.

Waldman, B. 1984. Kin recognition and sibling association among wood frog (*Rana sylvatica*) tadpoles. *Behav. Ecol. Sociobiol.* 14:171–180.

Waldschmidt, S. 1983. The effect of supplemental feeding on home range size and activity patterns in the lizard *Uta stansburiana*. *Oecologia* 57:1–5.

Waldschmidt, S., and C. R. Tracy. 1983. Interactions between a lizard and its thermal environment: implications for sprint performance and space utilization in the lizard *Uta stansburiana*. *Ecology* 64:476–484.

Waltz, E. C., and L. L. Wolf. 1984. By Jove!! Why do alternative mating tactics assume so many different forms? *Am. Nat.* 24:333–343.

Warner, R. R. 1984. Mating behavior and hermaphroditism in coral reef fishes. *Am. Sci.* 72:128–136.

———. 1991. The use of phenotypic plasticity in coral reef fishes as tests of theory in evolutionary ecology. In *The ecology of fishes on coral reefs*, edited by P. F. Sale, 387–398. Academic, San Diego.

Warner, R. R., and S. G. Hoffman. 1980. Population density and the economics of territorial defense in a coral reef fish. *Ecology* 61:772–780.

Watson, J. T., and E. Adkins-Regan. 1989. Testosterone implanted in the preoptic area of male Japanese quail must be aromatized to activate copulation. *Horm. Behav.* 23:432–447.

Wellborn, G. A. 1995. Determinants of reproductive success in freshwater amphipod species that experience different mortality regimes. *Anim. Behav.* 50:353–363.

Wells, K. D., and T. L. Taigen. 1986. The effect of social interactions on calling energetics in the gray treefrog (*Hyla versicolor*). *Behav. Ecol. Sociobiol.* 19:9–18.

Werner, D. I. 1978. On the biology of *Tropidurus delanonis*, Baur (Iguanidae). *Z. Tierpsychol.* 47:337–395.

West-Eberhard, M. J. 1979. Sexual selection, social competition, and evolution. *Proc. Am. Philos. Soc.* 123:222–234.

———. 1989. Phenotypic plasticity and the origins of diversity. *Annu. Rev. Ecol. Syst.* 20:249–278.

Westneat, D. F., P. W. Sherman, and M. L. Morton. 1990. The ecology and evolution of extra-pair copulations in birds. *Curr. Ornithol.* 7:330–369.

Whitfield, D. P. 1986. Plumage variability and territoriality in breeding turnstone *Arenaria interpres:* status signalling or individual recognition? *Anim. Behav.* 34:1471–1482.

———. 1987. Plumage variability, status signalling and individual recognition in avian flocks. *Trends Ecol. Evol.* 2:1318.

———. 1988. The social significance of plumage variability in wintering turnstone *Arenaria interpres. Anim. Behav.* 36:408–415.

Whiting, M. J. 1999. When to be neighbourly: differential agonistic responses in the lizard *Platysaurus broadleyi. Behav. Ecol. Sociobiol.* 46:210–214.

Whiting, M. J., and P. W. Bateman. 1999. Male preference for large females in the lizard *Platysaurus broadleyi* (Sauria: Cordylidae). *J. Herpetol.* 33:309–312.

Whiting, M. J., and J. M. Greeff. 1997. Facultative frugivory in the Cape flat lizard, *Platysaurus capensis* (Sauria: Cordylidae). *Copeia* 1997:811–818.

———. 1999. Use of heterospecific cues by the lizard *Platysaurus broadleyi* for food location. *Behav. Ecol. Sociobiol.* 45:420–423.

Whitman, C. O. 1919. The behavior of pigeons. *Publ. Carnegie Inst.* 257:1–161.

Wickler, W. 1957. Vergleichende verhaltenstudien an grundfishen. I. Bertra ge zür biologie, besonders zur ethologie von *Blenius fluviatilis* asso im vergleich zu einigen anderen bodenfischen. *Z. Tierpsychol.* 14:393–428.

Wiens, J. J. 1999. Phylogenetic evidence for multiple losses of a sexually selected character in phrynosomatid lizards. *Proc. Roy. Soc. Lond., Ser. B* 266:1529–1535.

———. 2000. Decoupled evolution of display morphology and display behaviour in phrynosomatid lizards. *Biol. J. Linn. Soc.* 70:597–612.

Wiens, J. J., and T. W. Reeder. 1997. Phylogeny of the spiny lizards (*Sceloporus*) based on molecular and morphological evidence. *Herpetol. Monogr.* 11:1–101.

Wiens, J. J., T. W. Reeder, and A. N. Montes de Oca. 1999. Molecular phylogenetics and evolution of sexual dichromatism among populations of the Yarrow's spiny lizard (*Sceloporus jarrovii*). *Evolution* 53:1884–1897.

Wikelski, M. 1994. Evolution of body size in the marine iguana (*Amblyrhynchus cristatus*): ultimate and proximate aspects. Ph.D. dissertation, Bielefeld University, Bielefeld, Germany.

Wikelski, M., C. Carbone, and F. Trillmich. 1996. Lekking in marine iguanas: female grouping and male reproductive strategies. *Anim. Behav.* 52:581–596.

Wilbur, H. M., D. I. Rubenstein, and L. Fairchild. 1978. Sexual selection in toads: the roles of female choice and male body size. *Evolution* 32:264–270.

Williams, E. E. 1969. The ecology of colonization as seen in the zoogeography of anoline lizards on small islands. *Quart. Rev. Biol.* 44:345–389.

———. 1972. The origin of faunas. Evolution of lizard congeners in a complex island fauna: a trial analysis. *Evol. Biol.* 6:47–89.

———. 1983. Ecomorphs, faunas, island size, and diverse end points in island radiations of *Anolis*. In *Lizard ecology: studies of a model organism,* edited by R. B. Huey, E. R. Pianka, and T. W. Schoener, 326–370. Harvard University Press, Cambridge, Mass.

Williams, E. E., and S. Rand. 1977. Species recognition, dewlap function, and faunal size. *Am. Zool.* 19:261–270.

Williams, G. C. 1966. *Adaptation and natural selection: a critique of some current evolutionary thought.* Princeton University Press, Princeton, N.J.

———. 1975. *Sex and evolution.* Princeton University Press, Princeton, N.J.

Williams, G. C., and D. C. Williams. 1957. Natural selection of individually harmful social adaptations among sibs with special reference to social insects. *Evolution* 11:32–39.

Wilson, B. S. 1991. Latitudinal variation in activity season mortality rates of the lizard *Uta stansburiana. Ecol. Monogr.* 61:393–414.

Wilson, E. O. 1975. *Sociobiology: the new synthesis.* Harvard University Press, Cambridge, Mass.

Wilson, J. D. 1992. A re-assessment of the significance of status signalling in populations of wild great tits, *Parus major. Anim. Behav.* 43:999–1009.

Wilson, R. S., and D. T. Booth. 1998. Effect of tail loss on reproductive output and its ecological significance in the skink *Eulamprus quoyii. J. Herpetol.* 32:128–131.

Wingfield, J. C. 1994. Hormone–behavior interactions and mating systems in male and female birds. In *The differences between the sexes,* edited by R. V. Short and E. Balaban, 303–330. Cambridge University Press, Cambridge, U.K.

Wingfield, J. C., and D. S. Farner. 1975. The determination of five steroids in avian plasma by radioimmunoassay and competitive protein-binding. *Steroids* 26:311–327.

Wingfield, J. C., and M. C. Moore. 1987. Hormonal, social and environmental factors in the reproductive biology of free-living male birds. In *Psychobiology of reproductive behavior: an evolutionary perspective,* edited by D. Crews, 149–175. Prentice-Hall, Englewood Cliffs, N.J.

Wittenberger, J. F. 1981. *Animal social behavior.* Duxbury, Boston.

———. 1983. Tactics of mate choice. In *Mate choice,* edited by P. Bateson, 3–32. Cambridge University Press, Cambridge, U.K.

Wood, R. A., K. A. Nagy, N. S. MacDonald, S. T. Wakakuwa, R. J. Beckman, and H. Kaaz. 1975. Determination of oxygen-18 in water contained in biological samples by charged particle activation. *Analyt. Chem.* 47:646–650.

Woodley, S. K., and M. C. Moore. 1999a. Female territorial aggression and steroid hormones in mountain spiny lizards. *Anim. Behav.* 57:1083–1089.

———. 1999b. Ovarian hormones influence territorial aggression in free-living female mountain spiny lizards. *Horm. Behav.* 35:205–214.

Wright, J. W. 1983. The evolution and biogeography of the lizards in the Galápagos Archipelago: evolutionary genetics of *Phyllodactylus* and *Tropidurus* populations. In *Patterns of evolution in Galápagos organisms,* edited by R. I. Bowman, M. Berson, and A. E. Leviton, 123–155. Pacific Division, AAAS, San Francisco.

Wright, S. J. 1981. Extinction-mediated competition: the *Anolis* lizards and insectivorous birds of the West Indies. *Am. Nat.* 117:181–192.

Wright, S. J., R. Kimsey, and C. J. Campbell. 1984. Mortality rates of insular *Anolis* lizards: a systematic effect of island area? *Am. Nat.* 123:134–142.

Yamane, A. 1998. Male reproductive tactics and reproductive success of the group-living feral cat (*Felis catus*). *Behav. Process.* 43:239–249.

Yasukawa, K. 1981. Male quality and female choice of mate in the red-winged blackbird (*Agelaius phoeniceus*). *Ecology* 62:922–929.

Ydenberg, R. C., and L. M. Dill. 1986. The economics of fleeing from predators. *Adv. Stud. Behav.* 16:229–249.

Yedlin, I. N., and G. W. Ferguson. 1973. Variations in aggressiveness of free-living male and female collared lizards, *Crotaphytus collaris. Herpetologica* 29:268–275.

Young, L. J., G. F. Lopreato, K. Horan, and D. Crews. 1994. Cloning and in situ hybridization analysis of estrogen receptor, progesterone receptor and androgen receptor expression in the brain of whiptail lizards (*Cnemidophorus uniparens* and *C. inornatus*). *J. Comp. Neurol.* 347:288–300.

Young, L. J., R. Nilsen, K. G. Waymire, G. R. MacGregor, and T. R. Insel. 1999. Increased affiliative response to vasopressin in mice expressing the V_{1a} receptor from a monogamous vole. *Nature* 400:766–768.

Young, L. J., Z. Wang, and T. R. Insel. 1997. Neuroendocrine bases of monogamy. *Trends Neurosci.* 21:71–75.

Zahavi, A. 1975. Mate selection—a selection for a handicap. *J. Theor. Biol.* 53:205–214.

———. 1977. The cost of honesty (further remarks on the handicap principle). *J. Theor. Biol.* 67:603–605.

Zamudio, K. R. 1998. The evolution of female-biased sexual size dimorphism: a population-level comparative study in horned lizards (*Phrynosoma*). *Evolution* 52:1821–1833.

Zamudio, K. R., and B. Sinervo. 2000. Polygyny, mate guarding, and posthumous fertilization as alternative male mating strategies. *Proc. Natl. Acad. Sci. USA* 97:14427–14432.

Zimmerer, E. J., and K. D. Kallman. 1988. The inheritance of vertical barring (aggression and appeasement signals) in the pygmy swordtail, *Xiphophorus nigrensis* (Poeciliidae, Teleostei). *Copeia* 1988:299–307.

Zimmermann, H., and E. Zimmermann. 1990. Behavioral systematics and zoogeography of the formation of species groups in dart-poison frogs (Anura, Dendrobatidae). *Newslett. Int. Soc. Study Dendrobatid Frogs* 3:75–116.

Zucker, N. 1994a. A dual status-signalling system: a matter of redundancy or differing roles? *Anim. Behav.* 47:15–22.

———. 1994b. Social influence on the use of a modifiable status signal. *Anim. Behav.* 48:1317–1324.

Zuk, M., T. S. Johnsen, and T. Maclarty. 1995. Endocrine-immune interactions, ornaments and mate choice in red jungle fowl. *Proc. Roy. Soc. Lond., Ser. B* 260:205–210.

Zuk, M., and G. R. Kolluru. 1998. Exploitation of sexual signals by predators and parasitoids. *Quart. Rev. Biol.* 73:415–438.

Zuk, M., R. Thornhill, J. D. Ligon, K. Johnson, S. Austad, S. H. Ligon, N. W. Thornhill, and C. Costin. 1990. The role of male ornaments and courtship behavior in female mate choice of red jungle fowl. *Am. Nat.* 136:459–473.

Zuri, I., and C. M. Bull. 2000. Reduced access to olfactory cues and home-range maintenance in the sleepy lizard (*Tiliqua rugosa*). *J. Zool. Lond.* 252:137–145.

Zykova, L. Y., and E. N. Panov. 1993. Notes on social organization and behavior of Khorosan agama, *Stellio erythrogaster*, in Badkhyz. *Zool. Zh.* 72:148–151.

Index